T0093042

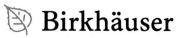

Applied and Numerical Harmonic Analysis

For further volumes:
http://www.springer.com/series/4968

Gitta Kutyniok • Demetrio Labate
Editors

Shearlets

Multiscale Analysis for Multivariate Data

 Birkhäuser

Editors
Gitta Kutyniok
Institut für Mathematik
Technische Universität Berlin
Berlin, Germany

Demetrio Labate
Department of Mathematics
University of Houston
Houston, TX, USA

ISBN 978-0-8176-8315-3   e-ISBN 978-0-8176-8316-0
DOI 10.1007/978-0-8176-8316-0
Springer New York Dordrecht Heidelberg London

Library of Congress Control Number: 2012932597

Mathematical Subject Classification (2010): 42C15, 42C40, 65T60, 68U10, 94A08

Printed on acid-free paper

Springer is part of Springer Science+Business Media
(www.birkhauser-science.com)

ANHA Series Preface

The *Applied and Numerical Harmonic Analysis (ANHA)* book series aims to provide the engineering, mathematical, and scientific communities with significant developments in harmonic analysis, ranging from abstract harmonic analysis to basic applications. The title of the series reflects the importance of applications and numerical implementation, but richness and relevance of applications and implementation depend fundamentally on the structure and depth of theoretical underpinnings. Thus, from our point of view, the interleaving of theory and applications and their creative symbiotic evolution is axiomatic.

Harmonic analysis is a wellspring of ideas and applicability that has flourished, developed, and deepened over time within many disciplines and by means of creative cross-fertilization with diverse areas. The intricate and fundamental relationship between harmonic analysis and fields such as signal processing, partial differential equations (PDEs), and image processing is reflected in our state-of-the-art *ANHA* series.

Our vision of modern harmonic analysis includes mathematical areas such as wavelet theory, Banach algebras, classical Fourier analysis, time–frequency analysis, and fractal geometry, as well as the diverse topics that impinge on them.

For example, wavelet theory can be considered an appropriate tool to deal with some basic problems in digital signal processing, speech and image processing, geophysics, pattern recognition, biomedical engineering, and turbulence. These areas implement the latest technology from sampling methods on surfaces to fast algorithms and computer vision methods. The underlying mathematics of wavelet theory depends not only on classical Fourier analysis, but also on ideas from abstract harmonic analysis, including von Neumann algebras and the affine group. This leads to a study of the Heisenberg group and its relationship to Gabor systems, and of the metaplectic group for a meaningful interaction of signal decomposition methods. The unifying influence of wavelet theory in the aforementioned topics illustrates the justification for providing a means for centralizing and disseminating information from the broader, but still focused, area of harmonic analysis. This will be a key role of *ANHA*. We intend to publish the scope and interaction that such a host of issues demands.

Along with our commitment to publish mathematically significant works at the frontiers of harmonic analysis, we have a comparably strong commitment to publish major advances in the following applicable topics in which harmonic analysis plays a substantial role:

Antenna theory	*Prediction theory*
Biomedical signal processing	*Radar applications*
Digital signal processing	*Sampling theory*
Fast algorithms	*Spectral estimation*
Gabor theory and applications	*Speech processing*
Image processing	*Time–frequency and*
Numerical partial differential equations	*time-scale analysis*
	Wavelet theory

The above point of view for the *ANHA* book series is inspired by the history of Fourier analysis itself, whose tentacles reach into so many fields.

In the last two centuries, Fourier analysis has had a major impact on the development of mathematics, on the understanding of many engineering and scientific phenomena, and on the solution of some of the most important problems in mathematics and the sciences. Historically, Fourier series were developed in the analysis of some of the classical PDEs of mathematical physics; these series were used to solve such equations. In order to understand Fourier series and the kinds of solutions they could represent, some of the most basic notions of analysis were defined, e.g., the concept of "function". Since the coefficients of Fourier series are integrals, it is no surprise that Riemann integrals were conceived to deal with uniqueness properties of trigonometric series. Cantors set theory was also developed because of such uniqueness questions.

A basic problem in Fourier analysis is to show how complicated phenomena, such as sound waves, can be described in terms of elementary harmonics. There are two aspects of this problem: first, to find, or even define properly, the harmonics or spectrum of a given phenomenon, e.g., the spectroscopy problem in optics; second, to determine which phenomena can be constructed from given classes of harmonics, as done, e.g., by the mechanical synthesizers in tidal analysis.

Fourier analysis is also the natural setting for many other problems in engineering, mathematics, and the sciences. For example, Wiener's Tauberian theorem in Fourier analysis not only characterizes the behavior of the prime numbers, but also provides the proper notion of spectrum for phenomena such as white light; this latter process leads to the Fourier analysis associated with correlation functions in filtering and prediction problems, and these problems, in turn, deal naturally with Hardy spaces in the theory of complex variables.

Nowadays, some of the theory of PDEs has given way to the study of Fourier integral operators. Problems in antenna theory are studied in terms of unimodular trigonometric polynomials. Applications of Fourier analysis abound in signal processing, whether with the fast Fourier transform (FFT), or filter design, or the adaptive modeling inherent in timefrequency-scale methods such as wavelet theory. The coherent states of mathematical physics are translated and modulated Fourier transforms, and these are used, in conjunctionwith the uncertainty principle, for dealing with signal reconstruction in communications theory.We are back to the raison d'être of the *ANHA* series!

University of Maryland   John J. Benedetto
College Park   Series Editor

Preface

The introduction of wavelets about 20 years ago has revolutionized applied mathematics, computer science, and engineering by providing a highly effective methodology for analyzing and processing univariate functions/signals containing singularities. However, wavelets do not perform equally well in the multivariate case due to the fact that they are capable of efficiently encoding only isotropic features. This limitation can be seen by observing that Besov spaces can be precisely characterized by decay properties of sequences of wavelet coefficients, but they are not capable of capturing those geometric features which could be associated with edges and other distributed singularities. Indeed, such geometric features are essential in the multivariate setting, since multivariate problems are typically governed by anisotropic phenomena such as singularities concentrated on lower dimensional embedded manifolds. To deal with this challenge, several approaches were proposed in the attempt to extend the benefits of the wavelet framework to higher dimensions, with the aim of introducing representation systems which could provide both optimally sparse approximations of anisotropic features and a unified treatment of the continuum and digital world. Among the various methodologies proposed, such as curvelets and contourlets, the shearlet system, which was introduced in 2005, stands out as the first and so far the only approach capable of satisfying this combination of requirements.

Today, various directions of research have been established in the theory of shearlets. These include, in particular, the theory of continuous shearlets—associated with a parameter set of continuous range—and its application to the analysis of distributions. Another direction is the theory of discrete shearlets—associated with a discrete parameter set—and their sparse approximation properties. Thanks to the fact that shearlets provide a unified treatment of the continuum and digital realm through the utilization of the shearing operator, digitalization and hence numerical realizations can be performed in a faithful manner, and this leads to very efficient algorithms. Building on these results, several shearlet-based algorithms were developed to address a range of problems in image and data processing.

This book is the first monograph devoted to shearlets. It is not only aimed at and accessible to a broad readership including graduate students and researchers in the areas of applied mathematics, computer science, and engineering, but it will also appeal to researchers working in any other field requiring highly efficient methodologies for the processing of multivariate data. Because of this fact, this volume can be used both as a state-of-the-art monograph on shearlets and advanced multiscale methods and as a textbook for graduate students.

This volume is organized into several tutorial-like chapters which cover the main aspects of theory and applications of shearlets and are written by the leading international experts in these areas. The first chapter provides a self-contained and comprehensive overview of the main results on shearlets and sets the basic notation and definitions which are used in the remainder of the book. The topics covered in the remaining chapters essentially follow the idea of going from the continuous setting, i.e., continuous shearlets and their microlocal properties, up to the discrete and digital setting, i.e., discrete shearlets, their digital realizations, and their applications. Each chapter is self-contained, which enables the reader to choose his/her own path through the book. Here is a brief outline of the content of each chapter.

The first chapter, written by the editors, provides an introduction and presents a self-contained overview of the main results on the theory and applications of shearlets. Starting with some background on frame theory and wavelets, it covers the definitions of continuous and discrete shearlets and the main results from the theory of shearlets, which are subsequently discussed in detail and expanded in the following chapters.

In the second chapter, Grohs focusses on the continuous shearlet transform. After making the reader familiar with concepts from microlocal analysis, he shows that the shearlet transform offers a simple and convenient way to characterize wavefront sets of distributions.

In the third chapter, Guo and Labate illustrate the ability of the continuous shearlet transform to characterize the set of singularities of multivariate functions and distributions. These properties set the groundwork for some of the imaging applications discussed in the eighth chapter.

In the fourth chapter, Dahlke et al. introduce the continuous shearlet transform for arbitrary space dimension. They further present the construction of smoothness spaces associated to shearlet representations and the analysis of their structural properties.

In the fifth chapter, Kutyniok et al. provide a comprehensive survey of the theory of sparse approximations of cartoon-like images using shearlets. Both the band-limited and the compactly supported shearlet frames are examined in this chapter.

In the sixth chapter, Sauer starts from the classical concepts of filterbanks and subband coding to present an entirely digital approach to shearlet multiresolution. This approach is not a discretization of the continuous transform, but is naturally connected to the filtering of digital data.

In the seventh chapter, Kutyniok et al. discuss the construction of digital realizations of the shearlet transform with a particular focus on a unified treatment of the continuum and digital realm. In particular, this chapter illustrates two distinct numerical implementations of the shearlet transform, one based on band-limited shearlets and the other based on compactly supported shearlets.

In the eighth chapter, Easley and Labate present the application of shearlets to several problems from imaging and data analysis to date. This includes the illustration of shearlet-based algorithms for image denoising, image enhancement, edge detection, image separation, deconvolution, and regularized reconstruction of Radon data. In all these applications, the ability of shearlet representations to handle anisotropic features efficiently is exploited in order to derive highly competitive numerical algorithms.

Finally, it is important to emphasize that the work presented in this volume would not have been possible without the interaction and discussions with many people during these years. We wish to thank the many students and researchers who over the years have given us insightful comments and suggestions, and helped this area of research to grow into its present form.

Berlin, Germany   Gitta Kutyniok
Houston, USA   Demetrio Labate

Contents

Contributors

Stephan Dahlke Philipps-Universität Marburg, FB12 Mathematik und Informatik, Hans-Meerwein Straße, Marburg, Germany

Glenn R. Easley System Planning Corporation, Arlington, VA, USA

Philipp Grohs ETH Zürich, Zürich, Switzerland

Kanghui Guo Department of Mathematics, Missouri State University, Springfield, MO, USA

Gitta Kutyniok Institut für Mathematik, Technische Universität Berlin, Berlin, Germany

Demetrio Labate Department of Mathematics, University of Houston, Houston, TX, USA

Jakob Lemvig Department of Mathematics, Technical University of Denmark, Lyngby, Denmark

Wang-Q Lim Institut für Mathematik, Technische Universität Berlin, Berlin, Germany

Tomas Sauer Lehrstuhl für Numerische Mathematik, Justus–Liebig–Universität Gießen, Gießen, Germany

Gabriele Steidl Universität Kaiserslautern, Fachbereich Mathematik, Kaiserslautern, Germany

Gerd Teschke Hochschule Neubrandenburg - University of Applied Sciences, Institute for Computational Mathematics in Science and Technology, Neubrandenburg, Germany

Xiaosheng Zhuang Institut für Mathematik, Technische Universität Berlin, Berlin, Germany

Introduction to Shearlets

Gitta Kutyniok and Demetrio Labate

Abstract Shearlets emerged in recent years among the most successful frameworks for the efficient representation of multidimensional data. Indeed, after it was recognized that traditional multiscale methods are not very efficient at capturing edges and other anisotropic features which frequently dominate multidimensional phenomena, several methods were introduced to overcome their limitations. The shearlet representation stands out since it offers a unique combination of some highly desirable properties: it has a single or finite set of generating functions, it provides optimally sparse representations for a large class of multidimensional data, it is possible to use compactly supported analyzing functions, it has fast algorithmic implementations and it allows a unified treatment of the continuum and digital realms. In this chapter, we present a self-contained overview of the main results concerning the theory and applications of shearlets.

Key words: Affine systems, Continuous wavelet transform, Image processing, Shearlets, Sparsity, Wavelets

1 Introduction

Scientists sometimes refer to the twenty-first century as the Age of Data. As a matter of fact, since technological advances make the acquisition of data easier and less expensive, we are coping today with a deluge of data including astronomical, medical, seismic, meteorological, and surveillance data, which require efficient analysis

G. Kutyniok
Institut für Mathematik, Technische Universität Berlin, 10623 Berlin, Germany
e-mail: kutyniok@math.tu-berlin.de

D. Labate
Department of Mathematics, University of Houston, Houston, TX 77204, USA
e-mail: dlabate@math.uh.edu

G. Kutyniok and D. Labate (eds.), *Shearlets: Multiscale Analysis for Multivariate Data*,
Applied and Numerical Harmonic Analysis, DOI 10.1007/978-0-8176-8316-0_1,
© Springer Science+Business Media, LLC 2012

1

and processing. The enormity of the challenge this poses is evidenced not only by the sheer amount of data but also by the diversity of data types and the variety of processing tasks which are required. To efficiently handle tasks ranging from feature analysis over classification to compression, highly sophisticated mathematical and computational methodologies are needed. From a mathematical standpoint data can be modeled, for example, as functions, distributions, point clouds, or graphs. Moreover, data can be classified by membership in one of the two categories: explicitly given data such as imaging or measurement data and implicitly given data such as solutions of differential or integral equations.

A fundamental property of virtually all data found in practical applications is that the relevant information which needs to be extracted or identified is sparse, i.e., data are typically highly correlated and the essential information lies on low-dimensional manifolds. This information can thus be captured, in principle, using just few terms in an appropriate dictionary. This observation is crucial not only for tasks such as data storage and transmission but also for feature extraction, classification, and other high-level tasks. Indeed, finding a dictionary which sparsely represents a certain data class entails the intimate understanding of its dominant features, which are typically associated with their geometric properties. This is closely related to the observation that virtually all multivariate data are typically dominated by anisotropic features such as singularities on lower dimensional embedded manifolds. This is exemplified, for instance, by edges in natural images or shock fronts in the solutions of transport equations. Hence, to efficiently analyze and process these data, it is of fundamental importance to discover and truly understand their *geometric structures*.

The subject of this volume is a recently introduced multiscale framework, the theory of *shearlets*, which allows optimal encoding of several classes of multivariate data through its ability to sparsely represent anisotropic features. As will be illustrated in the following, shearlets emerged as part of an extensive research activity developed during the last 10 years to create a new generation of analysis and processing tools for massive and higher dimensional data, which could go beyond the limitations of traditional Fourier and wavelet systems. One of the forerunners of this area of research is David L. Donoho, who observed that in higher dimensions traditional multiscale systems and wavelets ought to be replaced by a *Geometric Multiscale Analysis* in which multiscale analysis is adapted to intermediate-dimensional singularities. It is important to remark that many of the ideas which are at the core of this approach can be traced back to key results in harmonic analysis from the 1990s, such as Hart Smith's Hardy space for Fourier Integral Operators and Peter Jones' Analyst's Traveling Salesman theorem. Both results concern the higher dimensional setting, where geometric ideas are brought into play to discover "new architectures for decomposition, rearrangement, and reconstruction of operators and functions" [16].

This broader area of research is currently at the crossroads of applied mathematics, electrical engineering, and computer science, and has seen spectacular advances in recent years, resulting in highly sophisticated and efficient algorithms for image analysis and new paradigms for data compression and approximation. By presenting the theory and applications of shearlets obtained during the last 5 years, this book is also a journey into one of the most active and exciting areas of research in applied mathematics.

2 The Rise of Shearlets

2.1 The Role of Applied Harmonic Analysis

Applied harmonic analysis has established itself as the main area in applied mathematics focused on the efficient representation, analysis, and encoding of data. The primary object of this discipline is the process of "breaking into pieces" (this is the literal meaning of the Greek word *analysis*) to gain insight into an object. For example, given a class of data $\mathscr{C}$ in $L^2(\mathbb{R}^d)$, a collection of *analyzing* functions $(\varphi_i)_{i\in I} \subseteq L^2(\mathbb{R}^d)$ with I being a countable indexing set is sought such that, for all $f \in \mathscr{C}$, we have the expansion

$$f = \sum_{i\in I} c_i(f)\varphi_i. \tag{1}$$

This formula provides not only a decomposition for any element $f \in \mathscr{C}$ into a countable collection of linear measurements $(c_i(f))_{i\in I} \subseteq \ell^2(I)$, i.e., its *analysis*; it also illustrates the process of *synthesis*, where f is reconstructed from the expansion coefficients $(c_i(f))_{i\in I}$.

One major goal of applied harmonic analysis is the construction of special classes of analyzing elements which can best capture the most relevant information in a certain data class. Let us illustrate the two most successful types of analyzing systems in the one-dimensional setting. *Gabor systems* are designed to best represent the joint time–frequency content of data. In this case, the analyzing elements $(\varphi_i)_{i\in I}$ are obtained as translations and frequency shifts of a generating function $\varphi \in L^2(\mathbb{R})$ as follows:

$$\{\varphi_{p,q} = \varphi(\cdot - p)e^{2\pi i q\cdot} : p,q \in \mathbb{Z}\}.$$

In contrast to this approach, *wavelet systems* represent the data as associated with different location and resolution levels. In this case, the analyzing elements $(\varphi_i)_{i\in I}$ are obtained through the action of dilation and translation operators on a generating function $\psi \in L^2(\mathbb{R})$, called a *wavelet*, as:

$$\{\psi_{j,m} = 2^{j/2}\psi(2^j\cdot - m) : j,m \in \mathbb{Z}\}. \tag{2}$$

Given a prescribed class of data $\mathscr{C}$, one major objective is to design an analyzing system $(\varphi_i)_{i\in I}$ in such a way that, for each function $f \in \mathscr{C}$, the coefficient sequence $(c_i(f))_{i\in I}$ in (1) can be chosen to be *sparse*. In the situation of an infinite-dimensional Hilbert space—which is our focus here—the degree of sparsity is customarily measured as the decay rate of the error of best n-term approximation. Loosely speaking, this means that we can approximate any $f \in \mathscr{C}$ with high accuracy by using a coefficient sequence $(\tilde{c}_i(f))_{i \in I}$ containing very few nonzero entries. In the finite-dimensional setting, such a sequence is called *sparse*, and this explains the use of the term *sparse approximation*. Intuitively, if a function can be sparsely approximated, it is conceivable that "important" features can be detected by thresholding, i.e., by selecting the indices associated with the largest coefficients in absolute values, or that high compression rates can be achieved by storing only few large coefficients $c_i(f)$, see [19].

There is another fundamental phenomenon to observe here. If $(\varphi_i)_{i \in I}$ is an orthonormal basis, the coefficient sequence $(c_i(f))_{i \in I}$ in (1) is certainly uniquely determined. However, if we allow more freedom in the sense of choosing $(\varphi_i)_{i \in I}$ to form a frame—a redundant, yet stable system (see Sect. 3.3)—the sequences $(c_i(f))_{i \in I}$ might be chosen significantly sparser for each $f \in \mathscr{C}$. Thus, methodologies from *frame theory* will come into play, see Sect. 3.3 and [5, 7].

We can observe a close connection to yet another highly topical area. During the last 4 years, sparse recovery methodologies such as *Compressed Sensing* in particular have revolutionized the areas of applied mathematics, computer science, and electrical engineering by beating the traditional sampling theory limits, see [3, 23]. They exploit the fact that many types of signals can be represented using only a few nonvanishing coefficients when choosing a suitable basis or, more generally, a frame. Nonlinear optimization methods, such as ℓ_1 minimization, can then be employed to recover such signals from "very few" measurements under appropriate assumptions on the signal and on the basis or frame. These results can often be generalized to data which are merely sparsely approximated by a frame, thereby enabling compressed sensing methodologies for the situation we discussed above.

2.2 Wavelets and Beyond

The emergence of *wavelets* about 25 years ago represents a milestone in the development of efficient encoding of piecewise regular signals. The major reason for the spectacular success of wavelets consists not only in their ability to provide optimally sparse approximations of a large class of frequently occurring signals and to represent singularities much more efficiently than traditional Fourier methods, but also in the existence of fast algorithmic implementations which precisely digitalize the continuum domain transforms. The key property enabling such a unified treatment of the continuum and digital setting is a *Multiresolution Analysis*, which allows a direct transition between the realms of real variable functions and digital signals. This framework also combines very naturally with the theory of filter banks developed in the digital signal processing community. An additional aspect of the theory of wavelets which has contributed to its success is its rich mathematical structure, which allows one to design families of wavelets with various desirable properties expressed in terms of regularity, decay, or vanishing moments. As a consequence of all these properties, wavelets have literally revolutionized image and signal processing and produced a large number of very successful applications, including the algorithm of JPEG2000, the current standard for image compression. We refer the interested reader to [65] for more details about wavelets and their applications.

Despite their success, wavelets are not very effective when dealing with multivariate data. In fact, wavelet representations are optimal for approximating data with pointwise singularities only and cannot handle equally well distributed singularities such as singularities along curves. The intuitive reason for this is that wavelets are *isotropic* objects, being generated by isotropically dilating a single or finite set of generators. However, in dimensions two and higher, distributed discontinuities such as edges of surface boundaries are usually present or even dominant, and—as a result—wavelets are far from optimal in dealing with multivariate data.

The limitations of wavelets and traditional multiscale systems have stimulated a flurry of activity involving mathematicians, engineers, and applied scientists. Indeed, the need to introduce some form of directional sensitivity[1] in the wavelet framework was already recognized in the early filter bank literature, and several versions of "directional" wavelets were introduced, including the *steerable pyramid* by Simoncelli et al. [71], the *directional filter banks* by Bamberger and Smith [2], and the *2D directional wavelets* by Antoine et al. [1]. A more sophisticated approach was proposed more recently with the introduction of *complex wavelets* [44, 45]. However, even though they frequently outperform standard wavelets in applications, these methods do not provide optimally sparse approximations of multivariate data governed by anisotropic features. The fundamental reason for this failure is that these approaches are not truly multidimensional extensions of the wavelet approach.

The real breakthrough occurred with the introduction of *curvelets* by Candès and Donoho [4] in 2004, which was the first system providing optimally sparse approximations for a class of bivariate functions exhibiting anisotropic features. Curvelets form a pyramid of analyzing functions defined not only at various scales and locations as wavelets do, but also at various orientations, with the number of orientations increasing at finer scales. Another fundamental property is that their supports are highly anisotropic and become increasingly elongated at finer scales. Due to this anisotropy, curvelets are essentially as good as an adaptive representation system from the point of view of the ability to sparsely approximate images with edges. The two main drawbacks of the curvelet approach are that, firstly, this system is not singly generated, i.e., it is not derived from the action of countably many operators applied to a single (or finite set) of generating functions; secondly, its construction involves rotations and these operators do not preserve the digital lattice, which prevents a direct transition from the continuum to the digital setting.

Contourlets were introduced in 2005 by Do and Vetterli [14] as a purely discrete filter-bank version of the curvelet framework. This approach offers the advantage of

---

[1] It is important to recall that the importance of directional sensitivity in the efficient processing of natural images by the human brain has been a major finding in neuropsycological studies such as the work of Field and Olshausen [68], and a significant inspiration for some of the research developed in the harmonic analysis and image processing literature.

allowing a tree-structured filter bank implementation similar to the standard wavelet implementations which was exploited to obtain very efficient numerical algorithms. However, a proper continuum theory is missing in this approach.

In the same year, *shearlets* were introduced by Guo, Kutyniok, Labate, Lim, and Weiss in [30, 61]. This approach was derived within a larger class of affine-like systems—the so-called *composite wavelets* [39, 40, 41]—as a truly multivariate extension of the wavelet framework. One of the distinctive features of shearlets is the use of shearing to control directional selectivity, in contrast to rotation used by curvelets. This is a fundamentally different concept, since it allows shearlet systems to be derived from a single or finite set of generators, and it also ensures a unified treatment of the continuum and digital world due to the fact that the shear matrix preserves the integer lattice. Indeed, as will be extensively discussed in this volume, the shearlet representation offers a unique combination of the following list of desiderata:

- A single or a finite set of generating functions.
- Optimally sparse approximations of anisotropic features in multivariate data.
- Compactly supported analyzing elements.
- Fast algorithmic implementations.
- A unified treatment of the continuum and digital realms.
- Association with classical approximation spaces.

For completeness, it is important to recall yet another class of representation systems which are able to overcome the limitations of traditional wavelets and produce optimally efficient representations for a large class of images, namely the *bandelets* [70] and the *grouplets* [66]. Also in these methods, the idea is to take advantage of the geometry of the data. However, in this case, this is done *adaptively*, that is, by constructing a special data decomposition which is especially designed for each data set, rather than by using a fixed representation system as it is done using wavelets or shearlets. While one can achieve very efficient data decompositions using such an adaptive approach, this is usually numerically more intensive than using nonadaptive methods.

In the following sections, we will present a self-contained overview of the key results from the theory and applications of shearlets, focused primarily on the 2D setting. These results will be elaborated in much more detail in the various chapters of this volume, which will discuss both the continuum and digital aspects of shearlets. Before starting our overview, it will be useful to establish the notation adopted throughout this volume and to present some background material from harmonic analysis and wavelet theory.

3 Notation and Background Material

3.1 Fourier Analysis

The Fourier transform is the most fundamental tool in harmonic analysis. Before stating the definition, we remark that, in the following, vectors in $\mathbb{R}^d$ or $\mathscr{C}^d$ will always be understood as column vectors, and their inner product—as also the inner product in $L^2(\mathbb{R}^d)$—shall be denoted by $\langle \cdot, \cdot \rangle$. For a function $f \in L^1(\mathbb{R}^d)$, the *Fourier transform* of f is defined by

$$\hat{f}(\xi) = \int f(x)e^{-2\pi i \langle x, \xi \rangle} dx,$$

and f is called a *band-limited* function if its Fourier transform is compactly supported. The *inverse Fourier transform* of a function $g \in L^1(\mathbb{R}^d)$ is given as

$$\check{g}(x) = \int g(\xi)e^{2\pi i \langle x, \xi \rangle} d\xi.$$

If $f \in L^1(\mathbb{R}^d)$ with $\hat{f} \in L^1(\mathbb{R}^d)$, we have $f = (\hat{f})^\vee$, hence in this case—which is by far not the only possible case—the inverse Fourier transform is the "true" inverse. It is well known that this definition can be extended to $L^2(\mathbb{R}^d)$, and as usual, also these extensions will be denoted by $\hat{f}$ and $\check{g}$. By using this definition of the Fourier transform, the *Plancherel formula* for $f, g \in L^2(\mathbb{R}^n)$ reads

$$\langle f, g \rangle = \langle \hat{f}, \hat{g} \rangle,$$

and, in particular,

$$\|f\|_2 = \|\hat{f}\|_2.$$

We refer to [25] for additional background information on Fourier analysis.

3.2 Modeling of Signal Classes

In the continuum setting, the standard model of d-dimensional signals is the space of *square-integrable functions* on $\mathbb{R}^d$, denoted by $L^2(\mathbb{R}^d)$. However, this space also contains objects which are very far from natural images and data. Hence, it is convenient to introduce subclasses and subspaces which can better model the types of data encountered in applications. One approach for doing this consists in imposing some degree of regularity. Therefore, we consider the *continuous functions* $C(\mathbb{R}^d)$, the k-times continuously differentiable functions $C^k(\mathbb{R}^d)$, and the infinitely many-times continuously differentiable functions $C^\infty(\mathbb{R}^d)$, which are also referred to as *smooth functions*. Since images are compactly supported in nature, a notion for *compactly supported functions* is also required which will be indicated with the subscript 0, e.g., $C_0^\infty(\mathbb{R}^d)$.

Sometimes it is useful to consider curvilinear singularities such as edges in images as singularities of distributions, which requires the *space of distributions* $\mathcal{D}'(\mathbb{R}^d)$ as a model. For a distribution u, we say that $x \in \mathbb{R}^d$ is a *regular point* of u, if there exists a function $\phi \in C_0^\infty(U_x)$ with $\phi(x) \neq 0$ and U_x being a neighborhood of x. This implies $\phi u \in C_0^\infty(\mathbb{R}^d)$, which is equivalent to $(\phi u)^\wedge$ being rapidly decreasing.

The complement of the set of regular points of u is called the *singular support* of u and is denoted by sing supp(u). Notice that the singular support of u is a closed subset of supp(u).

The anisotropic nature of singularities on one- or multidimensional embedded manifolds becomes apparent through the notion of a wavefront set. For simplicity, we illustrate the two-dimensional case only. For a distribution u, a point $(x, s) \in \mathbb{R}^2 \times \mathbb{R}$ is a *regular directed point*, if there exist neighborhoods U_x of x and V_s of s as well as a function $\phi \in C_0^\infty(\mathbb{R}^2)$ satisfying $\phi|_{U_x} \equiv 1$ such that, for each $N > 0$, there exists a constant C_N with

$$|(u\phi)^\wedge(\eta)| \le C_N (1 + |\eta|)^{-N} \quad \text{for all } \eta = (\eta_1, \eta_2) \in \mathbb{R}^2 \text{ with } \frac{\eta_2}{\eta_1} \in V_s.$$

The complement in $\mathbb{R}^2 \times \mathbb{R}$ of the regular directed points of u is called the *wavefront set* of u and is denoted by $WF(u)$. Thus, the singular support describes the location of the set of singularities of u, and the wavefront set describes both the location and local perpendicular orientation of the singularity set.

Fig. 1 Natural images are governed by anisotropic structures

A class of functions, which is of particular interest in imaging sciences, is the class of so-called *cartoon-like images*. This class was introduced in [15] to provide a simplified model of natural images, which emphasizes anisotropic features, most notably edges, and is consistent with many models of the human visual system. Consider, for example, the photo displayed in Fig. 1. Since the image basically consists of smooth regions separated by edges, it is suggestive to use a model consisting of piecewise regular functions, such as the one illustrated in Fig. 2. For simplicity, the domain is set to be $[0, 1]^2$ and the regularity can be chosen to be C^2, leading to the following definition.

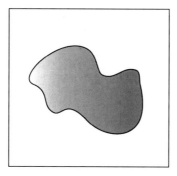

Fig. 2 Example of a cartoon-like image (function values represented using a gray scale map)

Definition 1. The *class* $\mathcal{E}^2(\mathbb{R}^2)$ *of cartoon-like images* is the set of functions $f :$ $\mathbb{R}^2 \to \mathbb{C}$ of the form

$$f = f_0 + f_1\chi_B,$$

where $B \subset [0,1]^2$ is a set with ∂B being a closed C^2-curve with bounded curvature and $f_i \in C^2(\mathbb{R}^2)$ are functions with $\mathrm{supp}\, f_i \subset [0,1]^2$ and $\|f_i\|_{C^2} \leq 1$ for each $i = 0,1$.

Let us finally mention that, in the digital setting, the usual models for d-dimensional signals are either functions on $\mathbb{Z}^d$ such as $\ell^2(\mathbb{Z}^d)$ or functions on $\{0,\dots,N-1\}^d$, sometimes denoted by $\mathbb{Z}_N^d$.

3.3 Frame Theory

When designing representation systems of functions, it is sometimes advantageous or unavoidable to go beyond the setting of orthonormal bases and consider redundant systems. The notion of a *frame*, originally introduced by Duffin and Schaeffer in [20] and later revived by Daubechies in [13], guarantees stability while allowing nonunique decompositions. Let us recall the basic definitions from frame theory in the setting of a general (real or complex) Hilbert space $\mathcal{H}$.

A sequence $(\varphi_i)_{i \in I}$ in $\mathcal{H}$ is called a *frame* for $\mathcal{H}$, if there exist constants $0 < A \leq B < \infty$ such that

$$A\|x\|^2 \leq \sum_{i \in I} |\langle x, \varphi_i \rangle|^2 \leq B\|x\|^2 \quad \text{for all } x \in \mathcal{H}.$$

The frame constants A and B are called *lower* and *upper frame bound*, respectively. The supremun over all A and the infimum over all B such that the frame inequalities hold are the *optimal frame bounds*. If A and B can be chosen with $A = B$, then the frame is called A-*tight*, and if $A = B = 1$ is possible, then $(\varphi_i)_{i \in I}$ is a *Parseval frame*. A frame is called *equal norm* if there exists some $c > 0$ such that $\|\varphi_i\| = c$ for all $i \in I$, and it is *unit norm* if $c = 1$.

Apart from providing redundant expansions, frames serve as an analysis tool. In fact, if $(\varphi_i)_{i \in I}$ in $\mathscr{H}$ is a frame for $\mathscr{H}$ it allows the analysis of data through the study of the associated *frame coefficients* $(\langle x, \varphi_i \rangle)_{i \in I}$, where the operator T defined by

$$T : \mathscr{H} \to \ell^2(I), \quad x \mapsto (\langle x, \varphi_i \rangle)_{i \in I}$$

is called the *analysis operator*. The adjoint T^* of the analysis operator is referred to as the *synthesis operator* and satisfies

$$T^* : \ell^2(I) \to \mathscr{H}, \quad ((c_i)_{i \in I}) \mapsto \sum_{i \in I} c_i \varphi_i.$$

The main operator associated with a frame, which provides a stable reconstruction process, is the *frame operator*

$$S = T^* T : \mathscr{H} \to \mathscr{H}, \quad x \mapsto \sum_{i \in I} \langle x, \varphi_i \rangle \varphi_i.$$

The operator S is a positive, self-adjoint invertible operator on $\mathscr{H}$ with $A \cdot I_{\mathscr{H}} \leq S \leq B \cdot I_{\mathscr{H}}$, where $I_{\mathscr{H}}$ denotes the identity operator on $\mathscr{H}$. In the case of a Parseval frame, this reduces to $S = I_{\mathscr{H}}$.

In general, a signal $x \in \mathscr{H}$ can be recovered from its frame coefficients through the reconstruction formula

$$x = \sum_{i \in I} \langle x, \varphi_i \rangle S^{-1} \varphi_i.$$

The sequence $(S^{-1} \varphi_i)_{i \in I}$, which can be shown to form a frame itself, is referred to as the *canonical dual frame*. Taking a different viewpoint and regarding a frame as a means for expansion in the system $(\varphi_i)_{i \in I}$, we observe that, for each vector $x \in \mathscr{H}$,

$$x = \sum_{i \in I} \langle x, S^{-1} \varphi_i \rangle \varphi_i.$$

If the frame $(\varphi_i)_{i \in I}$ does not constitute a basis, i.e., it is *redundant*, the coefficient sequence $(\langle x, S^{-1} \varphi_i \rangle)_{i \in I}$ of this expansion is certainly not unique. It is this property which then enables to derive much sparser expansions. It should also be noted that the sequence $(\langle x, S^{-1} \varphi_i \rangle)_{i \in I}$ has the distinct property of being the smallest in ℓ^2 norm of all expansion coefficient sequences.

For more details on frame theory, we refer the interested reader to [5, 7].

3.4 Wavelets

Wavelet analysis plays a central role in this volume since, as will be made more precise in the following, shearlets arise naturally from this general framework. Hence a full understanding of shearlets can only be derived through a sound understanding of wavelet theory.

We start by rewriting the definition of a *discrete wavelet system* in $L^2(\mathbb{R})$, stated at the beginning of the introduction in (2), as

$$\{\psi_{j,m} = D_2^{-j} T_m \psi = 2^{j/2} \psi(2^j \cdot -m) : j, m \in \mathbb{Z}\}, \qquad (3)$$

where $\psi \in L^2(\mathbb{R})$, D_2 is the dyadic *dilation operator* on $L^2(\mathbb{R})$ defined by

$$D_2 \psi(x) = 2^{-1/2} \psi(2^{-1}x), \qquad (4)$$

and T_t is the *translation operator* on $L^2(\mathbb{R})$, defined by

$$T_t \psi(x) = \psi(x - t), \qquad \text{for } t \in \mathbb{R}. \qquad (5)$$

The associated *Discrete Wavelet Transform* is then defined to be the mapping

$$L^2(\mathbb{R}) \ni f \mapsto \mathcal{W}_\psi f(j, m) = \langle f, \psi_{j,m} \rangle, \qquad j, m \in \mathbb{Z}.$$

If the system (3) is an orthonormal basis of $L^2(\mathbb{R})$, it is called an *orthonormal wavelet system*, and ψ is called a *wavelet*. Being a wavelet is by no means very restrictive and plenty of choices exist. In fact, it is possible to construct wavelets ψ which are *well localized*, in the sense that they have rapid decay both in the spatial and frequency domain, or which satisfy other regularity or decay requirements. Among the classical constructions, let us highlight the two most well known: the *Daubechies wavelets*, which have compact support and can be chosen to have high regularity, leading to good decay in the frequency domain; and the *Lemariè–Meyer wavelets*, which are band limited and C^∞ in the frequency domain, forcing rapid decay in the spatial domain. It should be emphasized that the localization properties of wavelet bases are among the major differences with respect to Fourier bases and play a fundamental role in their approximation properties, as we will show below.

In fact, there is a general machinery to construct orthonormal wavelet bases called *Multiresolution Analysis* (MRA). In dimension $d = 1$, this is defined as a sequence of closed subspaces $(V_j)_{j \in \mathbb{Z}}$ in $L^2(\mathbb{R})$ which satisfies the following properties:

(a) $\{0\} \subset \ldots \subset V_{-2} \subset V_{-1} \subset V_0 \subset V_1 \subset V_2 \subset \ldots \subset L^2(\mathbb{R}).$

(b) $\bigcap_{j \in \mathbb{Z}} V_j = \{0\} \quad \text{and} \quad \overline{\bigcup_{j \in \mathbb{Z}} V_j} = L^2(\mathbb{R}).$

(c) $f \in V_j \quad \text{if and only if} \quad D_2^{-1} f \in V_{j+1}.$

(d) There exists a $\phi \in L^2(\mathbb{R})$, called *scaling function*, such that $\{T_m \phi : m \in \mathbb{Z}\}$ is an orthonormal basis[2] for V_0.

This approach enables the decomposition of functions into different *resolution* levels associated with the so-called *wavelet spaces* W_j, $j \in \mathbb{Z}$. These spaces are defined by considering the orthogonal complements

$$W_j := V_{j+1} \ominus V_j, \qquad j \in \mathbb{Z}.$$

---

[2] This assumption can be replaced by the weaker assumption that $\{T_m \phi : m \in \mathbb{Z}\}$ is Riesz basis for the space V_0.

That is, a function $f_{j+1} \in V_{j+1}$ is decomposed as $f_{j+1} = f_j + g_j \in V_j \oplus W_j$, where f_j contains, roughly, the lower frequency component of f_{j+1} and g_j its higher frequency component. It follows that $L^2(\mathbb{R})$ can be broken up as a direct sum of wavelet spaces. Also, given an MRA, there always exists a function $\psi \in L^2(\mathbb{R})$ such that $\{\psi_{j,m} : j, m \in \mathbb{Z}\}$ is an orthonormal basis for $L^2(\mathbb{R})$. In fact, the MRA approach allows to introduce an alternative orthonormal basis involving both the wavelet and the scaling function, of the form

$$\{\phi_m = T_m\phi = \phi(\cdot - m) : m \in \mathbb{Z}\} \cup \{\psi_{j,m} : j \geq 0, m \in \mathbb{Z}\}.$$

In this case, the translates of the scaling function take care of the low-frequency region—the subspace $V_0 \subset L^2(\mathbb{R})$—and the wavelet terms of the high-frequency region—the complementary space $L^2(\mathbb{R}) \ominus V_0$. We refer to [65] for additional information about the theory of MRA.

The extension of wavelet theory to higher dimensions requires the introduction of some group theoretic tools. For this, it is useful to start by introducing the continuous *affine systems* of $L^2(\mathbb{R}^d)$, which are defined by

$$\left\{ \psi_{M,t} = T_t D_M^{-1} \psi = |\det M|^{1/2} \psi(M(\cdot - t)) : (M,t) \in G \times \mathbb{R}^d \right\}. \qquad (6)$$

In this definition, $\psi \in L^2(\mathbb{R}^d)$, G is a subset of $GL_d(\mathbb{R})$, the group of d-dimensional invertible matrices, D_M is the dilation operator on $L^2(\mathbb{R}^d)$, defined by

$$D_M\psi(x) = |\det M|^{-1/2}\psi(M^{-1}x), \qquad \text{for } M \in GL_d(\mathbb{R}), \qquad (7)$$

and T_t is the *translation operator* on $L^2(\mathbb{R}^d)$, defined by

$$T_t\psi(x) = \psi(x - t), \qquad \text{for } t \in \mathbb{R}^d. \qquad (8)$$

We now aim to derive conditions on ψ such that any $f \in L^2(\mathbb{R}^d)$ can be recovered from its coefficients $(\langle f, \psi_{M,t} \rangle)_{M,t}$. For this, we first equip the parameter set of (6) with a group structure by setting

$$(M,t) \cdot (M',t') = (MM', t + Mt').$$

The resulting group, typically denoted by $\mathscr{A}_d$, is the so-called *affine group on* $\mathbb{R}^d$. The mathematical structure of the affine systems becomes evident by observing that (6) can be generated by the action of the unitary representation $\pi_{(M,t)} = D_M T_t$ of $\mathscr{A}_d$ acting on $L^2(\mathbb{R}^d)$ (cf. [42] for details on the theory of group representations). Then the following result on reproducibility of functions in $L^2(\mathbb{R}^d)$ can be proven.

Theorem 1 ([29, 63]). *Retaining the notations introduced in this section, let* $d\mu$ *be a left-invariant Haar measure on* $G \subset GL_d(\mathbb{R})$, *and* $d\lambda$ *be a left Haar measure of* $\mathscr{A}_d$. *Further, suppose that* $\psi \in L^2(\mathbb{R}^d)$ *satisfies the admissibility condition*

$$\int_G |\hat{\psi}(M^T\xi)|^2 |\det M| \, d\mu(M) = 1.$$

Then any function $f \in L^2(\mathbb{R}^d)$ can be recovered via the reproducing formula

$$f = \int_{\mathscr{A}_d} \langle f, \psi_{M,t} \rangle \, \psi_{M,t} \, d\lambda(M,t),$$

interpreted weakly.

When the conditions of the above theorem are satisfied, $\psi \in L^2(\mathbb{R}^d)$ is called a *continuous wavelet*. The associated *Continuous Wavelet Transform* is defined to be the mapping

$$L^2(\mathbb{R}^d) \ni f \mapsto \mathscr{W}_\psi f(M,t) = \langle f, \psi_{M,t} \rangle, \quad (M,t) \in \mathscr{A}_d.$$

One interesting special case is obtained when the dilation group G has the form $G = \{aI_d : a > 0\}$, which corresponds to the case of *isotropic dilations*. In this case, the admissibility condition for ψ becomes

$$\int_{a>0} |\hat{\psi}(a\xi)|^2 \, \frac{da}{a} = 1,$$

and the *(isotropic) Continuous Wavelet Transform* is the mapping of $f \in L^2(\mathbb{R}^d)$ into

$$\mathscr{W}_\psi f(a,t) = a^{-d/2} \int_{\mathbb{R}^d} f(x) \overline{\psi(a^{-1}(x-t))} \, dx, \quad a > 0, t \in \mathbb{R}^d. \qquad (9)$$

Notice that the discrete wavelet systems (3) are obtained by discretizing the continuous affine systems (6) for $d = 1$, when choosing isotropic dilations with $G = \{2^j : j \in \mathbb{Z}\}$.

3.5 Wavelets for Multivariate Data and Their Limitations

The traditional theory of wavelets, which is based on the use of isotropic dilations, is essentially a one-dimensional theory. This can be illustrated by looking at the behavior of the isotropic Continuous Wavelet Transform of functions containing singularities. Indeed, consider a function or distribution f, which is regular everywhere except for a point singularity at x_0, and let us examine the behavior of $\mathscr{W}_\psi f(a,t)$, given by (9). Provided ψ is smooth, a direct computations shows that $\mathscr{W}_\psi f(a,t)$ has rapid asymptotic decay, as $a \to 0$, for all values of t, unless $t = x_0$. In this sense, the Continuous Wavelet Transform of f signals the location of the singularity through its asymptotic decay at fine scales. More generally, using this property, the Continuous Wavelet Transform can be used to characterize the singular support of a function or distribution [43].

However, due to its isotropic nature, the Continuous Wavelet Transform is unable to provide additional information about the geometry of the set of singularities of a function or distribution in terms of resolving the wavefront set. The key problem is that, although the isotropic wavelet transform has the advantage of simplicity, it lacks directional sensitivity and the ability to detect the geometry of f. The same phenomenon showing the limitation of the traditional wavelet framework can be illustrated using the Discrete Wavelet Transform.

Before doing this, let us recall the definition of *nonlinear approximation* and, in particular, the best N-term approximation, which is the proper notion of approximation in the context of wavelet bases. For a function $f \in L^2(\mathbb{R}^2)$, the *best N-term approximation* f_N of f with respect to a wavelet basis is obtained by approximating f from its N largest wavelet coefficients in magnitude—rather than from its "first" N which is the standard approach in linear Fourier approximations. Hence, denoting by Λ_N the index set corresponding to the N largest wavelet coefficients $|\langle f, \psi_\lambda \rangle|$ associated with some wavelet basis $(\psi_\lambda)_{\lambda \in \Lambda}$, the *best N-term approximation* of some $f \in L^2(\mathbb{R}^2)$ in $(\psi_\lambda)_{\lambda \in \Lambda}$ is defined as

$$f_N = \sum_{\lambda \in \Lambda_N} \langle f, \psi_\lambda \rangle \, \psi_\lambda.$$

If a function is expanded in a frame instead of a basis, the best N-term approximation can usually not be explicitly determined. A more detailed discussion of nonlinear approximation theory, encompassing the expansion in frames, is contained in chapter "Shearlets and Optimally Sparse Approximations" of this volume.

We can now present a simple heuristic argument, which highlights the limitations of traditional wavelet approximations with respect to more sophisticated multiscale methods—such as the shearlet framework—when aiming at optimally sparse approximations of cartoon-like images and other piecewise smooth functions on $\mathbb{R}^2$. Let f be a cartoon-like image (see Definition 1) containing a singularity along a smooth curve and $\{\psi_{j,m}\}$ be a standard wavelet basis of $L^2(\mathbb{R}^2)$. For j sufficiently large, the only significant wavelet coefficients $\langle f, \psi_{j,m} \rangle$ are those associated with the singularity. Since at scale 2^{-j}, each wavelet $\psi_{j,m}$ is supported or essentially supported inside a box of size $2^{-j} \times 2^{-j}$, there exist about 2^j elements of the wavelet basis overlapping the singularity curve. The associated wavelet coefficients can be controlled by

$$|\langle f, \psi_{j,m} \rangle| \leq \|f\|_\infty \|\psi_{j,m}\|_{L^1} \leq C 2^{-j}.$$

It follows that the Nth largest wavelet coefficient in magnitude, which we denote by $\langle f, \psi_{j,m} \rangle_{(N)}$, is bounded by $O(N^{-1})$. Thus, if f is approximated by its best N-term approximation f_N, the L^2 error obeys

$$\|f - f_N\|_{L^2}^2 \leq \sum_{\ell > N} |\langle f, \psi_{j,m} \rangle_{(\ell)}|^2 \leq C N^{-1}.$$

Indeed, this estimate can be proved rigorously and can be shown to be tight in the sense that there exist cartoon-like images for which the decay rate is also bounded below by $C N^{-1}$ for some constant $C > 0$ (cf. [65]).

However, the approximation rate $O(N^{-1})$ obtained using wavelet approximations is far from optimal for the class of cartoon-like images $\mathcal{E}^2(\mathbb{R}^2)$. Indeed, the following optimality result was proved in [15].

Theorem 2 ([15]). *Let $f \in \mathcal{E}^2(\mathbb{R}^2)$. There exists a constant C such that, for any N, a triangulation of $[0,1]^2$ with N triangles can be constructed so that the piecewise linear interpolation f_N of these triangles satisfies*

$$\|f - f_N\|_{L^2}^2 \leq CN^{-2}, \quad N \to \infty.$$

This result provides the optimal asymptotic decay rate of the nonlinear approximation error for objects in $\mathcal{E}^2(\mathbb{R}^2)$, in the sense that no other polynomial depth search algorithm[3] can yield a better rate. In fact, it shows that the adaptive triangle-based approximation of the image is as good as if the image had no singularities.

The approximation result from Theorem 2 provides a benchmark for optimally sparse approximation of two-dimensional data. Furthermore, the argument in the proof of Theorem 2, which uses adapted triangulations, suggests that analyzing elements with elongated and orientable supports are required to achieve optimally sparse approximations of piecewise smooth bivariate functions. Indeed, this observation is at the core of the construction of curvelets and shearlets. Notice, however, that, unlike the triangulation approximations in Theorem 2, curvelet and shearlet systems are nonadaptive. It is a remarkable fact that, even though they are nonadaptive, curvelet and shearlet representations are able to achieve (essentially) the same optimal approximation rate of Theorem 2. This result will be discussed below and, in more detail, in chapter "Shearlets and Optimally Sparse Approximations" of this volume.

4 Continuous Shearlet Systems

After discussing the limitations of wavelet systems in higher dimensions, we will now introduce shearlet systems as a general framework to overcome these limitations. We will first focus on *continuous* shearlet systems; discrete shearlet systems will be discussed next. As mentioned above, we restrict ourselves to the 2D case.

Before defining the system of shearlets in a formal way, let us introduce intuitively the ideas which are at the core of its construction. Our observations from the previous section suggest that, in order to achieve optimally sparse approximations of signals exhibiting anisotropic singularities such as cartoon-like images, the analyzing elements must consist of waveforms ranging over several scales, orientations, and locations with the ability to become very elongated. This requires a combination of an appropriate scaling operator to generate elements at different scales, an orthogonal operator to change their orientations, and a translation operator to displace these elements over the 2D plane.

---

[3] The role of the polynomial depth search condition is to limit how deep or how far down in the dictionary the algorithm is allowed to search. Without this condition, one could choose a countable dense set of $\mathcal{E}^2(\mathbb{R}^2)$ as a dictionary but this would make the search algorithm numerically impracticable. See a more detailed discussion in chapter on "Shearlets and Optimally Sparse Approximation" of this volume.

Since the scaling operator is required to generate waveforms with anisotropic support, we utilize the family of dilation operators D_{A_a}, $a > 0$, based on *parabolic scaling matrices* A_a of the form

$$A_a = \begin{pmatrix} a & 0 \\ 0 & a^{1/2} \end{pmatrix},$$

where the dilation operator is given by (7). This type of dilation corresponds to the so-called *parabolic scaling*, which has a long history in the harmonic analysis literature and can be traced back to the "second dyadic decomposition" from the theory of oscillatory integrals [24, 73] (see also the more recent work by Smith [72] on the decomposition of Fourier integral operators). It should be mentioned that, rather than A_a, the more general matrices $\mathrm{diag}(a, a^\alpha)$ with the parameter $\alpha \in (0,1)$ controlling the "degree of anisotropy" could be used. However, the value $\alpha = 1/2$ plays a special role in the discrete setting, i.e., when the parameters of the shearlet system are discretized. In fact, parabolic scaling is required in order to obtain optimally sparse approximations of cartoon-like images, since it is best adapted to the C^2-regularity of the curves of discontinuity in this model class. For simplicity, in the remainder of this chapter, we will only consider the case $\alpha = 1/2$, which is required for the sparsity results discussed below. For generalizations and extensions, we refer to chapters "Analysis and Identification of Multidimensional Singularities using the Continuous Shearlet Transform" and "Shearlets and Optimally Sparse Approximations" of this volume.

Next, we require an orthogonal transformation to change the orientations of the waveforms. The most obvious choice seems to be the rotation operator. However, rotations destroy the structure of the integer lattice $\mathbb{Z}^2$ whenever the rotation angle is different from 0, $\pm\frac{\pi}{2}$, $\pm\pi$, $\pm\frac{3\pi}{2}$. This issue becomes a serious problem for the transition from the continuum to the digital setting. As an alternative orthogonal transformation, we choose the shearing operator D_{S_s}, $s \in \mathbb{R}$, where the *shearing matrix* S_s is given by

$$S_s = \begin{pmatrix} 1 & s \\ 0 & 1 \end{pmatrix}.$$

The shearing matrix parameterizes the orientations using the variable s associated with the slopes rather than the angles, and has the advantage of leaving the integer lattice invariant, provided s is an integer.

Finally, for the translation operator we use the standard operator T_t given by (8).

Combining these three operators, we define continuous shearlet systems as follows.

Definition 2. For $\psi \in L^2(\mathbb{R}^2)$, the *continuous shearlet system* SH(ψ) is defined by

$$\mathrm{SH}(\psi) = \{\psi_{a,s,t} = T_t D_{A_a} D_{S_s} \psi : a > 0, s \in \mathbb{R}, t \in \mathbb{R}^2\}.$$

The next section will answer the question of how to choose a suitable generating function ψ so that the system SH(ψ) satisfies a reproducing formula for $L^2(\mathbb{R}^2)$.

4.1 Continuous Shearlet Systems and the Shearlet Group

One important structural property of the systems introduced in Definition 2 is their membership in the class of affine systems. Similar to the relation of wavelet systems to group representation theory discussed in Sect. 3.4, the theory of continuous shearlet systems can also be developed within the theory of unitary representations of the affine group and its generalizations [9].

To state this relation precisely, we define the so-called *shearlet group*, denoted by $\mathbb{S}$, as the semi-direct product

$$(\mathbb{R}^+ \times \mathbb{R}) \ltimes \mathbb{R}^2,$$

equipped with group multiplication given by

$$(a,s,t) \cdot (a',s',t') = (aa', s + s'\sqrt{a}, t + S_s A_a t').$$

A left-invariant Haar measure of this group is $\frac{da}{a^3} ds dt$. Letting the unitary representation $\sigma : \mathbb{S} \to \mathscr{U}(L^2(\mathbb{R}^2))$ be defined by

$$\sigma(a,s,t)\psi = T_t D_{A_a} D_{S_s} \psi,$$

where $\mathscr{U}(L^2(\mathbb{R}^2))$ denotes the group of unitary operators on $L^2(\mathbb{R}^2)$, a continuous shearlet system $SH(\psi)$ can be written as

$$SH(\psi) = \{\sigma(a,s,t)\psi : (a,s,t) \in \mathbb{S}\}.$$

The representation σ is unitary but not irreducible. If this additional property is desired, the shearlet group needs to be extended to $(\mathbb{R}^* \times \mathbb{R}) \ltimes \mathbb{R}^2$, where $\mathbb{R}^* = \mathbb{R} \setminus \{0\}$, yielding the continuous shearlet system

$$SH(\psi) = \{\sigma(a,s,t)\psi : a \in \mathbb{R}^*, s \in \mathbb{R}, t \in \mathbb{R}^2\}.$$

This point of view and its generalizations to higher dimensions will be examined in detail in chapter "Multivariate Shearlet Transform, Shearlet Coorbit Spaces and their Structural Properties" of this volume.

In the following, we provide an overview of the main results and definitions related to continuous shearlet systems for $L^2(\mathbb{R}^2)$.

4.2 The Continuous Shearlet Transform

Similar to the Continuous Wavelet Transform, the Continuous Shearlet Transform defines a mapping of $f \in L^2(\mathbb{R}^2)$ to the components of f associated with the elements of $\mathbb{S}$.

Definition 3. For $\psi \in L^2(\mathbb{R}^2)$, the *Continuous Shearlet Transform* of $f \in L^2(\mathbb{R}^2)$ is the mapping

$$L^2(\mathbb{R}^2) \ni f \to \mathscr{S}\mathscr{H}_\psi f(a,s,t) = \langle f, \sigma(a,s,t)\psi \rangle, \quad (a,s,t) \in \mathbb{S}.$$

Thus, $\mathscr{S}\mathscr{H}_\psi$ maps the function f to the coefficients $\mathscr{S}\mathscr{H}_\psi f(a,s,t)$ associated with the scale variable $a > 0$, the orientation variable $s \in \mathbb{R}$, and the location variable $t \in \mathbb{R}^2$.

Of particular importance are the conditions on ψ under which the Continuous Shearlet Transform is an isometry, since this is automatically associated with a reconstruction formula. For this, we define the notion of an *admissible* shearlet, also called *continuous shearlet*.

Definition 4. If $\psi \in L^2(\mathbb{R}^2)$ satisfies

$$\int_{\mathbb{R}^2} \frac{|\hat{\psi}(\xi_1,\xi_2)|^2}{\xi_1^2} \, d\xi_2 \, d\xi_1 < \infty,$$

it is called an *admissible shearlet*.

Notice that it is very easy to construct examples of admissible shearlets, including examples of admissible shearlets which are well localized. Essentially any function ψ such that $\hat{\psi}$ is compactly supported away from the origin is an admissible shearlet. Of particular importance is the following example, which is called *classical shearlet*. This was originally introduced in [39] and later slightly modified in [30, 61].

Definition 5. Let $\psi \in L^2(\mathbb{R}^2)$ be defined by

$$\hat{\psi}(\xi) = \hat{\psi}(\xi_1,\xi_2) = \hat{\psi}_1(\xi_1)\, \hat{\psi}_2\left(\tfrac{\xi_2}{\xi_1}\right),$$

where $\psi_1 \in L^2(\mathbb{R})$ is a discrete wavelet in the sense that it satisfies the discrete Calderón condition, given by

$$\sum_{j \in \mathbb{Z}} |\hat{\psi}_1(2^{-j}\xi)|^2 = 1 \quad \text{for a.e. } \xi \in \mathbb{R}, \tag{10}$$

with $\hat{\psi}_1 \in C^\infty(\mathbb{R})$ and $\operatorname{supp}\hat{\psi}_1 \subseteq [-\tfrac{1}{2}, -\tfrac{1}{16}] \cup [\tfrac{1}{16}, \tfrac{1}{2}]$, and $\psi_2 \in L^2(\mathbb{R})$ is a bump function in the sense that

$$\sum_{k=-1}^{1} |\hat{\psi}_2(\xi + k)|^2 = 1 \quad \text{for a.e. } \xi \in [-1,1], \tag{11}$$

satisfying $\hat{\psi}_2 \in C^\infty(\mathbb{R})$ and $\operatorname{supp}\hat{\psi}_2 \subseteq [-1,1]$. Then ψ is called a *classical shearlet*.

Thus, a classical shearlet ψ is a function which is wavelet-like along one axis and bump-like along another one. The frequency support of a classical shearlet is illustrated in Fig. 3a. Notice that there exist several choices of ψ_1 and ψ_2 satisfying conditions (10) and (11). One possible choice is to set ψ_1 to be a Lemariè–Meyer wavelet and $\hat{\psi}_2$ to be a spline (cf. [22, 31]).

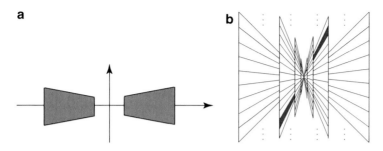

Support of the Fourier transform of a classical shearlet.

Fourier domain support of several elements of the shearlet system, for different values of a and s.

Fig. 3 Classical shearlets

The notion of admissible shearlets allows us to state sufficient conditions for a reconstruction formula in $L^2(\mathbb{R}^2)$.

Theorem 3 ([9]). *Let $\psi \in L^2(\mathbb{R}^2)$ be an admissible shearlet, and define*

$$C_\psi^+ = \int_0^\infty \int_{\mathbb{R}} \frac{|\hat{\psi}(\xi_1,\xi_2)|^2}{\xi_1^2} \, d\xi_2 \, d\xi_1 \quad and \quad C_\psi^- = \int_{-\infty}^0 \int_{\mathbb{R}} \frac{|\hat{\psi}(\xi_1,\xi_2)|^2}{\xi_1^2} \, d\xi_2 \, d\xi_1.$$

If $C_\psi^- = C_\psi^+ = 1$, then $\mathscr{SH}_\psi$ is an isometry.

Proof. By the Plancherel theorem, we obtain

$$\int_{\mathbb{S}} |\mathscr{SH}_\psi f(a,s,t)|^2 \frac{da}{a^3} \, ds \, dt$$

$$= \int_{\mathbb{S}} |f * \psi_{a,s,0}^*(t)|^2 \, dt \, ds \frac{da}{a^3}$$

$$= \int_0^\infty \int_{\mathbb{R}} \int_{\mathbb{R}^2} |\hat{f}(\xi)|^2 |\widehat{\psi_{a,s,0}^*}(\xi)|^2 \, d\xi \, ds \frac{da}{a^3}$$

$$= \int_0^\infty \int_{\mathbb{R}^2} \int_{\mathbb{R}} |\hat{f}(\xi)|^2 a^{-\frac{3}{2}} |\hat{\psi}(a\xi_1, \sqrt{a}(\xi_2 + s\xi_1))|^2 \, ds \, d\xi \, da,$$

where we used the notation $\psi^*(x) = \overline{\psi(-x)}$. By appropriate changes of variables,

$$\int_{\mathbb{S}} |\mathscr{SH}_\psi f(a,s,t)|^2 \frac{da}{a^3} \, ds \, dt$$

$$= \int_{\mathbb{R}} \int_0^\infty \int_0^\infty \int_{\mathbb{R}} |\hat{f}(\xi)|^2 a^{-2} \xi_1^{-1} |\hat{\psi}(a\xi_1, \omega_2)|^2 \, d\omega_2 \, da \, d\xi_1 d\xi_2$$

$$\quad - \int_{\mathbb{R}} \int_{-\infty}^0 \int_0^\infty \int_{\mathbb{R}} |\hat{f}(\xi)|^2 a^{-2} \xi_1^{-1} |\hat{\psi}(a\xi_1, \omega_2)|^2 \, d\omega_2 \, da \, d\xi_1 d\xi_2$$

$$= \int_{\mathbb{R}} \int_0^\infty |\hat{f}(\xi)|^2 \, d\xi_1 d\xi_2 \int_0^\infty \int_{\mathbb{R}} \frac{|\hat{\psi}(\omega_1, \omega_2)|^2}{\omega_1^2} \, d\omega_2 \, d\omega_1$$

$$\quad + \int_{\mathbb{R}} \int_{-\infty}^0 |\hat{f}(\xi)|^2 \, d\xi_1 d\xi_2 \int_{-\infty}^0 \int_{\mathbb{R}} \frac{|\hat{\psi}(\omega_1, \omega_2)|^2}{\omega_1^2} \, d\omega_2 \, d\omega_1.$$

The claim follows from here.   □

The classical shearlets, given in Definition 5, satisfy the hypothesis of admissibility, as the following result shows. The proof is straightforward; therefore we omit it.

Lemma 1 ([9]). *Let $\psi \in L^2(\mathbb{R}^2)$ be a classical shearlet. Retaining the notation from Theorem 3 we have $C_\psi^- = C_\psi^+ = 1$.*

4.3 Cone-Adapted Continuous Shearlet Systems

Although the continuous shearlet systems defined above exhibit an elegant group structure, they do have a directional bias, which is already recognizable in Fig. 3b. To illustrate the impact of this directional bias, consider a function or distribution which is mostly concentrated along the ξ_2 axis in the frequency domain. Then the energy of f is more and more concentrated in the shearlet components $\mathscr{SH}_\psi f(a,s,t)$ as $s \to \infty$. Hence, in the limiting case in which f is a delta distribution supported along the ξ_2 axis—the typical model for an edge along the x_1 axis in spatial domain—f can only be "detected" in the shearlet domain as $s \to \infty$. It is clear that this behavior can be a serious limitation for some applications.

One way to address this problem is to partition the Fourier domain into four cones, while separating the low-frequency region by cutting out a square centered around the origin. This yields a partition of the frequency plane as illustrated in Fig. 4. Notice that, within each cone, the shearing variable s is only allowed to vary over a finite range, hence producing elements whose orientations are distributed more uniformly.

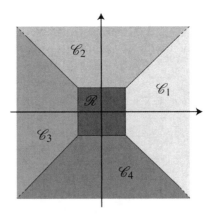

Fig. 4 Resolving the problem of biased treatment of directions by continuous shearlet systems. The frequency plane is partitioned into four cones $\mathscr{C}_i$, $i = 1,\ldots,4$, and the low-frequency box $\mathscr{R} = \{(\xi_1,\xi_2) : |\xi_1|,|\xi_2| \leq 1\}$

Thus, we define the following variant of continuous shearlet systems.

Definition 6. For $\phi, \psi, \tilde{\psi} \in L^2(\mathbb{R}^2)$, the *cone-adapted continuous shearlet system* $SH(\phi, \psi, \tilde{\psi})$ is defined by

$$SH(\phi, \psi, \tilde{\psi}) = \Phi(\phi) \cup \Psi(\psi) \cup \tilde{\Psi}(\tilde{\psi}),$$

where

$$\Phi(\phi) = \{\phi_t = \phi(\cdot - t) : t \in \mathbb{R}^2\},$$
$$\Psi(\psi) = \{\psi_{a,s,t} = a^{-\frac{3}{4}} \psi(A_a^{-1} S_s^{-1}(\cdot - t)) : a \in (0,1], |s| \leq 1 + a^{1/2}, t \in \mathbb{R}^2\},$$
$$\tilde{\Psi}(\tilde{\psi}) = \{\tilde{\psi}_{a,s,t} = a^{-\frac{3}{4}} \tilde{\psi}(\tilde{A}_a^{-1} S_s^{-T}(\cdot - t)) : a \in (0,1], |s| \leq 1 + a^{1/2}, t \in \mathbb{R}^2\},$$

and $\tilde{A}_a = \mathrm{diag}(a^{1/2}, a)$.

In the following, the function ϕ will be chosen to have compact frequency support near the origin, which ensures that the system $\Phi(\phi)$ is associated with the low-frequency region $\mathscr{R} = \{(\xi_1, \xi_2) : |\xi_1|, |\xi_2| \leq 1\}$. By choosing ψ to satisfy the conditions of Definition 5, the system $\Psi(\psi)$ is associated with the horizontal cones $\mathscr{C}_1 \cup \mathscr{C}_3 = \{(\xi_1, \xi_2) : |\xi_2/\xi_1| \leq 1, |\xi_1| > 1\}$. The shearlet $\tilde{\psi}$ can be chosen likewise with the roles of ξ_1 and ξ_2 reversed, i.e., $\tilde{\psi}(\xi_1, \xi_2) = \psi(\xi_2, \xi_1)$. Then the system $\tilde{\Psi}(\tilde{\psi})$ is associated with the vertical cones $\mathscr{C}_2 \cup \mathscr{C}_4 = \{(\xi_1, \xi_2) : |\xi_2/\xi_1| > 1, |\xi_2| > 1\}$.

4.4 The Cone-Adapted Continuous Shearlet Transform

Similar to the situation of continuous shearlet systems, an associated transform can be defined for cone-adapted continuous shearlet systems.

Definition 7. Set

$$\mathbb{S}_{\mathrm{cone}} = \{(a,s,t) : a \in (0,1], |s| \leq 1 + a^{1/2}, t \in \mathbb{R}^2\}.$$

Then, for $\phi, \psi, \tilde{\psi} \in L^2(\mathbb{R}^2)$, the *cone-adapted continuous shearlet transform* of $f \in L^2(\mathbb{R}^2)$ is the mapping

$$f \to \mathscr{SH}_{\phi,\psi,\tilde{\psi}} f(t', (a,s,t), (\tilde{a}, \tilde{s}, \tilde{t})) = (\langle f, \phi_{t'} \rangle, \langle f, \psi_{a,s,t} \rangle, \langle f, \tilde{\psi}_{\tilde{a}, \tilde{s}, \tilde{t}} \rangle),$$

where

$$(t', (a,s,t), (\tilde{a}, \tilde{s}, \tilde{t})) \in \mathbb{R}^2 \times \mathbb{S}_{\mathrm{cone}}^2.$$

Similar to the situation above, conditions on ψ, $\tilde{\psi}$, and ϕ can be formulated for which the mapping $\mathscr{SH}_{\phi,\psi,\tilde{\psi}}$ is an isometry. In fact, a similar argument to the one used in the proof of Theorem 3 yields the following result.

Theorem 4 ([52]). *Retaining the notation of Theorem 3, let* $\psi, \tilde{\psi} \in L^2(\mathbb{R}^2)$ *be admissible shearlets satisfying* $C_\psi^+ = C_\psi^- = 1$ *and* $C_{\tilde{\psi}}^+ = C_{\tilde{\psi}}^- = 1$, *respectively, and let* $\phi \in L^2(\mathbb{R}^2)$ *be such that, for a.e.* $\xi = (\xi_1, \xi_2) \in \mathbb{R}^2$,

$$|\hat{\phi}(\xi)|^2 + \chi_{\mathscr{C}_1 \cup \mathscr{C}_3}(\xi) \int_0^1 |\hat{\psi}_1(a\xi_1)|^2 \frac{da}{a} + \chi_{\mathscr{C}_2 \cup \mathscr{C}_4}(\xi) \int_0^1 |\hat{\psi}_1(a\xi_2)|^2 \frac{da}{a} = 1.$$

Then, for each $f \in L^2(\mathbb{R}^2)$,

$$\|f\|^2 = \int_{\mathbb{R}} |\langle f, T_t \phi \rangle|^2 dt + \int_{\mathbb{S}_{\text{cone}}} |\langle (\hat{f} \chi_{\mathscr{C}_1 \cup \mathscr{C}_3})^\vee, \psi_{a,s,t} \rangle|^2 \frac{da}{a^3} ds\, dt$$

$$+ \int_{\mathbb{S}_{\text{cone}}} |\langle (\hat{f} \chi_{\mathscr{C}_2 \cup \mathscr{C}_4})^\vee, \tilde{\psi}_{\tilde{a},\tilde{s},\tilde{t}} \rangle|^2 \frac{d\tilde{a}}{\tilde{a}^3} d\tilde{s}\, d\tilde{t}.$$

In this result, the function $\hat{\phi}$, $\hat{\psi}$, and $\hat{\tilde{\psi}}$ can in fact be chosen to be in $C_0^\infty(\mathbb{R}^2)$. In addition, the cone-adapted shearlet system can be designed so that the low-frequency and high-frequency parts are smoothly combined.

A more detailed analysis of the (cone-adapted) continuous shearlet transform and its generalizations can be found in [27] and in chapter "Shearlets and Microlocal Analysis" of this volume.

4.5 Microlocal Properties and Characterization of Singularities

As observed in Sect. 3.5, the Continuous Wavelet Transform is able to precisely characterize the singular support of functions and distributions. However, due to its isotropic nature, this approach fails to provide additional information about the geometry of the set of singularities in the sense of resolving the wavefront set.

In contrast to this behavior, the anisotropic shape of elements of a cone-adapted continuous shearlet system enables the Continuous Shearlet Transform to very precisely characterize the geometric properties of the set of singularities. For illustration purposes, let us examine the linear delta distribution $\mu_p(x_1, x_2) = \delta(x_1 + px_2)$, $p \in \mathbb{R}$, defined by

$$\langle \mu_p, f \rangle = \int_{\mathbb{R}} f(-px_2, x_2) dx_2,$$

as a simple model for a distributed singularity. For simplicity, we assume that $|p| \leq 1$. Letting ϕ be a scaling function and $\psi, \tilde{\psi}$ be classical shearlets, the asymptotic analysis of its cone-adapted continuous shearlet transform $\mathscr{SH}_{\phi,\psi,\tilde{\psi}}\mu_p$ shows that this transform precisely determines both the position and the orientation of the linear singularity by its decay behavior at fine scales. Specifically, we have the following result.

Proposition 1 ([52]). *Let* $t' \in \mathbb{R}^2$ *and* $(\tilde{a}, \tilde{s}, \tilde{t}) \in \mathbb{S}_{\text{cone}}$ *be a fixed value. For* $t_1 = -pt_2$ *and* $s = p$, *we have*

$$\mathscr{SH}_{\phi,\psi,\tilde{\psi}}\mu_p(t', (a,s,t), (\tilde{a},\tilde{s},\tilde{t})) \sim a^{-\frac{1}{4}} \quad \text{as } a \to 0.$$

In all other cases, $\mathscr{SH}_{\phi,\psi,\tilde{\psi}}\mu_p(t',(a,s,t),(\tilde{a},\tilde{s},\tilde{t}))$ decays rapidly as $a \to 0$; that is, for all $N \in \mathbb{N}$, there is a constant C_N such that

$$\mathscr{SH}_{\phi,\psi,\tilde{\psi}}\mu_p(t',(a,s,t),(\tilde{a},\tilde{s},\tilde{t})) \leq C_N\, a^N \quad as\ a \to 0.$$

In fact, it can be proven that the cone-adapted continuous shearlet transform precisely resolves the wavefront set for more general distributions [52, 26]. Furthermore, it can be used to provide a precise characterization of edge-discontinuities of functions of two variables. In particular, consider a function $f = \chi_B \subset L^2(\mathbb{R}^2)$, where $B \subset \mathbb{R}^2$ is a planar region with piecewise smooth boundary. Then $\mathscr{SH}_{\phi,\psi,\tilde{\psi}}f$ characterizes both the location and orientation of the boundary edge ∂B by its decay at fine scales [32, 38]. This property is very useful in applications which require the analysis or detection of edge discontinuities. For example, using these observations, a shearlet-based algorithm for edge detection and analysis was developed in [74], and related ideas were exploited to develop algorithms for the regularized inversion of the Radon transform in [6].

A more detailed discussion of these issues, including the extensions to higher dimensions, will be the content of chapters "Shearlets and Microlocal Analysis" and "Analysis and Identification of Multidimensional Singularities using the Continuous Shearlet Transform" of this volume.

5 Discrete Shearlet Systems

Starting from continuous shearlet systems defined in Definition 2, several discrete versions of shearlet systems can be constructed by an appropriate sampling of the continuous parameter set $\mathbb{S}$ or $\mathbb{S}_{\text{cone}}$. Various approaches have been suggested, aiming for discrete shearlet systems which preferably form an orthonormal basis or a tight frame for $L^2(\mathbb{R}^2)$.

One approach proposed in [8] and continued in [10] and [12] applies a powerful methodology called *coorbit theory*, which is used to derive different discretizations while ensuring frame properties. In fact, the regular shearlet frame which will be introduced in the next section can be derived using this machinery, and this approach will be further discussed in chapter "Multivariate Shearlet Transform, Shearlet Coorbit Spaces and their Structural Properties" of this volume. A different path, which also relies on the group properties of continuous shearlet systems, is taken in [50]. In this paper, a quantitative density measure for discrete subsets of the shearlet group $\mathbb{S}$ is introduced, adapted to its group multiplication, which is inspired by the well-known Beurling density for subsets of the Abelian group $\mathbb{R}^2$. These measures are shown to provide necessary conditions on the density of the sampling set for the existence of shearlet generators which yield a frame, thereby linking geometric properties of the sampling set to the frame properties of the resulting shearlet system. Notice, however, that the conditions derived using this approach are necessary but not sufficient. In a third approach [51], sufficient conditions are derived by studying the classical t_q-*equations* from the theory of wavelets. Recall that these equations are part of the sufficient conditions needed for an affine system to form a wavelet orthonormal basis or a tight frame (see [47] for a detailed discussion on this topic). Due to the close relationship between shearlet systems and affine systems discussed in Sect. 4.1, this ansatz can be transferred to the situation of cone-adapted continuous shearlet systems [49].

5.1 Discrete Shearlet Systems and Transforms

Discrete shearlet systems are formally defined by sampling continuous shearlet systems on a discrete subset of the shearlet group $\mathbb{S}$. This leads to the following definition.

Definition 8. Let $\psi \in L^2(\mathbb{R}^2)$ and $\Lambda \subseteq \mathbb{S}$. An *irregular discrete shearlet system* associated with ψ and Λ, denoted by $SH(\psi, \Lambda)$, is defined by

$$SH(\psi, \Lambda) = \{\psi_{a,s,t} = a^{-\frac{3}{4}} \psi(A_a^{-1} S_s^{-1}(\cdot - t)) : (a,s,t) \in \Lambda\}.$$

A *(regular) discrete shearlet system* associated with ψ, denoted by $SH(\psi)$, is defined by

$$SH(\psi) = \{\psi_{j,k,m} = 2^{\frac{3}{4}j} \psi(S_k A_{2^j} \cdot -m) : j,k \in \mathbb{Z}, m \in \mathbb{Z}^2\}.$$

Notice that the regular versions of discrete shearlet systems are derived from the irregular systems by choosing $\Lambda = \{(2^{-j}, -k, S_{-k}A_{2^{-j}}m) : j,k \in \mathbb{Z}, m \in \mathbb{Z}^2\}$. We also remark that, in the definition of a regular discrete shearlet system, the translation parameter is sometimes chosen to belong to $c_1\mathbb{Z} \times c_2\mathbb{Z}$ for some $(c_1, c_2) \in (\mathbb{R}^+)^2$. This provides some additional flexibility which is useful for some constructions.

Our goal is to apply shearlet systems as analysis and synthesis tools. Hence, it is of particular interest to examine the situation in which a discrete shearlet system $SH(\psi)$ forms a basis or, more generally, a frame. Similar to the wavelet case, we are particularly interested not only in finding generic generator functions ψ but also in selecting a generator ψ with special properties, e.g., regularity, vanishing moments, and compact support, so that the corresponding basis or frame of shearlets has satisfactory approximation properties. Particularly useful examples are the classical shearlets from Definition 5. As the following result shows, these shearlets generate shearlet Parseval frames for $L^2(\mathbb{R}^2)$.

Proposition 2. *Let* $\psi \in L^2(\mathbb{R}^2)$ *be a classical shearlet. Then* $SH(\psi)$ *is a Parseval frame for* $L^2(\mathbb{R}^2)$.

Proof. Using the properties of classical shearlets as stated in Definition 5, a direct computation gives that, for $a.e. \xi \in \mathbb{R}^2$,

$$\sum_{j \in \mathbb{Z}} \sum_{k \in \mathbb{Z}} |\hat{\psi}(S^T_{-k} A_{2^{-j}} \xi)|^2 = \sum_{j \in \mathbb{Z}} \sum_{k \in \mathbb{Z}} |\hat{\psi}_1(2^{-j} \xi_1)|^2 |\hat{\psi}_2(2^{j/2} \tfrac{\xi_2}{\xi_1} - k)|^2$$

$$= \sum_{j \in \mathbb{Z}} |\hat{\psi}_1(2^{-j} \xi_1)|^2 \sum_{k \in \mathbb{Z}} |\hat{\psi}_2(2^{j/2} \tfrac{\xi_2}{\xi_1} + k)|^2 = 1.$$

The claim follows immediately from this observation and the fact that $\operatorname{supp} \hat{\psi} \subset [-\frac{1}{2}, \frac{1}{2}]^2$. $\square$

Since a classical shearlet ψ is a well-localized function, Proposition 2 implies that there exit Parseval frames $SH(\psi)$ of well-localized discrete shearlets. The well localization property is critical for deriving superior approximation properties of shearlet systems and will be required for deriving optimally sparse approximations of cartoon-like images (cf. Sect. 5.4).

By removing the assumption that ψ is well localized in Definition 5, one can construct discrete shearlet systems which form not only tight frames but also orthonormal bases, as indicated in [39, 41]. This naturally raises the question whether well-localized shearlet orthonormal bases do exit. Unfortunately, the answer seems to be negative, according to the recent work in [48]. Thus, loosely speaking, a well-localized discrete shearlet system can form a frame or a tight frame but (most likely) not an orthonormal basis.

To achieve spatial domain localization, compactly supported discrete shearlet systems are required. It was recently shown that one can formulate sufficient conditions on ψ to generate a discrete shearlet frame of compactly supported functions with controllable frame bounds. This will be discussed in Sect. 5.3.

Finally, similar to the continuous case, we define a discrete shearlet transform as follows. We state this definition only for the regular case, with obvious extension to the irregular shearlet systems.

Definition 9. For $\psi \in L^2(\mathbb{R}^2)$, the *discrete shearlet transform* of $f \in L^2(\mathbb{R}^2)$ is the mapping defined by

$$f \mapsto \mathcal{SH}_\psi f(j,k,m) = \langle f, \psi_{j,k,m} \rangle, \quad (j,k,m) \in \mathbb{Z} \times \mathbb{Z} \times \mathbb{Z}^2.$$

Thus, $\mathcal{SH}_\psi$ maps the function f to the coefficients $\mathcal{SH}_\psi f(j,k,m)$ associated with the scale index j, the orientation index k, and the position index m.

5.2 Cone-Adapted Discrete Shearlet Systems and Transforms

Similar to the situation of continuous shearlet systems, discrete shearlet systems also suffer from a biased treatment of directions. As expected, this problem can be addressed by dividing the frequency plane into cones similar to Sect. 4.3. For the sake of generality, let us start by defining cone-adapted discrete shearlet systems with respect to an irregular parameter set.

Definition 10. Let $\phi, \psi, \tilde{\psi} \in L^2(\mathbb{R}^2)$, $\Delta \subset \mathbb{R}^2$, and $\Lambda, \tilde{\Lambda} \subset \mathbb{S}_{\text{cone}}$. Then the *irregular cone-adapted discrete shearlet system* $\text{SH}(\phi, \psi, \tilde{\psi}; \Delta, \Lambda, \tilde{\Lambda})$ is defined by

$$\text{SH}(\phi, \psi, \tilde{\psi}; \Delta, \Lambda, \tilde{\Lambda}) = \Phi(\phi; \Delta) \cup \Psi(\psi; \Lambda) \cup \tilde{\Psi}(\tilde{\psi}; \tilde{\Lambda}),$$

where

$$\Phi(\phi; \Delta) = \{\phi_t = \phi(\cdot - t) : t \in \Delta\},$$
$$\Psi(\psi; \Lambda) = \{\psi_{a,s,t} = a^{-\frac{3}{4}} \psi(A_a^{-1} S_s^{-1}(\cdot - t)) : (a,s,t) \in \Lambda\},$$
$$\tilde{\Psi}(\tilde{\psi}; \tilde{\Lambda}) = \{\tilde{\psi}_{a,s,t} = a^{-\frac{3}{4}} \tilde{\psi}(\tilde{A}_a^{-1} S_s^{-T}(\cdot - t)) : (a,s,t) \in \tilde{\Lambda}\}.$$

The regular variant of the cone-adapted discrete shearlet systems is much more frequently used. To allow more flexibility and enable changes to the density of the translation grid, we introduce a sampling factor $c = (c_1, c_2) \in (\mathbb{R}_+)^2$ in the translation index. Hence, we have the following definition.

Definition 11. For $\phi, \psi, \tilde{\psi} \in L^2(\mathbb{R}^2)$ and $c = (c_1, c_2) \in (\mathbb{R}_+)^2$, the *(regular) cone-adapted discrete shearlet system* $\text{SH}(\phi, \psi, \tilde{\psi}; c)$ is defined by

$$\text{SH}(\phi, \psi, \tilde{\psi}; c) = \Phi(\phi; c_1) \cup \Psi(\psi; c) \cup \tilde{\Psi}(\tilde{\psi}; c),$$

where

$$\Phi(\phi; c_1) = \{\phi_m = \phi(\cdot - c_1 m) : m \in \mathbb{Z}^2\},$$
$$\Psi(\psi; c) = \{\psi_{j,k,m} = 2^{\frac{3}{4}j} \psi(S_k A_{2^j} \cdot - M_c m) : j \geq 0, |k| \leq \lceil 2^{j/2} \rceil, m \in \mathbb{Z}^2\},$$
$$\tilde{\Psi}(\tilde{\psi}; c) = \{\tilde{\psi}_{j,k,m} = 2^{\frac{3}{4}j} \tilde{\psi}(S_k^T \tilde{A}_{2^j} \cdot - \tilde{M}_c m) : j \geq 0, |k| \leq \lceil 2^{j/2} \rceil, m \in \mathbb{Z}^2\},$$

with

$$M_c = \begin{pmatrix} c_1 & 0 \\ 0 & c_2 \end{pmatrix} \quad \text{and} \quad \tilde{M}_c = \begin{pmatrix} c_2 & 0 \\ 0 & c_1 \end{pmatrix}.$$

If $c = (1,1)$, the parameter c is omitted in the formulae above.

The generating functions ϕ will be referred to as *shearlet scaling functions* and the generating functions $\psi, \tilde{\psi}$ as *shearlet generators*. Notice that the system $\Phi(\phi; c_1)$ is associated with the low-frequency region, and the systems $\Psi(\psi; c)$ and $\tilde{\Psi}(\tilde{\psi}; c)$ are associated with the conic regions $\mathscr{C}_1 \cup \mathscr{C}_3$ and $\mathscr{C}_2 \cup \mathscr{C}_4$, respectively (cf. Fig. 4).

We already discussed the difficulties—or even the impossibility—of constructing a discrete shearlet orthonormal basis. Hence, one aims to derive Parseval frames. A first step toward this goal is the observation that a classical shearlet, according to Definition 5, is a shearlet generator of a Parseval frame for the subspace of $L^2(\mathbb{R}^2)$ of functions whose frequency support lies in the union of two cones $\mathscr{C}_1 \cup \mathscr{C}_3$.

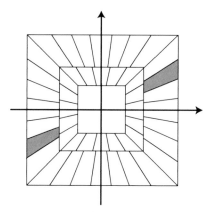

Fig. 5 Tiling of the frequency plane induced by a cone-adapted Parseval frame of shearlets

Theorem 5 ([30]). *Let* $\psi \in L^2(\mathbb{R}^2)$ *be a classical shearlet. Then the shearlet system*

$$\Psi(\psi) = \{\psi_{j,k,m} = 2^{\frac{3}{4}j}\psi(S_k A_{2^j} \cdot -m) : j \geq 0, |k| \leq \lceil 2^{j/2}\rceil, m \in \mathbb{Z}^2\}$$

is a Parseval frame for $L^2(\mathscr{C}_1 \cup \mathscr{C}_3)^\vee = \{f \in L^2(\mathbb{R}^2) : \operatorname{supp}\hat{f} \subset \mathscr{C}_1 \cup \mathscr{C}_3\}.$

Proof. Let ψ be a classical shearlet. Then (11) implies that, for any $j \geq 0$,

$$\sum_{|k| \leq \lceil 2^{j/2}\rceil} |\hat{\psi}_2(2^{j/2}\xi + k)|^2 = 1, \quad |\xi| \leq 1.$$

Thus, using this observation together with (10), a direct computation gives that, for a.e. $\xi = (\xi_1, \xi_2) \in \mathscr{C}_1 \cup \mathscr{C}_3$,

$$\sum_{j \geq 0} \sum_{|k| \leq \lceil 2^{j/2}\rceil} |\hat{\psi}(S^T_{-k} A_{2^{-j}}\xi)|^2 = \sum_{j \geq 0} \sum_{|k| \leq \lceil 2^{j/2}\rceil} |\hat{\psi}_1(2^{-j}\xi_1)|^2 |\hat{\psi}_2(2^{j/2}\tfrac{\xi_2}{\xi_1} - k)|^2$$

$$= \sum_{j \geq 0} |\hat{\psi}_1(2^{-j}\xi_1)|^2 \sum_{|k| \leq \lceil 2^{j/2}\rceil} |\hat{\psi}_2(2^{j/2}\tfrac{\xi_2}{\xi_1} + k)|^2 = 1.$$

The claim follows immediately from this observation and the fact that $\operatorname{supp}\hat{\psi} \subset [-\frac{1}{2}, \frac{1}{2}]^2$. $\quad\square$

It is clear that, if ψ is a replaced by $\tilde{\psi}$, a result very similar to Theorem 5 holds for the subspace of $L^2(\mathscr{C}_2 \cup \mathscr{C}_4)^\vee$. This indicates that one can build up a Parseval frame for the whole space $L^2(\mathbb{R}^2)$ by piecing together Parseval frames associated with different cones on the frequency domain together with a coarse scale system which takes care of the low-frequency region. Using this idea, we have the following result.

Theorem 6 ([30]). *Let $\psi \in L^2(\mathbb{R}^2)$ be a classical shearlet, and let $\phi \in L^2(\mathbb{R}^2)$ be chosen so that, for a.e. $\xi \in \mathbb{R}^2$,*

$$|\hat{\phi}(\xi)|^2 + \sum_{j \geq 0} \sum_{|k| \leq \lceil 2^{j/2} \rceil} |\hat{\psi}(S^T_{-k} A_{2^{-j}} \xi)|^2 \chi_C + \sum_{j \geq 0} \sum_{|k| \leq \lceil 2^{j/2} \rceil} |\hat{\tilde{\psi}}(S_{-k} \tilde{A}_{2^{-j}} \xi)|^2 \chi_{\tilde{C}} = 1.$$

Let $P_C \Psi(\psi)$ denote the set of shearlet elements in $\Psi(\psi)$ after projecting their Fourier transforms onto $C = \{(\xi_1, \xi_2) \in \mathbb{R}^2 : |\xi_2/\xi_1| \leq 1\}$, with a similar definition holding for $P_{\tilde{C}} \tilde{\Psi}(\tilde{\psi})$ where $\tilde{C} = \mathbb{R}^2 \setminus C$. Then, the modified cone-adapted discrete shearlet system $\Phi(\phi) \cup P_C \Psi(\psi) \cup P_{\tilde{C}} \tilde{\Psi}(\tilde{\psi})$ is a Parseval frame for $L^2(\mathbb{R}^2)$.

Notice that, despite its simplicity, the Parseval frame construction above has one drawback. When the cone-based shearlet systems are projected onto C and $\tilde{C}$, the shearlet elements overlapping the boundary lines $\xi_1 = \pm \xi_2$ in the frequency domain are cut so that the "boundary" shearlets lose their regularity properties. To avoid this problem, it is possible to redefine the "boundary" shearlets in such a way that their regularity is preserved. This requires to slightly modifying the definition of the classical shearlet. Then the boundary shearlets are obtained, essentially, by piecing together the shearlets overlapping the boundary lines $\xi_1 = \pm \xi_2$ which have been projected onto C and $\tilde{C}$. This modified construction yields smooth Parseval frames of band-limited shearlets and can be found in [37], where also the higher dimensional versions are discussed.

The tiling of the frequency plane induced by this Parseval frame of shearlets is illustrated in Fig. 5. The shearlet transform associated with regular cone-adapted discrete shearlet systems is defined as follows.

Definition 12. Set $\Lambda = \mathbb{N}_0 \times \{-\lceil 2^{j/2} \rceil, \dots, \lceil 2^{j/2} \rceil\} \times \mathbb{Z}^2$. For $\phi, \psi, \tilde{\psi} \in L^2(\mathbb{R}^2)$, the *cone-adapted discrete shearlet transform* of $f \in L^2(\mathbb{R}^2)$ is the mapping defined by

$$f \mapsto \mathscr{SH}_{\phi, \psi, \tilde{\psi}} f(m', (j, k, m), (\tilde{j}, \tilde{k}, \tilde{m})) = (\langle f, \phi_{m'} \rangle, \langle f, \psi_{j,k,m} \rangle, \langle f, \tilde{\psi}_{\tilde{j}, \tilde{k}, \tilde{m}} \rangle),$$

where

$$(m', (j, k, m), (\tilde{j}, \tilde{k}, \tilde{m})) \in \mathbb{Z}^2 \times \Lambda \times \Lambda.$$

5.3 Compactly Supported Shearlets

The shearlet systems generated by classical shearlets are band limited, i.e., they have compact support in the frequency domain and, hence, cannot be compactly supported in the spatial domain. Thus, a different approach is needed for the construction of compactly supported shearlet systems.

We start our discussion by examining sufficient conditions for the existence of cone-adapted discrete shearlet systems which are compactly supported and form a frame for $L^2(\mathbb{R}^2)$. These conditions can be derived by extending the classical t_q-equations from the theory of wavelets to this situation (cf. [47]). Before stating the main result, let us first introduce the following notation.

For functions $\phi, \psi, \tilde{\psi} \in L^2(\mathbb{R}^2)$, we define $\Theta : \mathbb{R}^2 \times \mathbb{R}^2 \to \mathbb{R}$ by

$$\Theta(\xi, \omega) = |\hat{\phi}(\xi)||\hat{\phi}(\xi + \omega)| + \Theta_1(\xi, \omega) + \Theta_2(\xi, \omega),$$

where

$$\Theta_1(\xi, \omega) = \sum_{j \geq 0} \sum_{|k| \leq \lceil 2^{j/2} \rceil} \left| \hat{\psi}(S_k^T A_{2^{-j}} \xi) \right| \left| \hat{\psi}(S_k^T A_{2^{-j}} \xi + \omega) \right|$$

and

$$\Theta_2(\xi, \omega) = \sum_{j \geq 0} \sum_{|k| \leq \lceil 2^{j/2} \rceil} \left| \hat{\tilde{\psi}}(S_k \tilde{A}_{2^{-j}} \xi) \right| \left| \hat{\tilde{\psi}}(S_k \tilde{A}_{2^{-j}} \xi + \omega) \right|.$$

Also, for $c = (c_1, c_2) \in (\mathbb{R}_+)^2$, let

$$R(c) = \sum_{m \in \mathbb{Z}^2 \setminus \{0\}} \left(\Gamma_0(c_1^{-1} m) \Gamma_0(-c_1^{-1} m) \right)^{\frac{1}{2}} + \left(\Gamma_1(M_c^{-1} m) \Gamma_1(-M_c^{-1} m) \right)^{\frac{1}{2}}$$

$$+ \left(\Gamma_2(\tilde{M}_c^{-1} m) \Gamma_2(-\tilde{M}_c^{-1} m) \right)^{\frac{1}{2}},$$

where

$$\Gamma_0(\omega) = \operatorname*{ess\,sup}_{\xi \in \mathbb{R}^2} |\hat{\phi}(\xi)||\hat{\phi}(\xi + \omega)| \quad \text{and} \quad \Gamma_i(\omega) = \operatorname*{ess\,sup}_{\xi \in \mathbb{R}^2} \Theta_i(\xi, \omega) \quad \text{for } i = 1, 2.$$

Using this notation, we can now state the following theorem from [49].

Theorem 7 ([49]). *Let* $\phi, \psi \in L^2(\mathbb{R}^2)$ *be such that*

$$\hat{\phi}(\xi_1, \xi_2) \leq C_1 \cdot \min\{1, |\xi_1|^{-\gamma}\} \cdot \min\{1, |\xi_2|^{-\gamma}\}$$

and

$$|\hat{\psi}(\xi_1, \xi_2)| \leq C_2 \cdot \min\{1, |\xi_1|^{\alpha}\} \cdot \min\{1, |\xi_1|^{-\gamma}\} \cdot \min\{1, |\xi_2|^{-\gamma}\},$$

for some positive constants $C_1, C_2 < \infty$ *and* $\alpha > \gamma > 3$. *Define* $\tilde{\psi}(x_1, x_2) = \psi(x_2, x_1)$, *and let* $L_{\inf}, L_{\sup}$ *be defined by*

$$L_{\inf} = \operatorname*{ess\,inf}_{\xi \in \mathbb{R}^2} \Theta(\xi, 0) \quad \text{and} \quad L_{\sup} = \operatorname*{ess\,sup}_{\xi \in \mathbb{R}^2} \Theta(\xi, 0).$$

Then there exists a sampling parameter $c = (c_1, c_2) \in (\mathbb{R}^+)^2$ *with* $c_1 = c_2$ *such that* $SH(\phi, \psi, \tilde{\psi}; c)$ *forms a frame for* $L^2(\mathbb{R}^2)$ *with frame bounds A and B satisfying*

$$0 < \frac{1}{|\det M_c|} [L_{\inf} - R(c)] \leq A \leq B \leq \frac{1}{|\det M_c|} [L_{\sup} + R(c)] < \infty.$$

It can be easily verified that the conditions imposed on ϕ and ψ by Theorem 7 are satisfied by many suitably chosen scaling functions and classical shearlets. In addition, one can construct various compactly supported *separable* shearlets that satisfy these conditions.

The difficulty, however, arises when aiming for compactly supported *separable* functions ϕ and ψ which ensure that the corresponding cone-adapted discrete shearlet system is a tight or almost tight frame. Separability is useful to achieve fast algorithmic implementations. In fact, it was shown in [49] that there exists a class of functions generating almost tight frames, which have (essentially) the form

$$\hat{\psi}(\xi) = m_1(4\xi_1)\hat{\phi}(\xi_1)\hat{\phi}(2\xi_2), \quad \xi = (\xi_1, \xi_2) \in \mathbb{R}^2,$$

where m_1 is a carefully chosen bandpass filter and ϕ an adaptively chosen scaling function. The proof of this fact is highly technical and will be omitted. We refer the reader to chapter "Shearlets and Optimally Sparse Approximations" of this volume and to [11, 53] for more details about compactly supported shearlets.

5.4 Sparse Approximations by Shearlets

One of the main motivations for the introduction of the shearlet framework is the derivation of optimally sparse approximations of multivariate functions. In Sect. 3.5, we presented a heuristic argument to justify why traditional wavelets are unable to take advantage of the geometry of typical functions of two variables. In fact, since traditional wavelets are not very efficient at dealing with anisotropic features, they do not provide optimally sparse approximations of images containing edges. As discussed above, shearlet systems are able to overcome these limitations.

Before stating the main results, it is enlightening to present a heuristic argument similar to the one used in Sect. 3.5, in order to describe how shearlet expansions are able to achieve optimally sparse approximations of cartoon-like images.

For this, consider a cartoon-like function f, and let $\mathrm{SH}(\phi, \psi, \tilde{\psi}; c)$ be a shearlet system. Since the elements of $\mathrm{SH}(\phi, \psi, \tilde{\psi}; c)$ are effectively or—in case of compactly supported elements—exactly supported inside a box of size $2^{-j/2} \times 2^{-j}$, it follows that at scale 2^{-j} there exist about $O(2^{j/2})$ such waveforms whose support is tangent to the curve of discontinuity. Similar to the wavelet case, for j sufficiently large, the shearlet elements associated with the smooth region of f, as well as the elements whose overlap with the curve of discontinuity is nontangential, yield negligible shearlet coefficients $\langle f, \psi_{j,k,m} \rangle$ (or $\langle f, \tilde{\psi}_{j,k,m} \rangle$). Each shearlet coefficient can be controlled by

$$|\langle f, \psi_{j,k,m} \rangle| \leq \|f\|_\infty \|\psi_{j,k,m}\|_{L^1} \leq C 2^{-3j/4},$$

similarly for $\langle f, \tilde{\psi}_{j,k,m} \rangle$. Using this estimate and the observation that there exist at most $O(2^{j/2})$ significant coefficients, we can conclude that the Nth largest shearlet coefficient, which we denote by $|s_N(f)|$, is bounded by $O(N^{-3/2})$. This implies that

$$\|f - f_N\|_{L^2}^2 \leq \sum_{\ell > N} |s_\ell(f)|^2 \leq C N^{-2},$$

where f_N denotes the N-term shearlet approximation using the N largest coefficients in the shearlets expansion. This is exactly the optimal decay rate for the upper bound stated in Theorem 2. Even though this is a simple heuristic argument, it provides an error rate which—up to a log-like factor—coincides exactly with what can be proved using a rigorous argument.

Indeed, the following result holds.

Theorem 8 ([31]). *Let $\Phi(\phi) \cup P_C \Psi(\psi) \cup P_{\tilde{C}} \tilde{\Psi}(\tilde{\psi})$ be a Parseval frame for $L^2(\mathbb{R}^2)$ as defined in Theorem 6, where $\psi \in L^2(\mathbb{R}^2)$ is a classical shearlet and $\hat{\phi} \in C_0^\infty(\mathbb{R}^2)$.*

Let $f \in \mathcal{E}^2(\mathbb{R}^2)$ and f_N be its nonlinear N-term approximation obtained by selecting the N largest coefficients in the expansion of f with respect to this shearlet system. Then there exists a constant $C > 0$, independent of f and N, such that

$$\|f - f_N\|_2^2 \le CN^{-2} (\log N)^3 \qquad as\ N \to \infty.$$

Since a log-like factor is negligible with respect to the other terms for large N, the optimal error decay rate is essentially achieved. It is remarkable that an approximation rate which is essentially as good as the one obtained using an adaptive construction can be achieved using a nonadaptive system,. The same approximation rate—with the same additional log-like factor—is obtained using a Parseval frame of curvelets, see [4].

Interestingly, the same error decay rate is also achieved using approximations based on compactly supported shearlet frames, as stated below.

Theorem 9 ([55]). *Let $\mathrm{SH}(\phi, \psi, \tilde{\psi}; c)$ be a frame for $L^2(\mathbb{R}^2)$, where $c > 0$, and $\phi, \psi, \tilde{\psi} \in L^2(\mathbb{R}^2)$ are compactly supported functions such that, for all $\xi = (\xi_1, \xi_2) \in \mathbb{R}^2$, the shearlet ψ satisfies*

(i) $|\hat{\psi}(\xi)| \le C_1 \min\{1, |\xi_1|^\alpha\} \min\{1, |\xi_1|^{-\gamma}\} \min\{1, |\xi_2|^{-\gamma}\}$ and

(ii) $\left| \frac{\partial}{\partial \xi_2} \hat{\psi}(\xi) \right| \le |h(\xi_1)| \left(1 + \frac{|\xi_2|}{|\xi_1|} \right)^{-\gamma}$,

where $\alpha > 5$, $\gamma \ge 4$, $h \in L^1(\mathbb{R})$, C_1 is a constant, and the shearlet $\tilde{\psi}$ satisfies (i) and (ii) with the roles of ξ_1 and ξ_2 reversed.

Let $f \in \mathcal{E}^2(\mathbb{R}^2)$ and f_N be its nonlinear N-term approximation obtained by selecting the N largest coefficients in the expansion of f with respect to the shearlet frame $\mathrm{SH}(\phi, \psi, \tilde{\psi}; c)$. Then there exists a constant $C > 0$, independent of f and N, such that

$$\|f - f_N\|_2^2 \le CN^{-2} (\log N)^3 \qquad as\ N \to \infty.$$

Conditions (i) and (ii) are rather mild conditions and might be regarded as a weak version of directional vanishing moment conditions.

The topic of sparse shearlet approximations, including extensions to higher dimensions, will be the main topic of chapter "Shearlets and Optimally Sparse Approximations" of this volume.

5.5 Shearlet Function Spaces

As already mentioned in Sect. 2.2, the study of the smoothness spaces associated with shearlet coefficients is particularly useful to thoroughly understand and take advantage of the approximation properties of shearlet representations. Intuitively, shearlet systems can be described as directional versions of wavelet systems. Hence, since wavelets are known to be naturally associated with Besov spaces (in the sense that Besov spaces are characterized by the decay of wavelet coefficients), it seems conceivable that shearlet systems could be effective at characterizing some anisotropic version of Besov spaces.

The theory of coorbit spaces was applied as a systematic approach toward the construction of *shearlet spaces* in the series of papers [8, 10, 11, 12]. This ansatz leads to the so-called *shearlet coorbit spaces*, which are associated with decay properties of shearlet coefficients of discrete shearlet frames. The main challenge then consists in relating these spaces to known function spaces such as Besov spaces and deriving appropriate embedding results. Chapter "Multivariate Shearlet Transform, Shearlet Coorbit Spaces and their Structural Properties" of this volume provides a thorough survey of this topic.

5.6 Extensions and Generalizations

A number of recent studies have focused on the construction of shearlet systems which are tailored to specific tasks or applications.

- *Shearlet on bounded domains.* Some applications such as the construction of numerical solvers of certain partial differential equations require systems defined on bounded domains. This could be a rectangle or, more generally, a polygonal-shaped domain. When shearlets are used for the expansion of functions—explicitly or implicitly given—defined on a bounded domain, the treatment of the boundary is crucial. One typical challenge is to set zero boundary conditions without destroying necessary (directional) vanishing moment conditions. A first attempt in this direction was undertaken in [57], but many challenges still remain.
- *Multidimensional extensions.* Many current high-impact applications such as, for example, the analysis of seismic or biological data require dealing with three-dimensional data. The computational challenges in this setting are much more demanding than in two dimensions, and sparse approximations are in great demand. Due to the simplicity of the mathematical structure of shearlets, their extensions to higher dimensions are very natural. Indeed, some basic ideas were already introduced in [41], where it was observed that there exist several ways to extend the shearing matrix to higher dimensions. A new construction yielding smooth Parseval frames of discrete shearlets in any dimensions was recently introduced in [37]. Several other results have also recently appeared, including

the extension of the optimally sparse approximation results and the analysis and detection of surface singularities [10, 33, 34, 35, 36, 54].

In three-dimensional data, different types of anisotropic features occur, namely, singularities on one-dimensional and two-dimensional manifolds. This situation is therefore very different from the situation in two dimensions, since anisotropic features of two different dimensions are involved. This is reflected in the following two main approaches to extend the parabolic scaling matrix:

$$\begin{pmatrix} 2^j & 0 & 0 \\ 0 & 2^{j/2} & 0 \\ 0 & 0 & 2^j \end{pmatrix} \quad \text{or} \quad \begin{pmatrix} 2^j & 0 & 0 \\ 0 & 2^{j/2} & 0 \\ 0 & 0 & 2^{j/2} \end{pmatrix}.$$

The first choice leads to needle-like shearlets, which are intuitively better suited to capture one-dimensional singularities. The second choice leads to plate-like shearlets, which are more suited to two-dimensional singularities. Intriguingly, both systems are needed if the goal is to distinguish these two types of singularities. However, for the construction of (nearly) optimally sparse approximations which extend the results of Sect. 5.4, it turns out that the plate-like shearlets are the right approach [34, 35, 36, 54].

These topics will be further discussed in chapters "Analysis and Identification of Multidimensional Singularities using the Continuous Shearlet Transform", "Multivariate Shearlet Transform, Shearlet Coorbit Spaces and their Structural Properties", and "Shearlets and Optimally Sparse Approximations" of this volume.

6 Algorithmic Implementations of the Shearlet Transform

One major feature of the shearlet approach is a unified treatment of the continuum and digital setting. The numerical implementations which have been developed in the literature aim—and succeed—to faithfully digitalize the discrete shearlet transform. This ensures that microlocal and approximation properties of shearlet expansions, which are proven in the continuum realm, can be carried over to the digital setting.

To date, several distinct numerical implementations of the discrete shearlet transform have been proposed [22, 46, 58, 60, 64] and some additional implementations were introduced to address specific applications such as edge detection [74]. Furthermore, several attempts were made to develop a multiresolution analysis similar to the one associated with wavelets, in an effort to develop MRA-based implementations [41, 46, 58]. We also refer to [28] for useful observations about shearlet-based numerical shearlet decompositions.

Finally, we remark that numerical shearlet algorithms are available and downloadable at the webpages www.math.uh.edu/~dlabate (associated with [22]) and www.ShearLab.org (associated with [60, 64]).

Let us next briefly describe the different approaches developed so far, by grouping these into two categories: the approaches which are Fourier-domain based and those which are spatial-domain based. All these topics will be discussed in much more detail in chapters "Shearlet Multiresolution and Multiple Refinement", "Digital Shearlet Transforms", and "Image Processing using Shearlets" of this volume.

6.1 Fourier-Based Implementations

The cone-adapted discrete shearlet transform provides a particular decomposition of the frequency plane into frequency regions associated with different scales and orientations, as illustrated in Fig. 5. Hence, a very natural and direct approach to a digitalization of the discrete shearlet transform is a *Fourier-based approach*, which aims to directly produce the same frequency tiling. This approach was adopted in the following two contributions.

- The first numerical implementation of the discrete shearlet transform was introduced in [22] as a cascade of a subband decomposition, based on the Laplacian Pyramid filter followed by a directional filtering stage which uses the Pseudo-Polar Discrete Fourier Transform.
- A different approach, which was introduced in [59, 60], consists of a carefully weighted Pseudo-Polar transform ensuring isometry followed by windowing and inverse FFT. This transform is associated with band-limited tight shearlet frames, thereby allowing the adjoint frame operator for reconstruction.

6.2 Spatial-Domain-Based Implementations

A *spatial domain approach* is a method where the filters associated with the transform are implemented by a convolution in the spatial domain. This approach is exploited from different viewpoints in the following four contributions.

- A numerical implementation of the discrete shearlet transform is presented in [22] where the directional filters are obtained as approximations of the inverse Fourier transforms of digitized band-limited window functions in the Fourier domain. With respect to the corresponding Fourier-based implementation also in [22], this alternative approach ensures that the filters have good spatial localization.
- In contrast to the method in [22], separable window functions—which allow compactly supported shearlets—are exploited in [64]. This algorithm enables the application of fast transforms separably along both axes even if the corresponding transform is not associated with a tight frame.
- Yet another approach is adopted in [58], which explores the theory of subdivision schemes, leading to an associated multiresolution analysis. The main idea here

is to adapt the construction of a multiresolution analysis for wavelets, which can also be regarded as being generated by subdivision schemes. This approach allows the possibility to obtain scaling functions "along the way."

- Related to [58], the approach developed in [46] introduces a general unitary extension principle, which—applied to the shearlet setting—determines the conditions on the filters needed for deriving a shearlet frame.

7 Shearlets in Applications

Shearlets were introduced to tackle a number of challenges in the representation and processing of multivariate data, and they have been successfully employed in several numerical applications. Let us briefly summarize the main applications below and refer to chapter "Image Processing using Shearlets" of this volume for a detailed overview.

- *Imaging Applications.* The sparsity of shearlet expansions is beneficial for various problems of data restoration and feature extraction. In particular, one class of imaging applications where shearlets have been proven very successful is image denoising problems and several shearlet-based image denoising algorithms were proposed, including those in [22, 64], which adapt wavelet thresholding to the shearlet setting, and the method in [21], which combines thresholding with minimization of bounded variation. Extensions to these ideas to video denoising were proposed in [62, 67]. Another class of imaging applications for which the microlocal properties of shearlets have been successfully exploited is the analysis and detection of edges [74].
- *Data Separation.* In several practical applications, it is important to separate data into their subcomponents. In astronomical imaging, it is very useful to separate stars from galaxies and in neurobiological imaging, spines from dendrites. In both cases, the goal is the separation of point- and curve-like structures. Using methodologies from sparse approximation and combining wavelet and shearlet expansions, a very effective method for data separation was developed in [17, 18, 56].
- *Inverse Problems.* Shearlet-based methods have also been applied to construct a regularized inversion algorithm for the Radon transform. This transform is at the basis of computerized tomography [6]. Similar ideas were also shown to be useful when dealing with more general classes of inverse problems, such as deblurring and deconvolution [69].

Acknowledgments

G.K. acknowledges support by Deutsche Forschungsgemeinschaft (DFG) Grant KU 1446/14 and DFG Grant SPP-1324 KU 1446/13 and support by the Einstein Foundation Berlin. D.L. acknowledges support by Grant DMS 1008900 and DMS (Career)

1005799. Any opinions, findings, and conclusions or recommendations expressed in this material are those of the author(s) and do not necessarily reflect the views of the National Science Foundation.

References

1. J. P. Antoine, P. Carrette, R. Murenzi, and B. Piette, *Image analysis with two-dimensional continuous wavelet transform*, Signal Process. **31** (1993), 241–272.
2. R. H. Bamberger and M. J. T. Smith, *A filter bank for the directional decomposition of images: theory and design*, IEEE Trans. Signal Process. **40** (1992), 882–893.
3. A. M. Bruckstein, D. L. Donoho, and M. Elad, *From sparse solutions of systems of equations to sparse modeling of signals and images*, SIAM Review **51** (2009), 34–81.
4. E. J. Candès and D. L. Donoho, *New tight frames of curvelets and optimal representations of objects with piecewise C^2 singularities*, Comm. Pure and Appl. Math. **56** (2004), 216–266.
5. P. G. Casazza and G. Kutyniok, *Finite Frames: Theory and Applications*, Birkhäuser, Boston, to appear.
6. F. Colonna, G. R. Easley, K. Guo, and D. Labate, *Radon transform inversion using the shearlet representation*, Appl. Comput. Harmon. Anal., **29**(2) (2010), 232–250 .
7. O. Christensen, *An Introduction to Frames and Riesz Bases*, Birkhäuser, Boston, 2003.
8. S. Dahlke, G. Kutyniok, G. Steidl, and G. Teschke, *Shearlet coorbit spaces and associated Banach frames*, Appl. Comput. Harmon. Anal. **27** (2009), 195–214.
9. S. Dahlke, G. Kutyniok, P. Maass, C. Sagiv, H.-G. Stark, and G. Teschke, *The uncertainty principle associated with the continuous shearlet transform*, Int. J. Wavelets Multiresolut. Inf. Process. **6** (2008), 157–181.
10. S. Dahlke, G. Steidl, and G. Teschke, *The continuous shearlet transform in arbitrary space dimensions*, J. Fourier Anal. Appl. **16** (2010), 340–364.
11. S. Dahlke, G. Steidl and G. Teschke, *Shearlet coorbit spaces: compactly supported analyzing shearlets, traces and embeddings*, to appear in J. Fourier Anal. Appl. **17** (2011), 1232–1255.
12. S. Dahlke and G. Teschke, *The continuous shearlet transform in higher dimensions: variations of a theme*, in Group Theory: Classes, Representation and Connections, and Applications, edited by C. W. Danellis, Math. Res. Develop., Nova Publishers, 2010, 167–175.
13. I. Daubechies, *Ten Lectures on Wavelets*, SIAM, Philadelphia, 1992.
14. M. N. Do and M. Vetterli, *The contourlet transform: an efficient directional multiresolution image representation*, IEEE Trans. Image Process. **14** (2005), 2091–2106.
15. D. L. Donoho, *Sparse components of images and optimal atomic decomposition*, Constr. Approx. **17** (2001), 353–382.
16. D. L. Donoho, *Emerging applications of geometric multiscale analysis*, Proceedings International Congress of Mathematicians **Vol. I** (2002), 209–233.
17. D. L. Donoho and G. Kutyniok, *Geometric separation using a wavelet-shearlet dictionary*, SampTA'09 (Marseille, France, 2009), Proc., 2009.
18. D. L. Donoho and G. Kutyniok, *Microlocal analysis of the geometric separation problem*, Comm. Pure Appl. Math., to appear.
19. D. L. Donoho, M. Vetterli, R. DeVore, and I. Daubechies, *Data compression and harmonic analysis*, IEEE Trans. Info. Theory **44** (1998), 2435–2476.
20. R.J. Duffin and A.C. Schaeffer, *A class of nonharmonic Fourier series*, Trans. Amer. Math. Soc. **72** (1952), 341–366.
21. G. R. Easley, D. Labate, and F. Colonna, *Shearlet based Total Variation for denoising*, IEEE Trans. Image Process. **18**(2) (2009), 260–268.
22. G. Easley, D. Labate, and W.-Q Lim, *Sparse directional image representations using the discrete shearlet transform*, Appl. Comput. Harmon. Anal. **25** (2008), 25–46.
23. M. Elad, *Sparse and Redundant Representations*, Springer, New York, 2010.

24. C. Fefferman, *A note on spherical summation multipliers*, Israel J. Math. **15** (1973), 44–52.
25. G. Folland, *Fourier Analysis and Its Applications,* American Mathematical Society, Rhode Island, 2009.
26. P. Grohs, *Continuous shearlet frames and resolution of the wavefront set*, Monatsh. Math. **164** (2011), 393–426.
27. P. Grohs, *Continuous shearlet tight frames*, J. Fourier Anal. Appl. **17** (2011), 506–518.
28. P. Grohs, Tree Approximation with anisotropic decompositions. Applied and Computational Harmonic Analysis (2011), to appear.
29. A. Grossmann, J. Morlet, and T. Paul, *Transforms associated to square integrable group representations I: General Results*, J. Math. Phys. **26** (1985), 2473–2479.
30. K. Guo, G. Kutyniok, and D. Labate, *Sparse multidimensional representations using anisotropic dilation and shear operators*, in Wavelets and Splines (Athens, GA, 2005), Nashboro Press, Nashville, TN, 2006, 189–201.
31. K. Guo and D. Labate, *Optimally sparse multidimensional representation using shearlets*, SIAM J. Math Anal. **39** (2007), 298–318.
32. K. Guo, and D. Labate, *Characterization and analysis of edges using the Continuous Shearlet Transform*, SIAM on Imaging Sciences **2** (2009), 959–986.
33. K. Guo, and D. Labate, *Analysis and detection of surface discontinuities using the 3D continuous shearlet transform*, Appl. Comput. Harmon. Anal. **30** (2011), 231–242.
34. K. Guo, and D. Labate, *Optimally sparse 3D approximations using shearlet representations*, Electronic Research Announcements in Mathematical Sciences **17** (2010), 126–138.
35. K. Guo, and D. Labate, *Optimally sparse shearlet approximations of 3D data* Proc. of SPIE Defense, Security, and Sensing (2011).
36. K. Guo, and D. Labate, *Optimally sparse representations of 3D Data with C^2 surface singularities using Parseval frames of shearlets*, SIAM J. Math Anal., to appear (2012).
37. K. Guo, and D. Labate, *The Construction of Smooth Parseval Frames of Shearlets*, preprint (2011).
38. K. Guo, D. Labate and W.-Q Lim, *Edge analysis and identification using the Continuous Shearlet Transform*, Appl. Comput. Harmon. Anal. **27** (2009), 24–46.
39. K. Guo, D. Labate, W.-Q Lim, G. Weiss, and E. Wilson, *Wavelets with composite dilations*, Electron. Res. Announc. Amer. Math. Soc. **10** (2004), 78–87.
40. K. Guo, D. Labate, W.-Q Lim, G. Weiss, and E. Wilson, *The theory of wavelets with composite dilations*, Harmonic analysis and applications, Appl. Numer. Harmon. Anal., Birkhäuser Boston, Boston, MA, 2006, 231–250.
41. K. Guo, W.-Q Lim, D. Labate, G. Weiss, and E. Wilson, *Wavelets with composite dilations and their MRA properties*, Appl. Comput. Harmon. Anal. **20** (2006), 220–236.
42. E. Hewitt and K.A. Ross, *Abstract Harmonic Analysis I, II*, Springer-Verlag, Berlin/ Heidelberg/New York, 1963.
43. M. Holschneider, *Wavelets. Analysis Tool*, Oxford University Press, Oxford, 1995.
44. N. Kingsbury, *Image processing with complex wavelets*, Phil. Trans. Royal Society London A, **357** (1999), 2543–2560.
45. N. Kingsbury, *Complex wavelets for shift invariant analysis and filtering of signals*, Appl. Computat. Harmon. Anal. **10** (2001), 234–253.
46. B. Han, G. Kutyniok, and Z. Shen. *Adaptive multiresolution analysis structures and shearlet systems*, SIAM J. Numer. Anal. **49** (2011), 1921–1946.
47. E. Hernandez and G. Weiss, *A First Course on Wavelets*, CRC, Boca Raton, FL, 1996
48. R. Houska, *The nonexistence of shearlet scaling functions*, to appear in Appl. Comput Harmon. Anal. **32** (2012), 28–44.
49. P. Kittipoom, G. Kutyniok, and W.-Q Lim, *Construction of compactly supported shearlet frames*, Constr. Approx. **35** (2012), 21–72.
50. P. Kittipoom, G. Kutyniok, and W.-Q Lim, *Irregular shearlet frames: geometry and approximation properties* J. Fourier Anal. Appl. **17** (2011), 604–639.
51. G. Kutyniok and D. Labate, *Construction of regular and irregular shearlets*, J. Wavelet Theory and Appl. **1** (2007), 1–10.

52. G. Kutyniok and D. Labate, *Resolution of the wavefront set using continuous shearlets*, Trans. Amer. Math. Soc. **361** (2009), 2719–2754.

53. G. Kutyniok, J. Lemvig, and W.-Q Lim, *Compactly supported shearlets*, in Approximation Theory XIII (San Antonio, TX, 2010), Springer Proc. Math. **13**, 163–186, Springer, 2012.

54. G. Kutyniok, J. Lemvig, and W.-Q Lim, *Compactly supported shearlet frames and optimally sparse approximations of functions in $L^2(\mathbb{R}^3)$ with piecewise C^2 singularities*, preprint.

55. G. Kutyniok and W.-Q Lim, *Compactly supported shearlets are optimally sparse*, J. Approx. Theory **163** (2011), 1564–1589.

56. G. Kutyniok and W.-Q Lim, *Image separation using wavelets and shearlets*, Curves and Surfaces (Avignon, France, 2010), Lecture Notes in Computer Science **6920**, Springer, 2012.

57. G. Kutyniok and W.-Q Lim, *Shearlets on bounded domains*, in Approximation Theory XIII (San Antonio, TX, 2010), Springer Proc. Math. **13**, 187–206, Springer, 2012.

58. G. Kutyniok and T. Sauer, *Adaptive directional subdivision schemes and shearlet multiresolution analysis*, SIAM J. Math. Anal. **41** (2009), 1436–1471.

59. G. Kutyniok, M. Shahram, and D. L. Donoho, *Development of a digital shearlet transform based on pseudo-polar FFT*, in Wavelets XIII, edited by V. K. Goyal, M. Papadakis, D. Van De Ville, SPIE Proc. **7446** (2008), SPIE, Bellingham, WA, 2009, 7446-12.

60. G. Kutyniok, M. Shahram, and X. Zhuang, *ShearLab: a rational design of a digital parabolic scaling algorithm*, preprint.

61. D. Labate, W.-Q Lim, G. Kutyniok, and G. Weiss. *Sparse multidimensional representation using shearlets*, in Wavelets XI, edited by M. Papadakis, A. F. Laine, and M. A. Unser, SPIE Proc. **5914** (2005), SPIE, Bellingham, WA, 2005, 254–262.

62. D. Labate and P. S. Negi, *3D Discrete Shearlet Transform and video denoising*, Wavelets XIV (San Diego, CA, 2011), SPIE Proc. (2011).

63. Laugesen, R. S., N. Weaver, G. Weiss, and E. Wilson, *A characterization of the higher dimensional groups associated with continuous wavelets*, J. Geom. Anal. **12** (2001), 89–102.

64. W.-Q Lim, *The discrete shearlet transform: a new directional transform and compactly supported shearlet frames*, IEEE Trans. Image Process. **19** (2010), 1166–1180.

65. Mallat, S., *A Wavelet Tour of Signal Processing*, Academic Press, San Diego 1998.

66. S. Mallat, *Geometrical Grouplets*, Appl. Comput. Harmon. Anal. **26** (2) (2009), 161–180.

67. P. S. Negi and D. Labate, *Video denoising and enhancement using the 3D Discrete Shearlet Transform*, to appear IEEE Trans. Image Process. (2012)

68. B. A. Olshausen, and D. J. Field, *Natural image statistics and efficient coding*, Network: Computation in Neural Systems **7** (1996), 333–339

69. V.M. Patel, G. Easley, D. M. Healy, *Shearlet-based deconvolution* IEEE Trans. Image Process. **18**(12) (2009), 2673-2685

70. E. L. Pennec and S. Mallat, *Sparse geometric image representations with bandelets*, IEEE Trans. Image Process. **14** (2005), 423–438.

71. E. P. Simoncelli, W. T. Freeman, E. H. Adelson, D. J. Heeger, *Shiftable multiscale transforms*, IEEE Trans. Inform. Theory **38** (1992), 587–607.

72. H. F. Smith, *A Hardy space for Fourier integral operators*, J. Geom. Anal. **8** (1998), 629–653.

73. Stein, E., *Harmonic Analysis: Real–Variable Mathods, Orthogonality and Oscillatory Integrals*, Princeton University Press, Princeton, 1993.

74. S. Yi, D. Labate, G. R. Easley, and H. Krim, *A shearlet approach to edge analysis and detection*, IEEE Trans. Image Process. **18** (2009), 929–941.

Shearlets and Microlocal Analysis

Philipp Grohs

Abstract Although wavelets are optimal for describing pointwise smoothness properties of univariate functions, they fail to efficiently characterize the subtle geometric phenomena of multidimensional singularities in high-dimensional functions. Mathematically these phenomena can be captured by the notion of the wavefront set which describes point- and direction-wise smoothness properties of tempered distributions. After familiarizing ourselves with the definition and basic properties of the wavefront set, we show that the shearlet transform offers a simple and convenient way to characterize the wavefront set in terms of the decay properties of the shearlet coefficients.

Key words: Microlocal analysis, Radon transform, Representation formulas, Wavefront set

1 Introduction

One of the main reasons for the popularity of the wavelet transform is its ability to characterize pointwise smoothness properties of functions. This property has proven to be extremely useful in both pure and applied mathematics. To give a random example we mention the beautiful work [16], where the pointwise smoothness of the Riemann function is studied with a precision that had not been achievable with other methods.

For multidimensional functions, however, pointwise smoothness does not fully capture the geometric features of the singularity set: it is also of interest in which direction the function is singular. A useful notion to capture this additional information

P. Grohs
ETH Zürich, Rämistraße 101, 8001 Zürich, Switzerland, e-mail: pgrohs@math.ethz.ch

G. Kutyniok and D. Labate (eds.), *Shearlets: Multiscale Analysis for Multivariate Data*,   39
Applied and Numerical Harmonic Analysis, DOI 10.1007/978-0-8176-8316-0_2,
© Springer Science+Business Media, LLC 2012

is the wavefront set which has been defined a literature reference to the introduction chapter of this book. It has its origins in the work of Lars Hörmander on the propagation of singularities of pseudodifferential operators [17, 21].

It turns out that the wavelet transform is unable to describe the wavefront set of a tempered distribution: even though in general the multidimensional wavelet transform *does* possess a directional parameter[1] [1], the fact that the degree of anisotropy of the wavelet elements does not change throughout different scales implies that microlocal phenomena occurring in frequency cones with small opening angles cannot be detected, compare also the discussions in [3].

The purpose of this chapter is to show that shearlets actually can describe directional smoothness properties of tempered distributions: it turns out that the wavefront set can be characterized as the point–direction pairs for which the shearlet coefficients are not of fast decay as the scale parameter tends to zero. Such a result is of great interest in both theory, since it provides a simple and elementary analysis tool to study refined notions of smoothness, and practice, where it is used for the detection of edges in images, compare chapter "Analysis and Identification of Multidimensional Singularities using the Continuous Shearlet Transform" in this volume. For real practical purposes such as analysis and classification of edges, still more refined results are needed [13, 14].

The first proof of this result has been given in [18] for "classical," bandlimited shearlets. In [9], an extension to general shearlet generators has been obtained.

1.1 Notation

Whenever possible we use the notation from chapter "Introduction to Shearlets" of this volume. We write $\mathscr{S}(\mathbb{R}^k)$ for the space of Schwartz functions [22] and $\mathscr{S}(\mathbb{R}^k)'$ for its dual w.r.t. to the canonical pairing $\langle \cdot, \cdot \rangle_{L^2(\mathbb{R}^k)}$, the space of tempered distributions. For $k = 2$, we simply write $\mathscr{S}$, $\mathscr{S}'$. We also denote

$$\mathscr{C} := \left\{ \xi \in \mathbb{R}^2 : |\xi_2|/|\xi_1| \leq 3/2 \right\}$$

and

$$\mathscr{C}' := \left\{ \xi \in \mathbb{R}^2 : |\xi_1|/|\xi_2| \leq 3/2 \right\},$$

the horizontal, resp. vertical frequency cone. For $A \subset \mathbb{R}^2$, we write χ_A for its indicator function, i.e.,

$$\chi_A(\xi) = \begin{cases} 1 & \text{if } \xi \in A \\ 0 & \text{if } \xi \notin A \end{cases}$$

The symbol $\mathbb{T}$ shall denote the one-dimensional torus.

---

[1] We thank J.-P. Antoine for pointing this out.

1.2 Getting to Know the Wavefront Set

Recall from Sect. 2.2 of chapter "Introduction to Shearlets" in this volume the definition of the wavefront set of a bivariate tempered distribution:

Definition 1. Let f be a tempered distribution on $\mathbb{R}^2$. We say that $t_0 \in \mathbb{R}^2$ is a *regular point* if there exists a neighborhood U_{t_0} of t_0 such that $\varphi f \in C^\infty$, where φ is a smooth cutoff function with $\varphi \equiv 1$ on U_{t_0}. The complement of the (open) set of regular points is called *singular support* of f and denoted

$$\text{sing supp}(f).$$

Furthermore, we call (t_0, s_0) a *regular directed point* if there exists a neighborhood U_{t_0} of t_0, a smooth cutoff function φ with $\varphi \equiv 1$ on U_{t_0}, and a neighborhood V_{s_0} of s_0 such that

$$(\varphi f)^\wedge(\eta) = O\big((1 + |\eta|)^{-N}\big) \quad \text{for all } \eta = (\eta_1, \eta_2) \text{ such that } \frac{\eta_2}{\eta_1} \in V_{s_0} \text{ and } N \in \mathbb{N}. \tag{1}$$

The *wavefront set* $\text{WF}(f)$ is the complement of the set of regular directed points.

Remark 1. The definition of the wavefront set as given in (1) excludes the case $s_0 = \infty$, or $\eta_1 = 0$. In order to avoid this problem, we can make the same definition with the coordinate directions reversed in this case. Alternatively, we can let the parameter s vary in the projective line $\mathbb{P}^1$. We will not explicitly state this in all places below but would like to remark that we can usually restrict our attention to $s \in [-1, 1]$, the other directions being handled by reversing the coordinate directions. ◇

The wavefront set is usually defined in the Fourier domain. An intuitive reason for this definition is as follows: let us assume that we are given a function with a singularity (think of a jump) in some direction. Then, if we zoom in on the singularity, all that remains are oscillations in the direction orthogonal to the singularity which corresponds to slow Fourier decay.

At first sight, this definition might not feel too natural, especially for readers with not much experience in Fourier analysis. Therefore, in order to get some feeling for this notion we first consider some examples for which we can immediately compute the wavefront sets.

Example 1. The Dirac distribution δ_t, defined by $\langle \delta_t, \varphi \rangle := \varphi(t)$ has singular support $\{t\}$. Clearly, at $x = t$ this distribution is nonregular in any direction. This is reflected by the nondecay of $\hat{\delta}_t = \exp(2\pi i \langle \cdot, t \rangle)$. It follows that $\text{WF}(\delta_t) \subset \{t\} \times \mathbb{R}$. On the other hand we have $\text{WF}(\delta_t) \supset \{t\} \times \mathbb{R}$ since δ_t is regular locally around any point $t' \neq t$. In summary, we obtain

$$\text{WF}(\delta_t) = \{t\} \times \mathbb{R}.$$

Example 2. The line distribution $\delta_{x_2=p+qx_1}$ defined by

$$\langle \delta_{x_2=p+qx_1}, \varphi \rangle := \int_{\mathbb{R}} \varphi(x_1, p+qx_1)dx_1$$

has singular support $\{(x_1,x_2) : x_2 = p + qx_1\}$. To describe the wavefront set of $\delta_{x_2=p+qx_1}$ we compute

$$\begin{aligned}
\hat{\delta}_{x_2=p+qx_1}(\xi) &= \langle \delta_{x_2=p+qx_1}, \exp(2\pi i \langle \xi, x \rangle) \rangle \\
&= \int_{\mathbb{R}} \exp(2\pi i(\xi_1 x_1 + \xi_2(p+qx_1)))\,dx_1 \\
&= e^{2\pi i p \xi_2} \int_{\mathbb{R}} \exp(2\pi i(\xi_1 + q\xi_2)x_1)\,dx_1 \\
&= e^{2\pi i p \xi_1}\,\delta_{\xi_1+q\xi_2=0}.
\end{aligned}$$

We remark that, despite the fact that the operations above do not seem to be well defined at first sight, it is possible to make them rigorous by noting that the equalities above are "in the sense of oscillatory integrals", compare [21]. Essentially, this means that the equality holds in a weak sense, which is appropriate since we are dealing with tempered distributions. It follows that $\hat{\delta}_{x_2=p+qx_1}(\xi)$ is of fast decay, except when $\xi_2/\xi_1 = -1/q$, and therefore

$$\mathrm{WF}(\delta_{x_2=p+qx_1}) = \{(x_1,x_2) : x_2 = p + qx_1\} \times \{-1/q\}.$$

Before we go to the next example we pause to introduce the Radon transform [7]. As we shall see below, it will serve us as a valuable tool in the proofs of the latter sections.

Definition 2. The Radon transform of a function f is defined by

$$\mathscr{R}f(u,s) := \int_{\mathbb{R}} f(u - sx_2, x_2)dx_2, \quad u, s \in \mathbb{R}, \tag{2}$$

whenever this expression makes sense.

Observe that our definition of the Radon transform differs from the most common one which parameterizes the directions in terms of the angle and not the slope as we do. It turns out that our definition is particularly well adapted to the mathematical structure of the shearlet transform. The next theorem already indicates that the Radon transform provides a useful tool in studying microlocal phenomena.

Theorem 1 (Projection Slice Theorem). *With $\omega \in \mathbb{R}$ and $\mathscr{F}_1$ denoting the univariate Fourier transform with respect to the first coordinate, we have the equality*

$$\mathscr{F}_1(\mathscr{R}f(u,s))(\omega) = \hat{f}(\omega(1,s)) \quad \forall s \in \mathbb{R}. \tag{3}$$

Proof.

$$\mathscr{F}_1(\mathscr{R}f(u,s))(\omega) = \int_{\mathbb{R}}\int_{\mathbb{R}} f(u - sx_2, x_2)e^{-2\pi i u\omega}dx_2 du$$

$$= \int_{\mathbb{R}}\int_{\mathbb{R}} f(\tilde{u}, x_2)e^{-2\pi i(\tilde{u} + sx_2)\omega}dx_2 d\tilde{u} = \hat{f}(\omega(1, s)).$$

□

By the Projection Slice Theorem, another way of stating that (t_0, s_0) is a regular directed point is that

$$\mathscr{F}_1(\mathscr{R}\varphi f(u,s))(\omega) = O(|\omega|^{-N}) \quad \text{and } s \in V_{s_0}, \text{ for all } N \in \mathbb{N}.$$

or in other words, that $\mathscr{R}\varphi f(u,s)$ is C^∞ in u around $s = s_0$. We can now consider the next example, the indicator function of the unit ball.

Example 3. We let $f = \chi_B$ with $B = \{(x_1, x_2) : x_1^2 + x_2^2 \leq 1\}$. Clearly we have

$$\text{sing supp}(f) = \partial B = \{(x_1, x_2) : x_1^2 + x_2^2 = 1\}.$$

In order to describe the wavefront set of f we pick a bump function φ around a point $t \in \partial B$ with $t_2/t_1 = s_0$ and look at the Radon transform

$$\mathscr{R}\varphi f(u,s) = \int_{\frac{us - \sqrt{1 + s^2 - u^2}}{1 + s^2}}^{\frac{us + \sqrt{1 + s^2 - u^2}}{1 + s^2}} \varphi(u - sx_2, x_2)dx_2. \tag{4}$$

This expression will always be zero unless

$$u \in [t_1 + st_2 - \varepsilon, t_1 + st_2 + \varepsilon]$$

with an arbitrarily small $\varepsilon > 0$ depending on the diameter of φ around t. To see this, let us assume that the function φ is supported in the set

$$(t_1 + [-\varepsilon, \varepsilon]) \times (t_2 + [-\varepsilon, \varepsilon]). \tag{5}$$

Therefore, for the expression above to be nonzero, we need to require that

$$x_2 \in t_2 + [-\varepsilon, \varepsilon]$$

which, by (5) implies that

$$u \in t_1 + st_2 + (1 + s)[-\varepsilon, \varepsilon].$$

By the definition of t we have

$$t_1 = \frac{1}{\sqrt{1 + s_0^2}}, \quad t_2 = \frac{s_0}{\sqrt{1 + s_0^2}},$$

and therefore u^2 will be close to

$$(t_1 + st_2)^2 = \frac{(1 + ss_0)^2}{1 + s_0^2}.$$

It follows that $u^2 - 1 - s^2$ is arbitrarily close to

$$\frac{(1 + ss_0)^2 - (1 + s^2)(1 + s_0^2)}{1 + s_0^2},$$

which is $\neq 0$ whenever $s \neq s_0$. But if $u^2 - 1 - s^2$ stays away from zero, by (4), the function $\mathscr{R}\varphi f$ is C^∞ and therefore (t,s) is a regular directed point for $s \neq s_0$. The same argument implies that $\mathscr{R}\varphi f$ is not smooth for $s = s_0$ and we arrive at

$$\mathrm{WF}(f) = \{(t,s) : t_1^2 + t_2^2 = 1, \ t_2 = st_1\}.$$

We hope that this last example convinced the reader that indeed the Radon transform is a useful tool for our purposes (compare [3, 18] where similar statements are shown using much less elementary tools such as Bessel functions and the method of stationary phase). It also gives a geometrical interpretation of the wavefront set: take a family of translated lines with a prescribed slope s and compute the integrals of f restricted to these lines. If these integrals do not vary smoothly with the translation parameter, then we have a point in the wavefront set.

1.2.1 Shearlets and the wavefront set

Now we will take a first look at the relation between the decay rate of the shearlet coefficients and directional regularity. Let us consider the classical bandlimited shearlet ψ described in the introduction at hand of the previous examples. First, we briefly recall the classical shearlet construction, see also Definition 1 from chapter "Introduction to Shearlets" of this volume. We pick functions $\psi_1, \psi_2 \in \mathscr{S}$ such that $\operatorname{supp} \hat{\psi}_1 \subset [-\frac{1}{2}, -\frac{1}{16}] \cup [\frac{1}{16}, \frac{1}{2}]$, $\operatorname{supp} \hat{\psi}_2 \subset [-1, 1]$ and define

$$\hat{\psi}(\xi) = \hat{\psi}_1(\xi_1)\hat{\psi}_2\left(\frac{\xi_2}{\xi_1}\right).$$

We would like to remark that the specific choices for the supports of the functions ψ_1, ψ_2 have no deeper meaning; the important thing is that $\hat{\psi}_1$ is supported away from zero (or in other words that ψ_1 is a wavelet) and $\hat{\psi}_2$ is supported around zero. In fact, all that is needed is a sufficient number of directional vanishing moments which imposes the crucial frequency localization property needed for the detection of anisotropic structures.

Definition 3. A function ψ possesses N directional vanishing moments in x_1-direction if

$$\frac{\hat{f}(\xi)}{\xi_1^N} \in L^2(\mathbb{R}^2).$$

Directional vanishing moments in other directions can be defined in a similar way.

For simplicity, in the present discussion we stick to the classical bandlimited construction and treat the general case in the following sections. We then have

$$\hat{\psi}_{a,s,t}(\xi) = a^{3/4} \exp(-2\pi i \langle t, \xi \rangle) \, \hat{\psi}\left(a\xi_1, a^{1/2}(\xi_2 - s\xi_1)\right)$$

$$= a^{3/4} \exp(-2\pi i \langle t, \xi \rangle) \, \hat{\psi}_1(a\xi_1) \hat{\psi}_2\left(a^{-1/2}\left(\frac{\xi_2}{\xi_1} - s\right)\right). \tag{6}$$

It follows that

$$\text{supp } \hat{\psi}_{a,s,t} \subset \left\{\xi \in \mathbb{R}^2 : \xi_1 \in \left[-\frac{1}{2a}, -\frac{1}{16a}\right] \cup \left[\frac{1}{16a}, \frac{1}{2a}\right], \left|\frac{\xi_2}{\xi_1} - s\right| \le \sqrt{a}\right\}. \tag{7}$$

The following two examples are inspired by the discussions in [18, Sect. 4].

Example 4. Let us consider again the distribution δ_0 and examine its shearlet coefficients. First consider the case $t = 0$ and look at

$$\langle \delta_0, \psi_{a,s,0} \rangle = \psi_{a,s,0}(0) = a^{-3/4} \psi(0).$$

On the other hand, for $t \ne 0$ we use the fact that $\psi \in \mathscr{S}$, which implies that

$$|\psi_{a,s,t}(0)| \lesssim a^{-3/4} \left(1 + \|A_a^{-1} S_s^{-1} t\|^2\right)^{-k}, \quad \forall k \in \mathbb{N}.$$

An elementary calculation shows that, e.g., for $s \in [-1, 1]$ and $t \ne 0$ this expression is of order $O(a^k)$ for all $k \in \mathbb{N}$. For other directions with slope greater than 1, we use the shearlet construction $\tilde{\Psi}(\tilde{\psi})$ for the vertical cone and apply the same analysis, see the introduction. From Example 1, it follows that the wavefront set of δ_0 is characterized precisely by the point–direction pairs for which the shearlet transform does not decay faster than any power of a:

Proposition 1. *If $t = 0$, we have*

$$\mathscr{SH}_\psi(\delta_0)(a,s,t) \sim a^{-3/4} \quad \text{as } a \to 0.$$

In all other cases $\mathscr{SH}_\psi(\delta_0)(a,s,t)$ decays rapidly as $a \to 0$.

Example 5. Here we study the behavior of the shearlet coefficients of the line singularity distribution $v = \delta_{x_2 = qx_1}$. As seen in Example 2 we have

$$\hat{v} = \hat{\delta}_{x_2 = qx_1} = \delta_{\xi_1 + q\xi_2 = 0},$$

and thus

$$\langle \psi_{a,s,t}, v \rangle = \langle \hat{\psi}_{a,s,t}, \hat{v} \rangle = \langle \hat{\psi}_{a,s,t}, \delta_{\xi_1 + q\xi_2 = 0} \rangle = \int_{\mathbb{R}} \hat{\psi}_{a,s,t}(-q\xi_2, \xi_2) d\xi_2.$$

Inserting (6) gives

$$\langle \psi_{a,s,t}, v \rangle = a^{3/4} \int_{\mathbb{R}} \exp\left(-2\pi i(-qt_1\xi_2 + t_2\xi_2)\right) \hat{\psi}_1(-aq\xi_2)\hat{\psi}_2$$
$$\times \left(a^{-1/2}\left(-\frac{1}{q} - s\right)\right) d\xi_2$$
$$= -\frac{a^{-1/4}}{q} \int_{\mathbb{R}} \exp\left(-2\pi i a^{-1}\left(t_1\xi_2 - \frac{1}{q}t_2\xi_2\right)\right) \hat{\psi}_1(\xi_2)\hat{\psi}_2$$
$$\times \left(a^{-1/2}\left(-\frac{1}{q} - s\right)\right) d\xi_2$$
$$= -\frac{a^{-1/4}}{q} \psi_1\left(a^{-1}\left(t_1 - \frac{1}{q}t_2\right)\right) \hat{\psi}_2\left(a^{-1/2}\left(-\frac{1}{q} - s\right)\right) \qquad (8)$$

Since $\psi_1 \in \mathscr{S}(\mathbb{R})$, we have

$$|\psi_1(x)| \le C_k \left(1 + x^2\right)^k, \quad \forall k \in \mathbb{N}$$

and thus

$$\psi_1\left(a^{-1}\left(t_1 - \frac{1}{q}t_2\right)\right) = O(a^k), \quad \forall k \in \mathbb{N},$$

whenever $t_1 \ne \frac{1}{q}t_2$. Let us now assume that $t_1 = \frac{1}{q}t_2$. We distinguish two cases, namely $s \ne -\frac{1}{q}$ and $s = -\frac{1}{q}$. First assume that $s \ne -\frac{1}{q}$. Then, for a sufficiently small, the expression

$$\hat{\psi}_2\left(a^{-1/2}\left(-\frac{1}{q} - s\right)\right)$$

will always be zero due to the support properties of $\hat{\psi}_2$. By (8) we again have the estimate

$$\langle v, \psi_{a,s,t} \rangle = O(a^k), \quad \forall k \in \mathbb{N}.$$

We are left with the final case $t_1 = \frac{1}{q}t_2$ and $s = -\frac{1}{q}$ where we have

$$\langle v, \psi_{a,s,t} \rangle = -\frac{a^{-1/4}}{q} \psi_1(0) \hat{\psi}_2(0) \sim a^{-1/4}.$$

Summarizing, we get

Proposition 2. *If $t_1 = \frac{1}{q}t_2$ and $s = -\frac{1}{q}$, we have*

$$\mathscr{S}\mathscr{H}_\psi(v)(a,s,t) \sim a^{-1/4}.$$

In all other cases $\mathscr{SH}_\psi(v)(a,s,t)$ decays rapidly as $a \to 0$. In other words, $WF(v)$ is given precisely by those indices (s,t) for which $\mathscr{SH}_\psi(v)$ does not decay rapidly with a.

Example 6. With a little more work, one can show the following result related to Example 3, compare [18, Sect. 4.4]. We let $f = \chi_B$ with B the two-dimensional unit ball. The following result (which we give here without proof) holds:

Proposition 3. *If $(t,s) \in WF(f)$, then we have*

$$\mathscr{SH}_\psi(f)(a,s,t) \sim a^{3/4}.$$

In all other cases $\mathscr{SH}_\psi(f)(a,s,t)$ decays rapidly as $a \to 0$.

The previous examples have shown that at least for very simple singularity distributions the wavefront set can be precisely described by the asymptotic behavior of the shearlet transform coefficients. It is the purpose of the remaining chapter to formulate and prove this fact in a more general setting. But first we take a look at a simpler transform:

1.2.2 Wavelets and the wavefront set

Also two-dimensional wavelets possess a directional parameter, and one might wonder whether the anisotropic scaling underlying the shearlet transform is really necessary. We would like to illustrate at hand of a simple example that it actually is. The discussion is inspired by [1] and [3, Sect. 6]. Similar to the one-dimensional setting, we can construct a two-dimensional wavelet transform by starting with a function $\psi \in \mathscr{S}$ satisfying the admissibility condition

$$\int_{\mathbb{R}^2} \frac{|\hat{\psi}(\xi)|^2}{|\xi|^2} d\xi < \infty.$$

Now we define functions $\psi_{a,\theta,t}$, $(a,\theta,t) \in \mathbb{R}_+ \times \mathbb{T} \times \mathbb{R}^2$ by

$$\psi_{a,\theta,t}(x) = a^{-1}\psi\left(\frac{R_\theta(x-t)}{a}\right),$$

R_θ denoting rotation by $\theta \in \mathbb{T}$. Define the two-dimensional wavelet transform of a tempered distribution $f \in \mathscr{S}'$ by

$$\mathscr{W}_\psi^{2D}(f)(a,\theta,t) := \langle f, \psi_{a,\theta,t} \rangle.$$

Using this definition it is possible to get, up to a constant, a representation formula which represents f as a superposition of the $\psi_{a,\theta,t}$'s with the corresponding coefficients given by $\mathscr{W}_\psi^{2D}(f)(a,\theta,t)$. We have

$$\int_{\mathbb{R}^2}\int_{\mathbb{T}}\int_{\mathbb{R}_+}|\mathscr{W}_\psi^{2D}(f)(a,\theta,t)|^2\frac{da}{a^3}d\theta dt = C_\psi^{2D}\int_{\mathbb{R}^2}|f(x)|^2dx,$$

with some constant C_ψ^{2D}, see [1, Proposition 2.2.1]. One possibility for the construction of ψ would be to start with a usual wavelet $\tilde{\psi}$ and define $\psi(x_1,x_2) := \tilde{\psi}(10x_1,x_2/10)$. This gives an anisotropic basis function ψ and the parameter θ in the two-dimensional wavelet transform gives a notion of directionality. A natural question to ask is whether this much simpler transform is able to characterize the wavefront set of a tempered distribution.

Example 7. Let us start with a simple point singularity δ_0. Then we have

$$\mathscr{W}_\psi^{2D}(a,\theta,t) = a^{-1}\psi\left(\frac{R_\theta(-t)}{a}\right) = O(a^k) \quad \forall k \in \mathbb{N},$$

whenever $t \neq 0$. If $t = 0$ we have

$$\mathscr{W}_\psi^{2D}(a,\theta,0) \sim a^{-1}.$$

It follows that the two-dimensional wavelet transform is able to describe the wavefront set of δ_0. ◊

Of course, in terms of describing anisotropic notions of regularity, the point distribution δ_0 is irrelevant. Let us see what happens if we analyze the simple line singularity v from Example 5.

Example 8. We consider the line distribution $v = \delta_{x_1=0}$ and its two-dimensional wavelet transform at the point $t = 0$ lying on the singularity line. We have

$$\langle v, \psi_{a,\theta,0}\rangle = \langle \hat{v}, \hat{\psi}_{a,\theta,0}\rangle = a\int_{\mathbb{R}}\hat{\psi}(a\cos(\theta)\xi_1, -a\sin(\theta)\xi_1)d\xi_1$$

$$= \int_{\mathbb{R}}\hat{\psi}(\cos(\theta)\xi_1, -\sin(\theta)\xi_1)d\xi_1 := A(\theta).$$

The function $A(\theta)$ varies smoothly with θ and in particular there is no way to sharply distinguish between the singularity direction corresponding to $\theta = 0$ and other directions nearby. In particular for θ near 0, the decay rate of $\mathscr{W}_\psi^{2D}(v)(a,\theta,0)$ is only $O(1)$. ◊

The previous example implies that for the description of truly anisotropic phenomena, the two-dimensional wavelet transform is not suitable:

Anisotropic scaling is necessary to describe anisotropic regularity!

Note that wavelets can characterize the singular support of a tempered distribution, see, e.g., [2] and the references therein.

1.3 Contributions

The main result that we would like to present is the fact that the wavefront set can be characterized by the magnitude of the shearlet coefficients as follows:

Theorem 2. *Let* $\psi \in \mathscr{S}$ *be a Schwartz function with infinitely many vanishing moments in* x_1-*direction. Let* f *be a tempered distribution and* $\mathscr{D} = \mathscr{D}_1 \cup \mathscr{D}_2$, *where* $\mathscr{D}_1 = \{(t_0, s_0) \in \mathbb{R}^2 \times [-1, 1] : \text{for } (s, t) \text{ in a neighborhood } U \text{ of } (s_0, t_0),$ $|\mathscr{SH}_\psi f(a, s, t)| = O(a^k) \text{ for all } k \in \mathbb{N}, \text{ with the implied constant uniform over } U\}$ *and* $\mathscr{D}_2 = \{(t_0, s_0) \in \mathbb{R}^2 \times (1, \infty] : \text{for } (1/s, t) \text{ in a neighborhood } U \text{ of } (s_0, t_0),$ $|\mathscr{SH}_{\tilde{\psi}} f(a, s, t)| = O(a^k) \text{ for all } k \in \mathbb{N}, \text{ with the implied constant uniform over } U\}$. *Then*

$$WF(f)^c = \mathscr{D}.$$

The proof of this result will require some preparations. In particular, we need to study continuous reconstruction formulas which allow to reconstruct an arbitrary function from its shearlet coefficients. For classical shearlet generators, such a formula is given by Theorem 3 from chapter "Introduction to Shearlets" in this volume. In Sect. 2, we develop analogous formulas for arbitrary shearlet generators. Then, using these representations, in Sect. 3 we prove our main result, Theorem 2. In addition, Fig. 2 provides an illustration of the result.

1.4 Other Ways to Characterize the Wavefront Set

The shearlet transform is not the only decomposition that is capable of characterizing the Wavefront Set. As an example, we mention the so-called FBI transform which is defined by

$$f \mapsto Tf(x, \xi, h) := \alpha_h \langle f, \exp\left(-2\pi i \|x - \cdot\|^2 / 2h\right) \exp\left(2\pi i \langle x - \cdot, \xi \rangle / h\right)\rangle,$$

where $x, y \in \mathbb{R}^2$ and h is a semiclassical parameter (see [21] for more information on semiclassical analysis) and α_h is some parameter. This transform can be interpreted as a semiclassical version of the Gabor transform [8] where the semiclassical Fourier transform is defined by

$$f \mapsto \hat{f}^h(\xi) := \int_{\mathbb{R}^2} f(x) \exp\left(2\pi i \langle x, \xi \rangle / h\right).$$

Heisenberg's uncertainty principle asserts that a time–frequency window must have area at least h. Therefore, by letting $h \to 0$ the time–frequency localization gets arbitrarily good which makes the FBI transform a useful tool in microlocal analysis. An important result is that the decay rate of $Tf(x, \xi, h)$ for $h \to 0$ determines whether the pair $(x, \xi_2/\xi_1)$ lies in the wavefront set of f [21].

Another transform which—being based on parabolic scaling—is much closer to the shearlet transform is the curvelet transform [3]. The curvelet transform is also capable of characterizing the wavefront set. Another transform based on parabolic scaling with analogous properties is the transform introduced by Hart Smith in [23].

2 Reproduction Formulas

A crucial role in the proof of Theorem 2 will be played by so-called reproduction formulas which allow to reconstruct an arbitrary function from its shearlet coefficients. The first such formula is given in [18] for classical shearlet generators and further studies can be found in [10]. We will follow this latter work in our exposition.

Example 9. To give some motivation, we mention the continuous wavelet transform which is defined by mapping a function f to its transform coefficients

$$\mathscr{W}_\psi f(a,b) := \langle f, \psi_{a,b} \rangle,$$

where

$$\psi_{a,b}(\cdot) := a^{-1/2} \psi\left(\frac{\cdot - b}{a}\right), \quad a,b \in \mathbb{R}.$$

It is well known that whenever the Calderòn condition

$$C_\psi^{\mathrm{wav}} := \int_\mathbb{R} \frac{|\hat{\psi}(\omega)|^2}{|\omega|} \, \mathrm{d}\omega < \infty$$

holds, we have the reconstruction formula

$$f = \frac{1}{C_\psi^{\mathrm{wav}}} \int_\mathbb{R} \int_\mathbb{R} \mathscr{W}_\psi(a,b) \overline{\psi_{a,b}} \frac{\mathrm{d}a}{a} \mathrm{d}b.$$

The measure $\frac{\mathrm{d}a}{a} \mathrm{d}b$ comes from the fact that the wavelet transform carries the structure of a group representation of the affine group for which this measure is the left Haar measure [15]. Another way to see why this measure is natural in the wavelet context is that the operations of dilation by a and translation by b map a unit square in (a,b)-space to a rectangle with volume a^{-1}. We also want to mention that it is not necessary to consider the wavelet transform over all scales a. Under some assumptions on ψ one can show that there exists a smooth function Φ such that

$$f = \frac{1}{C_\psi^{\mathrm{wav}}} \left(\int_\mathbb{R} \int_0^1 \mathscr{W}_\psi(a,b) \overline{\psi_{a,b}} \frac{\mathrm{d}a}{a} \mathrm{d}b + \int_\mathbb{R} \langle f, \Phi(\cdot - b) \rangle \overline{\Phi(\cdot - b)} \mathrm{d}b \right). \quad (9)$$

See [6] for more information on wavelets. ◇

We would like to find conditions for a formula similar to (9) to hold for the shearlet transform. In the case of the full shearlet transform where we have a group structure at hand, such a formula follows from standard arguments, see, e.g., chapter

"Multivariate Shearlet Transform, Shearlet Coorbit Spaces and their Structural Properties" in this volume.

Remark 2. The group structure provides us with the natural (left-) invariant measure for the shearlet transform: it is given by $\frac{da}{a^3}dsdt$. A heuristic explanation for the power of -3 in the density is the fact that this measure divides the parameter space into unit cells of side a by $\sqrt{a}$ in space (hence a factor $a^{-3/2}$), unit intervals of length $\sqrt{a}$ on the space of directions (hence, a factor $\sqrt{a}$), and finally a factor of a^{-1} since a is a scale parameter, see also [4]. ◇

In Example 9, we have seen an integral formula which is a C_ψ^{wav}-multiple of the identity. In the shearlet setting the corresponding constant arises in the following admissibility condition, compare also [5]. In the following, we will assume that ψ satisfies this condition. All the results regarding the resolution of the wavefront set also hold without this assumption, but in that case we would have to split the frequency domain into four half-cones depending on the signs of the coordinates ξ_1, ξ_2.

Definition 4. A function ψ is called *admissible* if

$$C_\psi = \int_{\mathbb{R}} \int_{-\infty}^0 \frac{|\hat{\psi}(\xi)|^2}{|\xi_1|^2} d\xi_1 d\xi_2 = \int_{\mathbb{R}} \int_0^\infty \frac{|\hat{\psi}(\xi)|^2}{|\xi_1|^2} d\xi_1 d\xi_2 < \infty. \tag{10}$$

For our purposes it is necessary that the directional parameter varies only in a compact set, otherwise the implicit constants in Theorem 2 would deteriorate. Therefore, we would like to find representations similar to (9) for the cone-adapted shearlet transform.

From now on, we will assume that the shearlet ψ possesses infinitely many directional vanishing moments, compare Definition 3. The main result is as follows:

Theorem 3. *We have the representation formula*

$$C_\psi f = \int_{\mathbb{R}^2} \int_{-2}^2 \int_0^1 \mathscr{S}\mathscr{H}_\psi f(a,s,t)\psi_{a,s,t} a^{-3} dadsdt + \int_{\mathbb{R}^2} \langle f, \Phi(\cdot - t)\rangle \Phi(\cdot - t)dt \tag{11}$$

which is valid for all $f \in L^2(\mathscr{C})^\vee$ with a smooth function Φ and C_ψ being the constant from the shearlet admissibility condition, see Definition 4. An analogous statement is true for the vertical cone $\mathscr{C}'$.

An important role in the proof of this theorem will be played by the function

$$\Delta_\psi(\xi) := \int_{-2}^2 \int_0^1 \left| \hat{\psi}\left(a\xi_1, a^{1/2}(\xi_2 - s\xi_1)\right) \right|^2 a^{-3/2} dads. \tag{12}$$

The reason for this fact is given in the next lemma:

Lemma 1. *The representation (11) holds if and only if*

$$\Delta_\psi(\xi) + |\hat{\Phi}(\xi)|^2 = C_\psi \quad \text{for all } \xi \in \mathscr{C}. \tag{13}$$

Proof. First we note that (11) is equivalent to

$$C_\psi^2 \|f\|_2^2 = \int_{\mathbb{R}^2} \int_{-2}^{2} \int_{0}^{1} |\langle f, \psi_{a,s,t}\rangle|^2 a^{-3} da\, ds\, dt$$
$$+ \int_{\mathbb{R}^2} |\langle f, \Phi(\cdot - t)\rangle|^2 dt. \tag{14}$$

This follows from polarization. Taking the Fourier transform of both sides in (14) yields

$$C_\psi \|\hat{f}\|_2^2 = \int_{\mathbb{R}^2} \int_{-2}^{2} \int_{0}^{1} |\langle \hat{f}, \hat{\psi}_{a,s,t}\rangle|^2 a^{-3} da\, ds\, dt$$
$$+ \int_{\mathbb{R}^2} |\langle \hat{f}, (\Phi(\cdot - t))^\wedge\rangle|^2 dt$$
$$= \int_{\mathbb{R}^2} \int_{-2}^{2} \int_{0}^{1} \langle \hat{f}, \hat{\psi}_{a,s,t}\rangle \overline{\langle \hat{f}, \hat{\psi}_{a,s,t}\rangle} a^{-3} da\, ds\, dt$$
$$+ \int_{\mathbb{R}^2} \langle \hat{f}, (\Phi(\cdot - t))^\wedge\rangle \overline{\langle \hat{f}, (\Phi(\cdot - t))^\wedge\rangle} dt$$

Plugging in the explicit formula for the Fourier transform lets us rewrite the above equation as follows:

$$= \int_{\mathbb{R}^2} \int_{-2}^{2} \int_{0}^{1} \int_{\mathbb{R}^2} \hat{f}(\xi) \overline{a^{3/4} e^{-2\pi i \langle t, \xi\rangle} \hat{\psi}(a\xi_1, a^{1/2}(\xi_2 - s\xi_1))} d\xi$$
$$\times \int_{\mathbb{R}^2} \overline{\hat{f}(\eta)} a^{3/4} e^{-2\pi i \langle t, \eta\rangle} \hat{\psi}(a\eta_1, a^{1/2}(\eta_2 - s\eta_1)) d\eta\, a^{-3} da\, ds\, dt$$
$$+ \int_{\mathbb{R}^2} \int_{\mathbb{R}^2} \hat{f}(\xi) \overline{e^{-2\pi i \langle t, \xi\rangle} \hat{\Phi}(\xi)} d\xi$$
$$\times \int_{\mathbb{R}^2} \overline{\hat{f}(\eta)} e^{-2\pi i \langle t, \eta\rangle} \hat{\Phi}(\eta) d\eta\, dt$$
$$= \int_{-2}^{2} \int_{0}^{1} \int_{\mathbb{R}^2} \int_{\mathbb{R}^2} \int_{\mathbb{R}^2} \exp(-2\pi i \langle \eta - \xi, t\rangle) \hat{f}(\xi) \overline{a^{3/4} \hat{\psi}(a\xi_1, a^{1/2}(\xi_2 - s\xi_1))}$$
$$\times \overline{\hat{f}(\eta)} a^{3/4} \hat{\psi}(a\eta_1, a^{1/2}(\eta_2 - s\eta_1)) d\eta\, d\xi\, dt\, a^{-3} da\, ds$$
$$+ \int_{\mathbb{R}^2} \int_{\mathbb{R}^2} \int_{\mathbb{R}^2} \exp(-2\pi i \langle \eta - \xi, t\rangle) \hat{f}(\xi) \overline{\hat{\Phi}(\xi)} \overline{\hat{f}(\eta)} \hat{\Phi}(\eta) d\xi\, d\eta\, dt.$$

An application of Parseval's formula yields

$$C_\psi \|\hat{f}\|_2^2 = \|\hat{f}\|_2^2 \left(\int_{-2}^{2} \int_{0}^{1} \left| \hat{\psi}\left(a\xi_1, a^{1/2}(\xi_2 - s\xi_1)\right) \right|^2 a^{-3/2} da\, ds + |\hat{\Phi}(\xi)|^2 \right).$$

This implies the statement.   □

Due to the previous lemma, the goal in proving Theorem 3 is to show that the (more precisely: any) function Φ defined by (13) is smooth. To this end, it suffices to show that

$$|\hat{\Phi}(\xi)|^2 = O(|\xi|^{-N}) \quad \text{for } \xi \in \mathscr{C}, \xi \to \infty.$$

Before we do this, we would like to understand the function Δ_ψ better. It turns out that if we allow to integrate over $\mathbb{R} \times \mathbb{R}_+$ instead of $[-2,2] \times [0,1]$, the integral is equal to the admissibility constant C_ψ.

Lemma 2. *We have*

$$C_\psi = \int_{\mathbb{R}} \int_{\mathbb{R}_+} \left| \hat{\psi}\left(a\xi_1, a^{1/2}(\xi_2 - s\xi_1)\right) \right|^2 a^{-3/2} da\, ds. \tag{15}$$

Proof. We make the substitution $\eta_1(a,s) = -a\xi_1, \eta_2(a,s) = a^{1/2}(\xi_2 - s\xi_1)$. The Jacobian of this substitution equals $a^{1/2}\xi_1^2 = a^{1/2}(\eta_1/a)^2 = a^{-3/2}\eta_1^2$ which shows the desired result. $\square$

Now, we can prove the Fourier decay of Φ.

Lemma 3. *We have*

$$|\hat{\Phi}(\xi)|^2 = O(|\xi|^{-N}), \quad \text{for all } N \in \mathbb{N} \text{ and } |\xi_2|/|\xi_1| \leq 3/2. \tag{16}$$

Proof. By Lemma 2, we have that

$$|\hat{\Phi}(\xi)|^2 = \left(\int_{a \in \mathbb{R}_+, |s| > 2} |\hat{\psi}(a\xi_1, \sqrt{a}(\xi_2 - s\xi_1))|^2 a^{-3/2} da\, ds \right.$$

$$\left. + \int_{a > 1, |s| < 2} |\hat{\psi}(a\xi_1, \sqrt{a}(\xi_2 - s\xi_1))|^2 a^{-3/2} da\, ds \right).$$

We will analyze these two integrals separately, starting with the second one. Due to the smoothness of ψ and the fact that s only varies in a compact set, we can estimate

$$\int_{a > 1, |s| > 2} |\hat{\psi}(a\xi_1, \sqrt{a}(\xi_2 - s\xi_1))|^2 a^{-3/2} da\, ds \lesssim \int_{a > 1} (a|\xi_1|)^{-N} a^{-3/2}$$

$$\lesssim |\xi_1|^{-N} \lesssim |\xi|^{-N}.$$

The last inequality follows since we can always estimate $|\xi_1|^{-1}$ by $|\xi|^{-1}$ due to the fact that $|\xi_2|/|\xi_1| \leq 3/2$. We turn to the estimation of

$$\int_{a \in \mathbb{R}_+, |s| > 2} |\hat{\psi}(a\xi_1, \sqrt{a}(\xi_2 - s\xi_1))|^2 a^{-3/2} da\, ds.$$

First, we treat the case $a > 1$ by estimating

$$\int_{a>1,|s|>2} |\hat{\psi}(a\xi_1, \sqrt{a}(\xi_2 - s\xi_1))|^2 a^{-3/2} dads \lesssim \int_{a>1,|s|>2} a^{-N}(\xi_2 - s\xi_1)^{-2N}$$
$$\times a^{-3/2} dads$$
$$= \int_{a>1,|s|>2} |\xi_1|^{-2N} a^{-N} |\xi_2/\xi_1 - s|^{-2N}$$
$$\times a^{-3/2} dads$$
$$\leq \int_{a>1,|s|>2} |\xi_1|^{-2N} a^{-N} |3/2 - |s||^{-2N}$$
$$\times a^{-3/2} dads \lesssim |\xi|^{-N}.$$

Now we come to the last case where we will utilize the fact that ψ possesses infinitely many moments as well as the smoothness of ψ in the second coordinate.

$$\int_{a<1,\ |s|>2} |\hat{\psi}(a\xi_1, \sqrt{a}(\xi_2 - s\xi_1))|^2 a^{-3/2} dads$$
$$\lesssim \int_{a<1,\ |s|>2} a^M |\xi_1|^M a^{-L} |\xi_2 - s\xi_1|^{-2L}$$
$$\times a^{-3/2} dads$$
$$\leq \int_{a<1,\ |s|>2} a^M |\xi_1|^{M-2L} a^{-L} |3/2 - |s||^{-2L}$$
$$\times a^{-3/2} dads$$

for any L, M, in particular for $L = N + 2$ and $M = L + 4$ which gives that

$$\int_{a<1,\ |s|>2} |\hat{\psi}(a\xi_1, \sqrt{a}(\xi_2 - s\xi_1))|^2 a^{-3/2} dads \lesssim |\xi|^{-N}.$$

Summing up all three estimates proves the lemma. □

Now we have collected all the necessary ingredients to prove Theorem 3.

Proof (of Theorem 3). By Lemma 1, all we need to show is that any Φ defined by (13) is smooth. But this is established by Lemma 3. □

Remark 3. The assumptions in Theorem 3 can be weakened considerably, see [10]. In that paper, it is also shown that it not possible to obtain useful representation formulas without first projecting to a frequency cone. In [12] slightly different continuous representation formulas are considered which are called *atomic decompositions*, see also [2, 23] where similar constructions are introduced for the curvelet transform. ◇

3 Resolution of the Wavefront Set

In this section we prove our main result, Theorem 2. The proof turns out to be rather long but nevertheless quite elementary. Intuitively it is not too surprising that the shearlet transform is capable of resolving the wavefront set since every shearlet element only interacts with frequency content which is contained in a cone that gets narrower as the scale increases. The difficult part is to overcome the technical details in making this intuition rigorous. To this end, the Radon transform will turn out to be a valuable tool.

We divide this section into three parts. In the first part, we prove one half of Theorem 2, namely, the fast decay of the shearlet coefficients corresponding to a regular directed point. This turns out to be the easier part. To prove the converse statement, we need to study the notion of the wavefront set a little more in the second part before we can tackle the full proof of Theorem 2 in the third part.

In the results that we present here, the choice of parabolic scaling is not essential—it could be replaced by any anisotropic scaling with corresponding dilation matrix diag $(a, a^\delta), 0 < \delta < 1$.

3.1 A Direct Theorem

We start by proving one half of Theorem 2, namely we show that if we are given a regular directed point of f, then only the parameter pair (s, t) corresponding to this point and direction can possibly have a large interaction with f. Such statements are usually called direct theorems (or also Jackson theorems).

Remark 4. The corresponding result for the wavelet case states that if a univariate function is smooth in a point then the wavelet coefficients of f associated with the location of that point decay fast with the scale, provided that the underlying wavelet has sufficiently many vanishing moments. The proofs in the wavelet case are considerably simpler, see e.g., [20]. ◊

Theorem 4 (Direct Theorem). *Assume that $f \in \mathscr{S}'$ and that (t_0, s_0) is a regular directed point of f. Let $\psi \in \mathscr{S}$ be a test function with infinitely many directional vanishing moments. Then there exists a neighborhood U_{t_0} of t_0 and V_{s_0} of s such that we have the decay estimate*

$$\mathscr{S}\mathscr{H}_\psi f(a, s, t) = O(a^N) \text{ for all } N \in \mathbb{N}. \tag{17}$$

Proof. In the proof, we will denote by N an unspecified and arbitrarily large integer. We can without loss of generality assume that f is already localized around t_0, i.e., $f = \varphi f$ where φ is the cutoff function from the definition of the wavefront set which equals 1 around t_0. More specifically, we prove that

$$\langle (1 - \varphi)f, \psi_{a,s,t} \rangle = O(a^N). \tag{18}$$

Since we have assumed that ψ is in the Schwartz class, we have for any $P > 0$ that

$$|\psi(x)| \lesssim (1 + |x|)^{-P} \tag{19}$$

By definition we have

$$\psi_{a,s,t}(x_1, x_2) = a^{-3/4} \psi \left(\frac{(x_1 - t_1) + s(x_2 - t_2)}{a}, \frac{x_2 - t_2}{a^{1/2}} \right). \tag{20}$$

Now we note that in computing the inner product (18) we can assume that $|x - t| > \delta$ for some $\delta > 0$ and t in a small neighborhood U_{t_0} of t_0, since $(1 - \Phi)f = 0$ around t_0. By (19), we estimate

$$
\begin{aligned}
|\psi_{a,s,t}(x)| &\lesssim a^{-3/4} \left(1 + \left| \begin{pmatrix} a^{-1} & sa^{-1} \\ 0 & a^{-1/2} \end{pmatrix} (x - t) \right| \right)^{-P} \\
&\leq a^{-3/4} \left(1 + \left\| \begin{pmatrix} a & -sa^{1/2} \\ 0 & a^{1/2} \end{pmatrix} \right\|^{-1} |x - t| \right)^{-P} \\
&\lesssim a^{-3/4} \left(1 + C(s) a^{-1/2} |x - t| \right)^{-P} = O \left(a^{-3/4 + P/2} |x - t|^{-P} \right)
\end{aligned}
$$

for $|x - t| > \delta$ and $C(s) = \left(1 + \frac{s^2}{2} + (s^2 + \frac{s^2}{4})^{1/2} \right)^{1/2}$ (compare [18, Lemma 5.2]).

Let us for now assume that f is a slowly growing function (i.e., a function with at most polynomial growth). Then we can estimate

$$
\begin{aligned}
\langle (1 - \varphi)f, \psi_{a,s,t} \rangle &\lesssim a^{-3/4 + P/2} \int_{|x - t| \geq \delta} |x - t|^{-P} |1 - \varphi(x_1, x_2)| |f(x_1, x_2)| dx_1 dx_2 \\
&= O(a^N), \tag{21}
\end{aligned}
$$

for $t \in U_{t_0}$ and P large enough, which yields (18). For a general tempered distribution f, we use the fact that f can be written as a finite superposition of terms of the form $D^\beta g$, where g has slow growth, D denotes the total differential and $\beta \in \mathbb{N}^2$ [22]. Then we can use integration by parts together with the fact that also the derivatives of ψ obey the decay property (19) to arrive at the general case.

Now, assuming that $f = \varphi f$ is localized, we go on to estimate the shearlet coefficients $|\langle f, \psi_{a,s,t} \rangle|$. To do this, we utilize the Fourier transform. Furthermore, we assume that $f \in L^2(\mathbb{R}^2)$. The general case can again be handled by repeated partial integrations, at the expense of some (finitely many) powers of a. First note that the Fourier transform of $\psi_{a,s,t}$ is given by

$$\hat{\psi}_{a,s,t}(\xi) = a^{3/4} \exp(-2\pi i \langle t, \xi \rangle) \hat{\psi} \left(a\xi_1, a^{1/2}(\xi_2 - s\xi_1) \right). \tag{22}$$

Now, pick $\frac{1}{2} < \alpha < 1$ and write

$$|\langle f, \psi_{a,s,t}\rangle| = |\langle \hat{f}, \hat{\psi}_{a,s,t}\rangle| \leq a^{3/4} \int_{\mathbb{R}^2} |\hat{f}(\xi_1,\xi_2)| \left|\hat{\psi}\left(a\xi_1, a^{1/2}(\xi_2 - s\xi_1)\right)\right| d\xi$$

$$= \underbrace{a^{3/4} \int_{|\xi_1|<a^{-\alpha}}}_{A} + \underbrace{a^{3/4} \int_{|\xi_1|>a^{-\alpha}}}_{B} . \tag{23}$$

Since ψ possesses M moments in the x_1 direction which means that

$$\hat{\psi}(\xi_1,\xi_2) = \xi_1^M \hat{\theta}(\xi_1,\xi_2)$$

with some $\theta \in L^2(\mathbb{R}^2)$, we can estimate A as

$$A = a^{3/4} \int_{|\xi_1|<a^{-\alpha}} |\hat{f}(\xi_1,\xi_2)| \left|\hat{\psi}\left(a\xi_1, a^{1/2}(\xi_2 - s\xi_1)\right)\right| d\xi$$

$$= a^{3/4} \int_{|\xi_1|<a^{-\alpha}} a^M |\xi_1|^M |\hat{f}(\xi_1,\xi_2)| \left|\hat{\theta}\left(a\xi_1, a^{1/2}(\xi_2 - s\xi_1)\right)\right| d\xi$$

$$\leq a^{M(1-\alpha)} a^{3/4} \int_{|\xi_1|<a^{-\alpha}} |\hat{f}(\xi_1,\xi_2)| \left|\hat{\theta}\left(a\xi_1, a^{1/2}(\xi_2 - s\xi_1)\right)\right| d\xi$$

$$\leq a^{(1-\alpha)M} \langle |\hat{f}|, |\hat{\theta}_{a,s,t}|\rangle \leq a^{(1-\alpha)M} \|\hat{f}\|_2 \|\hat{\theta}_{a,s,t}\|_2 = a^{(1-\alpha)M} \|f\|_2 \|\theta\|_2$$

$$= O\left(a^N\right) \tag{24}$$

with M large enough. In order to estimate B, we make the following substitution:

$$\begin{pmatrix} a & 0 \\ -a^{1/2}s & a^{1/2} \end{pmatrix} \begin{pmatrix} \xi_1 \\ \xi_2 \end{pmatrix} = \begin{pmatrix} \tilde{\xi}_1 \\ \tilde{\xi}_2 \end{pmatrix}, \quad d\xi_1 d\xi_2 = a^{-3/2} d\tilde{\xi}_1 d\tilde{\xi}_2.$$

Then

$$B = a^{-3/4} \int_{\frac{|\tilde{\xi}_1|}{a}>a^{-\alpha}} \left|\hat{f}\left(\frac{\tilde{\xi}_1}{a}, \frac{s}{a}\tilde{\xi}_1 + a^{-1/2}\tilde{\xi}_2\right)\right| \left|\hat{\psi}\left(\tilde{\xi}_1, \tilde{\xi}_2\right)\right| d\tilde{\xi}. \tag{25}$$

Now we shall use that (t_0, s_0) is a regular directed point of f. This means that there is a neighborhood $(s_0 - \varepsilon, s_0 + \varepsilon)$ such that

$$\hat{f}(\eta_1, \eta_2) \lesssim (1 + |\eta|)^{-R} \quad \text{for all } \frac{\eta_2}{\eta_1} \in (s_0 - \varepsilon, s_0 + \varepsilon). \tag{26}$$

Looking at (25), we now consider $\frac{\eta_2}{\eta_1}$ with

$$\eta_1 := \frac{\tilde{\xi}_1}{a}, \quad \eta_2 := \frac{s}{a}\tilde{\xi}_1 + a^{-1/2}\tilde{\xi}_2 \quad \text{and} \quad \frac{\tilde{\xi}_1}{a} > a^{-\alpha}.$$

and get the estimate

$$s - a^{\alpha - 1/2}\tilde{\xi}_2 \leq \frac{\eta_2}{\eta_1} = s + a^{-1/2}\tilde{\xi}_2 \frac{a}{\tilde{\xi}_1} \leq s + a^{\alpha - 1/2}\tilde{\xi}_2. \tag{27}$$

By (26), we have that

$$\left| \hat{f}\left(\frac{\tilde{\xi}_1}{a}, \frac{s}{a}\tilde{\xi}_1 + a^{-1/2}\tilde{\xi}_2 \right) \right| \lesssim \left(1 + \frac{|\tilde{\xi}_1|}{a} \right)^{-R} \tag{28}$$

for s in a neighborhood V_{s_0} of s_0, $\frac{|\tilde{\xi}_1|}{a} > a^{-\alpha}$ and $|\tilde{\xi}_2| < \varepsilon' a^{1/2-\alpha}$ for some $\varepsilon' < \varepsilon$. Now we first split the integral B according to

$$B = a^{-3/4} \int_{|\tilde{\xi}_1|/a \geq a^{-\alpha}} \left| \hat{f}\left(\tilde{\xi}_1/a, \frac{s}{a}\tilde{\xi}_1 + a^{-1/2}\tilde{\xi}_2 \right) \right| \left| \hat{\psi}\left(\tilde{\xi}_1, \tilde{\xi}_2 \right) \right| d\tilde{\xi}_1 d\tilde{\xi}_2$$

$$= \underbrace{a^{-3/4} \int_{|\tilde{\xi}_1|/a \geq a^{-\alpha}, \, |\tilde{\xi}_2| < \varepsilon' a^{1/2-\alpha}}}_{B_1} + \underbrace{a^{-3/4} \int_{|\tilde{\xi}_1|/a \geq a^{-\alpha} \, |\tilde{\xi}_2| > \varepsilon' a^{1/2-\alpha}}}_{B_2} \tag{29}$$

By (28), we can estimate B_1 according to

$$B_1 = O\left(a^{\alpha R - 3/4} \|\hat{\psi}\|_1 \right) = O\left(a^N \right) \tag{30}$$

whenever R is large enough.

It only remains to estimate B_2. For this, we will use the fact that $\frac{\partial^L}{\partial x_2^L}\psi \in L^2(\mathbb{R}^2)$. This implies that

$$B_2 \leq a^{-3/4} \int_{|\tilde{\xi}_1|/a \geq a^{-\alpha} \, |\tilde{\xi}_2| > \varepsilon' a^{1/2-\alpha}} \left| \hat{f}\left(\tilde{\xi}_1/a, \frac{s}{a}\tilde{\xi}_1 + a^{-1/2}\tilde{\xi}_2 \right) \hat{\psi}(\tilde{\xi}_1, \tilde{\xi}_2) \right| d\tilde{\xi}_1 d\tilde{\xi}_2$$

$$= a^{-3/4} \int \left| \hat{f}\left(\tilde{\xi}_1/a, \frac{s}{a}\tilde{\xi}_1 + a^{-1/2}\tilde{\xi}_2 \right) \tilde{\xi}_2^{-L} \left(\frac{\partial^L}{\partial x_2^L}\psi \right)^{\wedge} (\tilde{\xi}_1, \tilde{\xi}_2) \right| d\tilde{\xi}_1 d\tilde{\xi}_2$$

$$\leq (\varepsilon')^{-L} a^{-3/4 + (\alpha - 1/2)L} \tag{31}$$

$$\times \int_{\mathbb{R}^2} \left| \hat{f}\left(\tilde{\xi}_1/a, \frac{s}{a}\tilde{\xi}_1 + a^{-1/2}\tilde{\xi}_2 \right) \right| \left| \left(\frac{\partial^L}{\partial x_2^L}\psi \right)^{\wedge} (\tilde{\xi}_1, \tilde{\xi}_2) \right| d\tilde{\xi}_1 d\tilde{\xi}_2$$

$$= (\varepsilon')^{-L} a^{(\alpha - 1/2)L} \left| \left\langle |\hat{f}|, \left| \left(\frac{\partial^L}{\partial x_2^L}\psi_{a,s,t} \right)^{\wedge} \right| \right\rangle \right|$$

$$\leq (\varepsilon')^{-L} a^{(\alpha - 1/2)L} \|f\|_2 \left\| \frac{\partial^L}{\partial x_2^L}\psi \right\|_2 = O\left(a^N \right). \tag{32}$$

Putting together the estimates (21), (24), (30), and (32) we finally arrive at the desired conclusion. □

Remark 5. Observe that in the proof of the direct theorem, it is nowhere essential that we have parabolic scaling of a in the first and $a^{1/2}$ in the second coordinate. All the results that we present in this chapter hold equally well for any anisotropic scaling of a in the first coordinate and a^δ in the second coordinate where $0 < \delta < 1$ is arbitrary, see also the discussion at the end of [18]. This stands in contrast to the results on Fourier integral operators [12] and sparse approximation of cartoon images [11, 19], where the parabolic scaling plays an essential role. $\diamond$

3.2 Properties of the Wavefront Set

Here we prove to basic results related to the wavefront set. The first result concerns its well definedness. Recall that the definition of a regular directed point involves a localization by a bump function. The first thing we need to show is that the property of being a regular directed point does not depend on the choice of such a function. The second result concerns the frequency side and states that a point–direction pair comprises a regular directed point of f if and only if it is a regular directed point of the frequency projection of f onto a cone containing the direction of the point–direction pair. These results might seem obvious but they need to be proven, nevertheless.

We start with the first statement.

Lemma 4. *Assume that (t_0, s_0) is a regular directed point of f and φ is a test function. Then, (t_0, s_0) is a regular directed point of φf.*

Proof. Assume that (t_0, s_0) is a regular directed point of f and let ξ be such that $\xi_2/\xi_1 = s_0$ (if $\xi_1 = 0$ we need to reverse the coordinate directions, compare Remark 1). Then we can write $\xi = t e_0$ where e_0 denotes the unit vector with slope s_0 and t proportional to $|\xi|$. What we want to show is that

$$\widehat{\varphi f}(t e_0) = O\left(|t|^{-N}\right).$$

Since pointwise multiplication transforms into convolution in the Fourier domain, this is equivalent to

$$\hat{\varphi} * \hat{f}(t e_0) = \int_{\mathbb{R}^2} \hat{f}(t e_0 - \xi)\hat{\varphi}(\xi)\mathrm{d}\xi = O\left(|t|^{-N}\right). \tag{33}$$

Since (t_0, s_0) is a regular directed point, by definition there exists $0 < \delta < 1$ such that $t e_0 + B_\delta$ is still contained in the frequency cone with slopes $s \in V_{s_0}$ for all $t \in \mathbb{R}$. Here, B_δ denotes the unit ball in $\mathbb{R}^2$ with radius δ around the origin. After picking δ, we can split the integral in (33) into

$$\int_{|\xi|<\delta t} \hat{f}(t e_0 - \xi)\hat{\varphi}(\xi)\mathrm{d}\xi + \int_{|\xi|>\delta t} \hat{f}(t e_0 - \xi)\hat{\varphi}(\xi)\mathrm{d}\xi.$$

We start by estimating the first term. By assumption we then have that

$$\hat{f}(te_0 - \xi) = O\left(|te_0 - \xi|^{-N}\right) = O\left(|t|^{-N}\right),$$

and this suffices to establish that

$$\int_{|\xi|<\delta t} \hat{f}(te_0 - \xi)\hat{\varphi}(\xi)d\xi = O\left(|t|^{-N}\right).$$

Now the second term. As before in the proof of Theorem 4, we assume that $\hat{f}$ is a slowly growing function. Again, this is no restriction since any tempered distribution is a finite sum of derivatives of slowly growing functions. To get rid of the derivatives, we simply do some integrations by parts in the integral (33) and shift them to $\hat{\varphi}$. Since $\hat{\varphi}$ is still a test function, this does not do any harm. Now we can establish the second part as by estimating

$$\int_{|\xi|>\delta t} \hat{f}(te_0 - \xi)\hat{\varphi}(\xi)d\xi \lesssim \int_{|\xi|>\delta t} |te_0 - \xi|^L |\xi|^{-M} d\xi$$

$$\lesssim \int_{|\xi|>\delta t} |t|^L |\xi|^L |\xi|^{-M} d\xi$$

with M arbitrary and L the (finite) order of growth of $\hat{f}$. Picking M sufficiently large and using the fact that $|\xi| \gtrsim |t|$ we arrive at the desired estimate. □

The second basic result that we want to establish is that a frequency projection onto a cone does not affect the set of regular directed points.

Lemma 5. *Assume that (t_0, s_0) is a regular directed point of f. Let $\mathscr{C}_0$ be a cone containing the slope s_0. Then (t_0, s_0) is a regular directed point of $\hat{P}_{\mathscr{C}_0} f$, where $\hat{P}_{\mathscr{C}_0}$ denotes the frequency projection of f onto the frequency cone $\mathscr{C}_0$. The converse also holds true.*

Proof. To show this, we first assume that (t_0, s_0) is a regular directed point of f. By definition, we then can pick a bump function φ such that φf has fast Fourier decay in a frequency cone around s_0, i.e.,

$$\hat{\varphi} * \hat{f}(\xi) = O\left(|\xi|^{-N}\right), \ \xi_2/\xi_1 \in V_{s_0}.$$

By shrinking the neighborhood V_{s_0} of s_0 if necessary, we can assume without loss of generality that for some small $\delta > 0$ we have the inclusion (see Fig. 1b)

$$\left\{\eta + B_\delta|\eta| : \ \eta_2/\eta_1 \in V_{s_0}\right\} \subset \mathscr{C}_0. \tag{34}$$

The inclusion (34) implies that

$$\xi \in \mathscr{C}_0^c \Rightarrow |\eta - \xi| > \delta|\eta| \ \eta_2/\eta_1 \in V_{s_0}. \tag{35}$$

Write

$$\hat{\varphi} * \hat{f}(\eta) = \int_{\mathbb{R}^2} \chi_{\mathscr{C}_0} \hat{f}(\eta - \xi) \hat{\varphi}(\xi) \mathrm{d}\xi + \int_{\mathbb{R}^2} \chi_{\mathscr{C}_0^c} \hat{f}(\eta - \xi) \hat{\varphi}(\xi) \mathrm{d}\xi.$$

The statement is proven if we can show that

$$\int_{\mathbb{R}^2} \chi_{\mathscr{C}_0^c} \hat{f}(\eta - \xi) \hat{\varphi}(\xi) \mathrm{d}\xi = O\left(|\eta|^{-N}\right), \text{ for } \eta_2/\eta_1 \in V_{s_0}. \tag{36}$$

But this follows by writing (36) as

$$\int_{\mathbb{R}^2} \chi_{\mathscr{C}_0^c} \hat{f}(\xi) \hat{\varphi}(\eta - \xi) \mathrm{d}\xi$$

and using (35) together with the fact that

$$\hat{\varphi}(\eta - \xi) = O\left(|\eta - \xi|^{-N}\right).$$

□

The last result in particular implies that in order to study the Wavefront Set of a tempered distribution f we can restrict ourselves to studying the Wavefront Sets of the two frequency projections $\hat{P}_{\mathscr{C}} f$, $\hat{P}_{\mathscr{C}'} f$ separately. This also holds true for the shearlet coefficients of a tempered distribution:

Lemma 6. *Assume that f is a tempered distribution. Let (t_0, s_0) be a point–direction pair and $\mathscr{C}_0$ a frequency cone around the direction with slope s_0. Then we have the equivalence*

$$\mathscr{SH}_\psi f(a, s_0, t_0) = O\left(a^N\right) \Leftrightarrow \mathscr{SH}_\psi \left(\hat{P}_{\mathscr{C}_0} f\right)(a, s_0, t_0) = O\left(a^N\right).$$

Proof. By linearity of the shearlet transform, we have

$$\mathscr{SH}_\psi f(a, s_0, t_0) - \mathscr{SH}_\psi \left(\hat{P}_{\mathscr{C}_0} f\right)(a, s_0, t_0) = \mathscr{SH}_\psi \left(\hat{P}_{\mathscr{C}_0^c} f\right)(a, s_0, t_0).$$

But clearly (t_0, s_0) is a regular directed point of $\hat{P}_{\mathscr{C}_0^c} f$. Therefore, by Theorem 4 we can establish that

$$\mathscr{SH}_\psi \left(\hat{P}_{\mathscr{C}_0^c} f\right)(a, s_0, t_0) = O\left(a^N\right)$$

which proves the statement. □

3.3 Proof of the Main Result

We are almost ready to tackle the second half of Theorem 2. First we need the following localization lemma.

Lemma 7. *Consider a tempered distribution f and a smooth bump function φ which is supported in a small neighborhood V_{t_0} of $t_0 \in \mathbb{R}^2$. Let U_{t_0} be another neighborhood of t_0 with $V_{t_0} \subset\subset U_{t_0}{}^2$. Consider the function*

$$g(x) = \int_{t \in U_{t_0}^c,\, s \in [-2,2],\, a \in [0,1]} \langle f, \psi_{a,s,t} \rangle \varphi(x) \psi_{a,s,t}(x) a^{-3} \,da\,ds\,dt.$$

Then

$$\hat{g}(\xi) = O\left(|\xi|^{-N}\right), \quad \xi \in \mathscr{C}. \tag{37}$$

Proof. Consider for $s \in [-1,1]$ the Radon transform

$$I(u) := \mathscr{R}g(u,s).$$

By the projection slice theorem we need to show that

$$I^{(N)}(u) := \left(\frac{d}{du}\right)^N I \in L^1(\mathbb{R}) \tag{38}$$

which implies that

$$\omega^N \hat{I}(\omega) = \omega^N \hat{g}(\omega, s\omega) \lesssim 1,$$

and therefore since $|s| \le 1$, this implies (37). By the product rule, $I^{(N)}$ can be written as a sum of terms of the form

$$\int_{t \in U(t_0)^c,\, s \in [-2,2],\, a \in [0,1]} \langle f, \psi_{a,s,t} \rangle \int_{\mathbb{R}} \left(\frac{d}{dx_1}\right)^{N-j} \varphi(u - sx, s) a^{-j}$$

$$\left(\left(\frac{d}{dx_1}\right)^{N-j} \psi\right)_{a,s,t} (u - sx, x) \,dx\, a^{-3}\,da\,ds\,dt.$$

By the support properties of φ, the points $y := (u - sx, x)$ must lie in $V(t_0)$ for this expression to be nonzero. With the same argument as in the beginning of the proof of Theorem 4, leading to (18), we can establish that

$$\left(\left(\frac{d}{dx_1}\right)^{N-j} \psi\right)_{a,s,t} (y) = O\left(a^N |y - t|^{-N}\right). \tag{39}$$

Since we can assume that $y \in V_{t_0}$ and $t \in U_{t_0}^c$, we obtain (see Fig. 1a)

$$|y - t| \gtrsim |t - t_0|, \tag{40}$$

which, together with (39), establishes the desired claim. Note that by Fubini's theorem the application of the Radon transform is justified a posteriori. □

---

[2] With the notation $A \subset\subset B$, we mean that a tube of diameter δ around A is still contained in B for some $\delta > 0$.

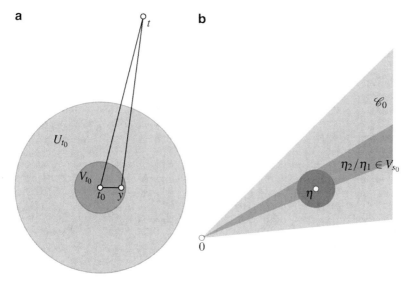

Fig. 1 (**a**) Illustration of the proof of (40). (**b**) Illustration of (34)

Theorem 5. *Assume that $f \in \mathscr{S}'$ is a tempered distribution and that for $(s_0, t_0) \in [-1, 1] \times \mathbb{R}^2$ we have in a neighborhood U of (s_0, t_0), $|\mathscr{S}\mathscr{H}_\psi f(a, s, t)| = O(a^N)$ for all $N \in \mathbb{N}$, with the implied constant uniform over U. Then (s_0, t_0) is a regular directed point of f. An analogous result holds for $\frac{1}{s_0} \in [-1, 1]$ and the shearlet $\tilde{\psi}$ for the vertical cone $\mathscr{C}'$.*

Proof. First we assume without loss of generality that f has its Fourier transform supported in $\mathscr{C}$. Otherwise we continue with the frequency projection $\hat{P}_{\mathscr{C}} f$ and invoke Lemmas 5 and 6 to arrive at the theorem.

By Theorem 3, we can represent f as

$$f = \int_{\mathbb{R}^2} \int_{-2}^{2} \int_0^1 \mathscr{S}\mathscr{H}_\psi f(a, s, t) \psi_{a,s,t} a^{-3} \, da \, ds \, dt + \int_{\mathbb{R}^2} \langle f, \Phi(\cdot - t) \rangle \Phi(\cdot - t) \, dt,$$

modulo an irrelevant constant. A further simplification can be obtained by noting that $\int_{\mathbb{R}^2} \langle f, \Phi(\cdot - t) \rangle \Phi(\cdot - t) \, dt$ is always smooth, since

$$\left(\int_{\mathbb{R}^2} \langle f, \Phi(\cdot - t) \rangle \Phi(\cdot - t) \, dt \right)^{\wedge} (\xi) = \hat{f}(\xi) |\hat{\Phi}(\xi)|^2 = O\left(|\xi|^{-N} \right)$$

by Lemma 3 (this holds if $\hat{f}$ is a slowly growing function, the general case is handled by integration-by-parts as usual). Therefore, all we need to show is that (t_0, s_0) is a regular directed point of

$$\int_{\mathbb{R}^2} \int_{-2}^{2} \int_0^1 \mathscr{S}\mathscr{H}_\psi f(a, s, t) \psi_{a,s,t} a^{-3} \, da \, ds \, dt.$$

To this end, we multiply this expression by a smooth bump function φ localized around t_0 and note that by Lemma 7 we actually only need to show that (t_0, s_0) is a regular directed point of

$$h := \int_{U_{t_0}} \int_{-2}^{2} \int_{0}^{1} \mathscr{S}\mathscr{H}_\psi f(a,s,t) \varphi \psi_{a,s,t} a^{-3} \, \mathrm{d}a\mathrm{d}s\mathrm{d}t,$$

where U_{t_0} is a compact neighborhood of t_0. To show this, we will establish that

$$I^{(N)}(u) \in L^1(\mathbb{R}),$$

where

$$I(u) := \mathscr{R}h(u, s_0).$$

With the same computations as in the proof of Lemma 7, we see that $I^{(N)}$ consists of terms of the form

$$\int_{t \in U_{t_0},\, s \in [-2,2],\, a \in [0,1]} \langle f, \psi_{a,s,t} \rangle \int_{\mathbb{R}} \left(\frac{\mathrm{d}}{\mathrm{d}x_1} \right)^{N-j} \varphi(u - s_0 x, s) a^{-j}$$
$$\left(\left(\frac{\mathrm{d}}{\mathrm{d}x_1} \right)^{N-j} \psi \right)_{a,s,t} (u - s_0 x, x) \mathrm{d}x a^{-3} \, \mathrm{d}a\mathrm{d}s\mathrm{d}t.$$

By making U_{t_0} (and the support of φ) sufficiently small, we can establish the existence of $\varepsilon > 0$ such that for all $t \in U_{t_0}$ and $s \in [s_0 - \varepsilon, s_0 + \varepsilon]$ we have $(s,t) \in U$. We now split the above integral according to $s \in [s_0 - \varepsilon, s_0 + \varepsilon]$ and $|s - s_0| > \varepsilon$. For the first part, we invoke the fast decay of the shearlet coefficients of f for $(s,t) \in U$ to see that

$$\int_{t \in U_{t_0},\, s \in [s_0-\varepsilon,s_0+\varepsilon],\, a \in [0,1]} \langle f, \psi_{a,s,t} \rangle \int_{\mathbb{R}} \left(\frac{\mathrm{d}}{\mathrm{d}x_1} \right)^{N-j} \varphi(u - s_0 x, s) a^{-j}$$
$$\left(\left(\frac{\mathrm{d}}{\mathrm{d}x_1} \right)^{N-j} \psi \right)_{a,s,t} (u - s_0 x, x) \mathrm{d}x a^{-3} \, \mathrm{d}a\mathrm{d}s\mathrm{d}t = O(1). \quad (41)$$

In order to handle the case $|s - s_0| > \varepsilon$, we note that the corresponding integral can be written as

$$\int_{t \in U_{t_0},\, s \in [s_0-\varepsilon,s_0+\varepsilon]^c,\, a \in [0,1]} \langle f, \psi_{a,s,t} \rangle a^{-j}$$
$$\times \mathscr{R} \left(\left(\frac{\mathrm{d}}{\mathrm{d}x_1} \right)^{N-j} \varphi \left(\left(\frac{\mathrm{d}}{\mathrm{d}x_1} \right)^{N-j} \psi \right)_{a,s,t} \right) (u, s_0) a^{-3} \, \mathrm{d}a\mathrm{d}s\mathrm{d}t. \quad (42)$$

Note that we can write

$$\mathscr{R}\left(\left(\left(\frac{d}{dx_1}\right)^{N-j}\varphi\left(\left(\frac{d}{dx_1}\right)^{N-j}\psi\right)\right)_{a,s,t}\right)(u,s_0) = \langle\tilde{\delta}_{u,s_0}, \theta_{a,s,t}\rangle, \qquad (43)$$

where

$$\theta := \left(\frac{d}{dx_1}\right)^{N-j}\psi$$

and

$$\tilde{\delta}_{u,s_0} := \left(\frac{d}{dx_1}\right)^{N-j}\varphi\delta_{x_1=u-s_0x_2}.$$

The Wavefront Set of $\tilde{\delta}_{u,s_0}$ is given by

$$\{(x_1,x_2,s) : x_1 = u - s_0x_2, \ s = s_0\},$$

as can be seen from the computations in Example 2. Since the function θ satisfies the assumptions of Theorem 4, we can apply this result and obtain that

$$\langle\tilde{\delta}_{u,s_0}, \theta_{a,s,t}\rangle = \mathscr{S}\mathscr{H}_\theta\tilde{\delta}_{u,s_0}(a,s,t) = O(a^N).$$

By (43), this implies that also the expression (43) is bounded. Together with (41), this proves that $I^{(N)}$ is bounded and therefore in L_1 since it is compactly supported. This proves the theorem. The argument for the dual cone follows from obvious modifications. $\square$

Putting together Theorems 4 and 5, we have finally proved Theorem 2:

Corollary 1. *Theorem 2 holds true.*

Remark 6. It is possible to weaken the assumptions in Theorem 2 considerably if one is only interested in determining directional regularity of a finite order, as opposed to our definition where Fourier decay of arbitrary order is asked in the definition of a regular directed point. In that case, only finitely many vanishing moments and only finite smoothness of ψ are required. The details are given in [9]. $\lozenge$

Remark 7. The shearlet transform (and similar related transform like for instance the curvelet transform) are able to characterize even finer notions of microlocal smoothness. We say that that $f \in \mathscr{S}'$ belongs to the microlocal Sobolev space $H^\alpha(t_0,s_0)$ if there exists a smooth bump function $\varphi \in \mathscr{S}$ around t_0 and a frequency cone $\mathscr{C}_{s_0}$ around s_0 such that

$$\int_{\mathscr{C}_{s_0}} |\xi|^{2\alpha}|(\varphi f)^\wedge(\xi)|^2 d\xi < \infty.$$

Define the shearlet square function

$$S_2^\alpha(f)(t,s) := \left(\int_0^1 |\mathscr{S}\mathscr{H}_\psi(f)(a,s,t)| \, a^{-2\alpha}\frac{da}{a^3}\right)^{1/2}.$$

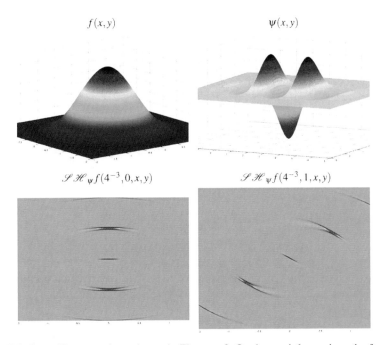

Fig. 2 This figure illustrates the main result, Theorem 2. On the *top left*, we show the function f to be analyzed—a rotated quadratic B-spline curve which has curvature discontinuities at integer points. Therefore, the wavefront set of f consists of the origin with all directions attached and the tangents of concentric circles with integer radius. The *top right* shows the analyzing shearlet which is of tensor product type. The two figures at the bottom show the magnitudes of the shearlet coefficients corresponding to two different directions—the horizontal direction on the *bottom left* and the diagonal direction with slope 1 on the *bottom right*. It is evident from these pictures that only the parameters corresponding to points of the wavefront set have nonnegligible shearlet coefficients

Then it holds that

$$f \in H^\alpha(t_0, s_0) \Leftrightarrow S_2^\alpha(f) \in L^2(\mathcal{N}),$$

for a neighborhood $\mathcal{N}$ of (t_0, s_0). The proof is similar to the proof of the corresponding curvelet result [3] and will be given elsewhere. ◇

Acknowledgments

This work was supported by the European Research Council under grant ERC AdG 247277.

References

1. J.-P. Antoine, R. Murenzi, P. Vandergheynst, and S. T. Ali. *Two-Dimensional Wavelets and their Relatives*. Cambridge University Press, 2004.
2. E. Candes and L. Demanet. The curvelet representation of wave propagators is optimally sparse. *Communications in Pure and Applied Mathematics*, 58:1472–1528, 2004.
3. E. Candes and D. Donoho. Continuous curvelet transform: I. resolution of the wavefront set. *Applied and Computational Harmonic Analysis*, 19(2):162–197, 2005.
4. E. Candes and D. Donoho. Continuous curvelet transform: II. discretization and frames. *Applied and Computational Harmonic Analysis*, 19(2):198–222, 2005.
5. S. Dahlke, G. Kutyniok, P. Maass, C. Sagiv, H.-G. Stark, and G. Teschke. The uncertainty principle associated with the continuous shearlet transform. *International Journal of Wavelets Multiresolution and Information Processing*, 6(2):157, 2008.
6. I. Daubechies. *Ten Lectures on Wavelets*. SIAM, 1992.
7. S. R. Deans. *The Radon Transform and Some of Its Applications*. John Wiley and Sons, 1983.
8. K. Gröchenig. *Foundations of Time-Frequency Analysis*. Birkhäuser, 2001.
9. P. Grohs. Continuous shearlet frames and resolution of the wavefront set. *Monatshefte für Mathematik*, 164(4):393–426, 2011.
10. P. Grohs. Continuous shearlet tight frames. *Journal of Fourier Analysis and Applications*, 17(3):506–518, 2011.
11. K. Guo and D. Labate. Optimally sparse multidimensional representation using shearlets. *SIAM journal on mathematical analysis*, 39(1):298–318, 2008.
12. K. Guo and D. Labate. Representation of Fourier integral operators using shearlets. *Journal of Fourier Analysis and Applications*, 14(3):327–371, 2008.
13. K. Guo and D. Labate. Characterization and analysis of edges using the continuous shearlet transform. *SIAM journal on Imaging Sciences*, 2:959–986, 2009.
14. K. Guo, D. Labate, and W.-Q. Lim. Edge analysis and identification using the continuous shearlet transform. *Applied and Computational Harmonic Analysis*, 27(1):24–46, 2009.
15. E. Hewitt and K. Ross. *Abstract Harmonic Analysis I*. Springer, 1979.
16. M. Holschneider and P. Tchamitchian. Pointwise analysis of Riemann's "nondifferentiable" function. *Inventiones Mathematicae*, 105:157–175, 1991.
17. L. Hörmander. *The Analysis of linear Partial Differential Operators*. Springer, 1983.
18. G. Kutyniok and D. Labate. Resolution of the wavefront set using continuous shearlets. *Transactions of the American Mathematical Society*, 361:2719–2754, 2009.
19. G. Kutyniok and W.-Q. Lim. Compactly supported shearlets are optimally sparse. Technical report, 2010.
20. S. Mallat. *A wavelet tour of signal processing*. Academic Press, 2003.
21. A. Martinez. *An Introduction to Semiclassical and Microlocal Analysis*. Springer, 2002.
22. W. Rudin. *Functional Analysis*. Mc Graw-Hill, 1991.
23. H. Smith. A Hardy space for Fourier integral operators. *Journal of Geometric Analysis*, 8:629–653, 1998.

Analysis and Identification of Multidimensional Singularities Using the Continuous Shearlet Transform

Kanghui Guo and Demetrio Labate

Abstract In this chapter, we illustrate the properties of the continuous shearlet transform with respect to its ability to describe the set of singularities of multidimensional functions and distributions. This is of particular interest since singularities and other irregular structures typically carry the most essential information in multidimensional phenomena. Consider, for example, the edges of natural images or the moving fronts in the solutions of transport equations. In the following, we show that the continuous shearlet transform provides a precise geometrical characterization of the singularity sets of multidimensional functions and precisely characterizes the boundaries of 2D and 3D regions through its asymptotic decay at fine scales. These properties go far beyond the continuous wavelet transform and other classical methods, and set the groundwork for very competitive algorithms for edge detection and feature extraction of 2D and 3D data.

Key words: Analysis of singularities, Continuous wavelet transform, Shearlets, sparsity, Wavefront set, Wavelets

1 Introduction

A well-known property of the continuous wavelet transform is its special ability to identify the singularities of functions and distributions. This property is a manifestation of the locality of the wavelet transform, which is extremely sensitive to local regularity structures. Specifically, if f is a function which is smooth apart from a

K. Guo
Department of Mathematics, Missouri State University, Springfield, MO 65804, USA
e-mail: KanghuiGuo@MissouriState.edu

D. Labate
Department of Mathematics, University of Houston, Houston, TX 77204, USA
e-mail: dlabate@math.uh.edu

G. Kutyniok and D. Labate (eds.), *Shearlets: Multiscale Analysis for Multivariate Data,*   69
Applied and Numerical Harmonic Analysis, DOI 10.1007/978-0-8176-8316-0_3,
© Springer Science+Business Media, LLC 2012

discontinuity at a point x_0, then the continuous wavelet transform of f, denoted by $\mathscr{W}_\psi f(a,t)$, exhibits very rapid asymptotic decay as a approaches 0, unless t is near x_0 [14, 16]. The complement of the set of locations where $\mathscr{W}_\psi f(a,t)$ has rapid asymptotic decay, as $a \to 0$, is called the *singular support* of f, and it corresponds, essentially, to the set where f is not regular. Indeed, an even finer analysis of the local regularity properties of f, expressed in terms of local Lipschitz regularity, can be performed using the wavelet transform. This shows that there is a precise correspondence between the Lipschitz exponent α of f at a point x_0 (where α measures the regularity type) and the asymptotic behavior of $\mathscr{W}_\psi f(a,x_0)$ as $a \to 0$ (see [10, 11]). This is in contrast with the traditional Fourier analysis which is only sensitive to global regularity properties and cannot be used to measure the regularity of a function f at a specific location.

However, despite all its good properties, the wavelet transform is unable to provide additional information about the *geometry* of the set of singularities of f. In many situations, such as in the study of the propagation of singularities associated with PDEs or in image-processing applications such as edge detection and image restoration and enhancement, it is extremely important not only to identify the location of singularities, but also to capture their geometrical information, such as the orientation and the curvature of discontinuity curves. We will show that, to this purpose, directional multiscale methods such as the shearlet transform provide, in a certain sense, the most effective solution, thanks to their ability to combine the microlocal properties of the wavelet transform and a sharp sensitivity for directional information.

In fact, it was shown that the continuous curvelet and shearlet transforms can be used to characterize the *wavefront set* of functions and distributions [1, 4, 12]. Furthermore, as will be discussed in this chapter, the continuous shearlet transform can be used to provide a very precise geometric description of the set of singularities, together with the analysis of the singularity type. Not only do these results demonstrate that the continuous shearlet transform is a highly effective microanalysis tool, going far beyond the traditional wavelet framework; they also set the theoretical groundwork for very competitive numerical algorithms for edge analysis and detection, which will be discussed in chapter "Image Processing using Shearlets" of this volume (see also the work in [18]).

In order to illustrate the properties of the shearlet transform with respect to the analysis of singularities, let us start by presenting a few simple examples. This will introduce some general concepts which will be further elaborated in the rest of the chapter.

1.1 Example: Line Singularity

In many signal and image-processing applications, there is a particular interest in functions which are *well localized*, that is, they have rapid decay both in $\mathbb{R}^n$ and in the Fourier domain. Since functions which have rapid decay must have highly regular Fourier transform and vice versa, it follows that well-localized functions must also exhibit good decay both in $\mathbb{R}^n$ and in the Fourier domain. Hence, one way to obtain well-localized functions is to define them in the Fourier domain so that they have both compact support and high regularity. In particular, we are interested in affine families of functions on $\mathbb{R}^n$ of the form $\psi_{M,t}(x) = |\det M|^{-1/2} \psi(M^{-1}(x-t))$, where $t \in \mathbb{R}^n$ and $M \in GL_n(\mathbb{R})$. Indeed we have the following observation stating, essentially, that if $\hat{\psi}$ has compact support and high regularity, then the functions $\psi_{M,t}$ are well localized.

Proposition 1. *Suppose that $\psi \in L^2(\mathbb{R}^n)$ is such that $\hat{\psi} \in C_c^\infty(R)$, where $R = \text{supp } \hat{\psi} \subset \mathbb{R}^n$. Then, for each $k \in \mathbb{N}$, there is a constant $C_k > 0$ such that, for any $x \in \mathbb{R}^n$, we have*

$$|\psi_{M,t}(x)| \le C_k |\det M|^{-\frac{1}{2}} (1 + |M^{-1}(x-t)|^2)^{-k}.$$

In particular, $C_k = k\, m(R) \left(\|\hat{\psi}\|_\infty + \|\triangle^k \hat{\psi}\|_\infty \right)$, where $\triangle = \sum_{i=1}^n \frac{\partial^2}{\partial \xi_i^2}$ is the frequency domain Laplacian operator and $m(R)$ is the Lebesgue measure of R.

Proof. From the definition of the Fourier transform, it follows that, for every $x \in \mathbb{R}^n$,

$$|\psi(x)| \le m(R) \|\hat{\psi}\|_\infty. \tag{1}$$

An integration by parts shows that

$$\int_R \triangle \hat{\psi}(\xi) e^{2\pi i \langle \xi, x \rangle} \, d\xi = -(2\pi)^2 |x|^2 \, \psi(x)$$

and thus, for every $x \in \mathbb{R}^n$,

$$(2\pi |x|)^{2k} |\psi(x)| \le m(R) \|\triangle^k \hat{\psi}\|_\infty. \tag{2}$$

Using (1) and (2), we have

$$\left(1 + (2\pi |x|)^{2k}\right) |\psi(x)| \le m(R) \left(\|\hat{\psi}\|_\infty + \|\triangle^k \hat{\psi}\|_\infty \right). \tag{3}$$

Observe that, for each $k \in \mathbb{N}$,

$$(1 + |x|^2)^k \le \left(1 + (2\pi)^2 |x|^2\right)^k \le k \left(1 + (2\pi |x|)^{2k}\right).$$

Using this last inequality and (3), we have that for each $x \in \mathbb{R}^n$

$$|\psi(x)| \le k\, m(R) (1 + |x|^2)^{-k} \left(\|\hat{\psi}\|_\infty + \|\triangle^k \hat{\psi}\|_\infty \right).$$

The proof is completed using a simple change of variables.   □

It is clear that Proposition 1 applies, in particular, to the continuous shearlet systems $\{\psi_{a,s,t} = \psi_{M_{as},t}\}$, where $M_{as} = \begin{pmatrix} a & -a^{1/2} s \\ 0 & a^{1/2} \end{pmatrix}$, for $a > 0$, $s \in \mathbb{R}$, $t \in \mathbb{R}^2$ (provided that ψ satisfies the assumptions of the proposition).

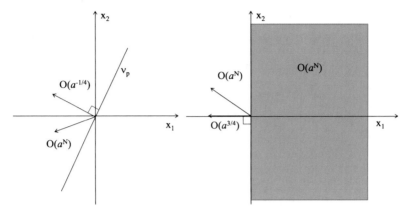

Fig. 1 Linear singularity. *Left*: The continuous shearlet transform of the linear delta distribution v_p has rapid asymptotic decay except when the location variable t is on the support of v_p and the shearing variable s corresponds to the normal direction at t; in this case $\mathcal{SH}_\psi v_p(a,s,t) \sim a^{-1/4}$, as $a \to 0$. *Right*: The continuous shearlet transform of the Heaviside function has rapid asymptotic decay except when the location variable t is on the line of $x_1 = 0$ and the shearing variable s corresponds to the normal direction to this line; in this case $\mathcal{SH}_\psi v_p(a,s,t) \sim a^{3/4}$, as $a \to 0$

In the following, we will assume that ψ, the generator of the continuous shearlet system, is a *classical shearlet*, according to the definition in chapter "Introduction to Shearlets" of this volume. That is, for $\xi = (\xi_1, \xi_2) \in \mathbb{R}^2$, $\xi_2 \neq 0$, we have

$$\hat{\psi}(\xi) = \hat{\psi}_1(\xi_1)\,\hat{\psi}_2\left(\frac{\xi_2}{\xi_1}\right)$$

where

- $\hat{\psi}_1 \in C_c^\infty(\mathbb{R})$, $\operatorname{supp}\hat{\psi}_1 \subset [-2, -\frac{1}{2}] \cup [\frac{1}{2}, 2]$, and it satisfies the Calderòn condition

$$\int_0^\infty |\hat{\psi}_1(a\omega)|^2\,\frac{da}{a} = 1, \text{ for a.e. } \omega \in \mathbb{R}. \tag{4}$$

- $\hat{\psi}_2 \in C_c^\infty(\mathbb{R})$, $\operatorname{supp}\hat{\psi}_2 \subset [-\frac{\sqrt{2}}{4}, \frac{\sqrt{2}}{4}]$, and $\|\psi_2\|_2 = 1$. $\tag{5}$

Since $\hat{\psi} \in C_c^\infty(\mathbb{R}^2)$, it follows that $\psi \in \mathscr{S}(\mathbb{R}^2)$ and, therefore, the continuous shearlet transform of f, denoted by

$$\mathcal{SH}_\psi f(a,s,t) = \langle f, \psi_{a,s,t}\rangle, \quad (a,s,t) \in \mathbb{R}_+ \times \mathbb{R}, \times \mathbb{R}^2,$$

is well defined for all tempered distributions $f \in \mathscr{S}'$. Hence, we can examine the continuous shearlet transform of the delta distribution supported along the line $x_1 = -px_2$, which we denote by $v_p(x_1, x_2) = \delta(x_1 + px_2)$, $p \in \mathbb{R}$ and is defined by

$$\langle v_p, f\rangle = \int_{\mathbb{R}} f(-px_2, x_2)\,dx_2.$$

The following simple result from [12] shows that $\mathcal{SH}_\psi v_p(a,s,t)$, the continuous shearlet transform of v_p, has rapid asymptotic decay, as $a \to 0$, for all values of

s and t, except when t is on the singularity line and s corresponds to the normal orientation to the singularity line (see Fig. 1). Here and in the following, by *rapid asymptotic decay*, we mean that, given any $N \in \mathbb{N}$, there is a $C_N > 0$ such that $|\mathscr{SH}_\psi v_p(a,s,p)| \leq C_N a^N$, as $a \to 0$.

Proposition 2. *If $t_1 = -pt_2$ and $s = p$, we have*

$$\lim_{a \to 0} a^{\frac{1}{4}} \mathscr{SH}_\psi v_p(a,s,t) \neq 0.$$

In all other cases,

$$\lim_{a \to 0} a^{-N} \mathscr{SH}_\psi v_p(a,s,t) = 0, \quad \text{for all } N > 0.$$

Proof. Recall that the Fourier transform of v_p is given by

$$\hat{v}_p(\xi_1, \xi_2) = \int \int \delta(x_1 + px_2) e^{-2\pi i \langle \xi, x \rangle} dx_2 dx_1$$

$$= \int e^{-2\pi i x_2(\xi_2 - p\xi_1)} dx_2 = \delta(\xi_2 - p\xi_1) = v_{(-\frac{1}{p})}(\xi_1, \xi_2).$$

That is, the Fourier transform of the linear delta on $\mathbb{R}^2$ is another linear delta on $\mathbb{R}^2$, where the slope $-\frac{1}{p}$ is replaced by the slope p. Hence, a direct computation gives:

$$\mathscr{SH}_\psi v_p(a,s,t) = \langle \hat{v}_p, \hat{\psi}_{a,s,t} \rangle$$

$$= \int_{\mathbb{R}} \hat{\psi}_{a,s,t}(\xi_1, p\xi_1) d\xi_1$$

$$= a^{\frac{3}{4}} \int_{\mathbb{R}} \hat{\psi}(a\xi_1, \sqrt{a}p\xi_1 - \sqrt{a}s\xi_1) e^{-2\pi i \xi_1(t_1 + pt_2)} d\xi_1$$

$$= a^{-\frac{1}{4}} \int_{\mathbb{R}} \hat{\psi}(\xi_1, a^{-\frac{1}{2}}p\xi_1 - a^{-\frac{1}{2}}s\xi_1) e^{-2\pi i a^{-1}\xi_1(t_1 + pt_2)} d\xi_1$$

$$= a^{-\frac{1}{4}} \int_{\mathbb{R}} \hat{\psi}_1(\xi_1) \hat{\psi}_2(a^{-\frac{1}{2}}(p-s)) e^{-2\pi i a^{-1}\xi_1(t_1 + pt_2)} d\xi_1$$

$$= a^{-\frac{1}{4}} \hat{\psi}_2(a^{-\frac{1}{2}}(p-s)) \psi_1(-a^{-1}(t_1 + pt_2)).$$

Recall that $\hat{\psi}_2$ is compactly supported in the interval $[-1,1]$. It follows that, if $s \neq p$, then $\hat{\psi}_2(a^{-1/2}(p-s)) \to 0$ as $a \to 0$ since $|p-s| > \sqrt{a}$ for a sufficiently small. Thus, $\lim_{a \to 0} \mathscr{SH}_\psi v_p(a,s,t) = \lim_{a \to 0} \langle \hat{v}_p, \hat{\psi}_{a,s,t} \rangle = 0$ for $s \neq p$. On the other hand, if $t_1 = -pt_2$ and $s = p$, then $\hat{\psi}_2(a^{-1/2}(p-s)) = \hat{\psi}_2(0) \neq 0$, and

$$\langle \hat{v}_p, \hat{\psi}_{a,s,t} \rangle = a^{-\frac{1}{4}} \hat{\psi}_2(a^{-\frac{1}{2}}(p-s)) \psi_1(0) \sim a^{-\frac{1}{4}}, \quad \text{as } a \to 0.$$

Finally, if $t_1 \neq -pt_2$, Proposition 1 implies that, for all $N \in \mathbb{N}$,

$$\langle \hat{v}_p, \hat{\psi}_{a,s,t} \rangle$$
$$\leq a^{-\frac{1}{4}} \hat{\psi}_2(a^{-\frac{1}{2}}(p-s)) |\psi_1(a^{-1}(t_1 + pt_2))|$$
$$\leq C_N a^{-\frac{1}{4}} \hat{\psi}_2(a^{-\frac{1}{2}}(p-s)) (1 + a^{-2}(t_1 + pt_2)^2)^{-N} \sim a^{2N-\frac{1}{4}}, \quad \text{as } a \to 0. \quad \square$$

As another example of line singularity, we will consider the Heaviside function $H(x_1,x_2) = \chi_{x_1>0}(x_1,x_2)$. Also in this case, as the following result shows, the continuous shearlet transform has rapid asymptotic decay, as $a \to 0$, for all values of s and t, except when t is on the singularity line and s corresponds to the normal orientation to the singularity line (see Fig. 1). Notice that, using an appropriate change of variables, this result can be extended to deal with step discontinuities along lines with arbitrary orientations.

Proposition 3. *If $t = (0,t_2)$ and $s = 0$, we have*

$$\lim_{a \to 0} a^{-\frac{3}{4}} \mathscr{SH}_\psi H(a,0,(0,t_2)) \neq 0.$$

In all other cases,

$$\lim_{a \to 0} a^{-N} \mathscr{SH}_\psi H(a,s,t) = 0, \quad \text{for all } N > 0.$$

Proof. Observe that $\frac{\partial}{\partial x_1} H = \delta_1$, where δ_1 is the delta distribution defined by

$$\langle \delta_1, \phi \rangle = \int \phi(0,x_2)\,dx_2,$$

and ϕ is a function in the Schwartz class $\mathscr{S}(\mathbb{R}^2)$ (notice that here we use the notation of the inner product $\langle,\rangle$ to denote the functional on $\mathscr{S}$). Hence

$$\widehat{H}(\xi_1,\xi_2) = (2\pi i\xi_1)^{-1}\hat{\delta}_1(\xi_1,\xi_2),$$

where $\hat{\delta}_1$ is the distribution obeying

$$\langle \hat{\delta}_1, \hat{\phi} \rangle = \int \hat{\phi}(\xi_1,0)\,d\xi_1.$$

The continuous shearlet transform of H can now be expressed as

$$\begin{aligned}
\mathscr{SH}_\psi H(a,s,t) &= \langle H, \psi_{a,s,t} \rangle \\
&= \int_{\mathbb{R}^2} (2\pi i\xi_1)^{-1}\hat{\delta}_1(\xi)\,\overline{\hat{\psi}_{a,s,t}}(\xi)\,d\xi \\
&= \int_{\mathbb{R}} (2\pi i\xi_1)^{-1}\,\overline{\hat{\psi}_{a,s,t}}(\xi_1,0)\,d\xi_1 \\
&= \int_{\mathbb{R}} \frac{a^{3/4}}{2\pi i\xi_1}\,\overline{\hat{\psi}_1}(a\xi_1)\,\overline{\hat{\psi}_2}(a^{-\frac{1}{2}}s)\,e^{2\pi i\xi_1 t_1}\,d\xi_1 \\
&= \frac{a^{3/4}}{2\pi i}\,\overline{\hat{\psi}_2}(a^{-\frac{1}{2}}s)\int_{\mathbb{R}} \overline{\hat{\psi}_1}(u)\,e^{2\pi i u\frac{t_1}{a}}\,\frac{du}{u},
\end{aligned}$$

where t_1 is the first component of $t \in \mathbb{R}^2$.

Notice that, by the properties of ψ_1, the function $\tilde{\psi}_1(v) = \int_{\mathbb{R}} \overline{\hat{\psi}_1(u)} e^{2\pi i u v} \frac{du}{u}$ decays rapidly, asymptotically, as $v \to \infty$. Hence, if $t_1 \neq 0$, it follows that $\tilde{\psi}_1(\frac{t_1}{a})$ decays rapidly, asymptotically, as $a \to 0$, and, as a result, $\mathscr{SH}_\psi H(a,s,t)$ also has rapid decay as $a \to 0$. Similarly, by the support conditions of $\hat{\psi}_2$, if $s \neq 0$ it follows that the function $\hat{\psi}_2(a^{-\frac{1}{2}}s)$ approaches 0 as $a \to 0$. Finally, if $t_1 = s = 0$, then

$$a^{-3/4} \mathscr{SH}_\psi H(a, 0, (0, t_2)) = \frac{1}{2\pi i} \overline{\hat{\psi}_2(0)} \int_{\mathbb{R}} \overline{\hat{\psi}_1(u)} \frac{du}{u} \neq 0. \quad \square$$

1.2 General Singularities

The two examples presented above show that the continuous shearlet transform describes, through its asymptotic decay at fine scales, both the location and the orientation of delta-type and step-type singularities along lines. As will be discussed below, this result holds in much greater generality. In particular, *away from the singularities*, the continuous shearlet transform has always rapid asymptotic decay. *At the singularity*, as the examples suggest, the behavior depends (1) on singularity type and (2) on the geometry of the singularity set.

For the dependence on the singularity type, both examples show that nonrapid decay occurs at the singularity, in the normal orientation. However, for the step singularity, we found that, as $a \to 0$, $\mathscr{SH}_\psi H(a, 0, (0, t_2)) \sim a^{3/4}$ (slow decay), whereas for the delta singularity we found $\mathscr{SH}_\psi V_p(a, s, t) \sim a^{-1/4}$ (increase). This shows that the sensitivity on the singularity type is consistent with the wavelet analysis as presented in [10, 11]. Recall, in fact, that if $f \in L^2(\mathbb{R})$ is uniformly Lipschitz α in a neighborhood of t and $\tilde{\psi}$ is a nice wavelet, the continuous wavelet transform of f satisfies

$$\mathscr{W}_{\tilde{\psi}} f(a, t) \leq C a^{\alpha + 1/2},$$

which shows that the decay is controlled by the regularity of f at t. This analysis extends to the case where f has a jump or delta singularity at t, corresponding to $\alpha = 0$ and $\alpha = -1$, respectively. This yields that $\mathscr{W}_{\tilde{\psi}} f(a, t)$ has slow decay $a^{1/2}$, if t is a jump discontinuity, while it increases as $a^{-1/2}$ if t is a delta-type singularity (cf. [15]).

That the qualitative behaviors of the decay of the continuous shearlet transform at the singularity is similar to the continuous wavelet transform should not be surprising since the continuous shearlet transform preserves the microlocal features of the continuous wavelet transform. Additional observations about this aspect are found in [5, 12].

On the other hand, the ability of the continuous shearlet transform to detect the geometry of the singularity set goes far beyond the continuous wavelet transform and is its most distinctive feature. As a particular manifestation of this ability, we will show that the continuous shearlet transform provides a very general and elegant characterization of step discontinuities along 2D piecewise smooth curves, which

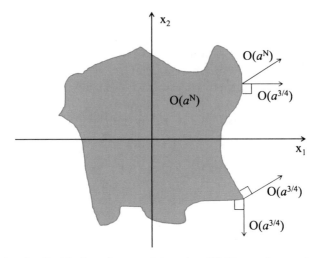

Fig. 2 General region S with piecewise smooth boundary ∂S. The continuous shearlet transform of $B = \chi_S$ has rapid asymptotic decay except when the location variable t is on ∂S and the shearing variable s corresponds to the normal direction at t; in this case $\mathscr{SH}_\psi B(a,s,t) \sim a^{3/4}$, as $a \to 0$. The same slow decay rate occurs at the corner points, for normal orientations

can summarized as follows (see [6, 7]). Let $B = \chi_S$, where $S \subset \mathbb{R}^2$ and its boundary ∂S is a piecewise smooth curve.

- If $t \notin \partial S$, then $\mathscr{SH}_\psi B(a,s,t)$ has rapid asymptotic decay, as $a \to 0$, for each $s \in \mathbb{R}$.
- If $t \in \partial S$ and ∂S is smooth near t, then $\mathscr{SH}_\psi B(a,s,t)$ has rapid asymptotic decay, as $a \to 0$, for each $s \in \mathbb{R}$ unless $s = s_0$ is the normal orientation to ∂S at p. In this last case, $\mathscr{SH}_\psi B(a,s_0,t) \sim a^{\frac{3}{4}}$, as $a \to 0$.
- If t is a corner point of ∂S and $s = s_0$, $s = s_1$ are the normal orientations to ∂S at t, then $\mathscr{SH}_\psi B(a,s_0,t), \mathscr{SH}_\psi B(a,s_1,t) \sim a^{\frac{3}{4}}$, as $a \to 0$. For all other orientations, the asymptotic decay of $\mathscr{SH}_\psi B(a,s,t)$ is faster (even if not necessarily "rapid").

This behavior is illustrated in Fig. 2.

The proof of this result and its generalization to higher dimensions will be the main content of the sections below.

2 Analysis of Step Singularities (2D)

In this section, we describe how the shearlet framework is employed to provide a characterization of the step singularities associated with functions of the form $B = \chi_S$, where S is a bounded region in $\mathbb{R}^2$, and the boundary set ∂S is piecewise smooth.

Before presenting this result, let us discuss the properties of the continuous shearlet transform that are needed for this analysis. As already noticed in chapter "Introduction to Shearlets" of this volume, the continuous shearlet transform exhibits a directional bias which is manifested, in particular, by the fact that the shearlet detection of the line singularity given by Proposition 2 does not cover the case where the singularity line is $t_2 = 0$. In this case, in fact, to be able to capture the correct orientation, the shearing variable should be taken in the limit value $s \to \infty$. As also discussed in chapter "Introduction to Shearlets" of this volume, this limitation can be overcome by using the cone-adapted continuous shearlet systems.

Hence, for $\xi = (\xi_1, \xi_2) \in \mathbb{R}^2$, we let $\psi^{(h)}, \psi^{(v)}$ be defined by

$$\hat{\psi}^{(h)}(\xi_1, \xi_2) = \hat{\psi}_1(\xi_1)\,\hat{\psi}_2\left(\tfrac{\xi_2}{\xi_1}\right), \quad \hat{\psi}^{(h)}(\xi_1, \xi_2) = \hat{\psi}_1(\xi_2)\,\hat{\psi}_2\left(\tfrac{\xi_1}{\xi_2}\right),$$

and, for

$$M_{as} = \begin{pmatrix} a & -a^{1/2}s \\ 0 & a^{1/2} \end{pmatrix}, \quad N_{as} = \begin{pmatrix} a^{1/2} & 0 \\ -a^{1/2}s & a \end{pmatrix},$$

we define the systems of "horizontal" and "vertical" continuous shearlets by

$$\{\psi_{a,s,t}^{(h)} = |\det M_{as}|^{-\frac{1}{2}}\psi^{(h)}(M_{as}^{-1}(x-t)) : 0 < a \le \tfrac{1}{4}, -\tfrac{3}{2} \le s \le \tfrac{3}{2}, t \in \mathbb{R}^2\},$$
$$\{\psi_{a,s,t}^{(v)} = |\det N_{as}|^{-\frac{1}{2}}\psi^{(h)}(N_{as}^{-1}(x-t)) : 0 < a \le \tfrac{1}{4}, -\tfrac{3}{2} \le s \le \tfrac{3}{2}, t \in \mathbb{R}^2\}.$$

Notice that, in the new definition, the shearing variable is now allowed to vary over a compact interval only. In fact, as also described in chapter "Introduction to Shearlets" of this volume, each system of continuous shearlets spans only a subspace of $L^2(\mathbb{R}^2)$, namely, the spaces $L^2(\mathscr{P}^{(h)})^\vee$ and $L^2(\mathscr{P}^{(v)})^\vee$, where $\mathscr{P}^{(h)}$ and $\mathscr{P}^{(v)}$ are the horizontal and vertical cones in the frequency domain, given by

$$\mathscr{P}^{(h)} = \{(\xi_1, \xi_2) \in \mathbb{R}^2 : |\xi_1| \ge 2 \text{ and } |\tfrac{\xi_2}{\xi_1}| \le 1\}$$
$$\mathscr{P}^{(v)} = \{(\xi_1, \xi_2) \in \mathbb{R}^2 : |\xi_1| \ge 2 \text{ and } |\tfrac{\xi_2}{\xi_1}| > 1\}.$$

Hence, we will define a continuous shearlet transform which uses as analyzing elements either the "horizontal" or the "vertical" continuous shearlet system. That is, for $0 < a \le 1/4$, $s \in \mathbb{R}$, $t \in \mathbb{R}^2$, we define the *(fine-scale) cone-adapted continuous shearlet transform* as the mapping from $f \in L^2(\mathbb{R}^2 \setminus [-2,2]^2)^\vee$ into $\mathscr{SH}_\psi f$ which is given by

$$\mathscr{SH}_\psi f(a,s,t) = \begin{cases} \mathscr{SH}_\psi^{(h)} f(a,s,t) = \langle f, \psi_{a,s,t}^{(h)} \rangle, & \text{if } |s| \le 1 \\ \mathscr{SH}_\psi^{(v)} f(a, \tfrac{1}{s}, t) = \langle f, \psi_{a,s,t}^{(v)} \rangle, & \text{if } |s| > 1. \end{cases}$$

The term *fine scale* refers to the fact that this shearlet transform is only defined for values $0 < a \le 1/4$. In fact, for the analysis of the boundaries of planar regions, we will only be interested in deriving asymptotic estimates as a approaches 0.

Finally, we will still assume the shearlet generators $\psi^{(h)}, \psi^{(v)}$ are well-localized functions. However a few additional assumptions with respect to Sect. 1.1 are needed. For completeness, we summarize below the complete set of assumptions on the functions ψ_1, ψ_2. We assume that:

- $\hat{\psi}_1 \in C_c^{\infty}(\mathbb{R})$, supp $\hat{\psi}_1 \subset [-2, -\frac{1}{2}] \cup [\frac{1}{2}, 2]$, is odd, nonnegative on $[\frac{1}{2}, 2]$ and it satisfies $\int_0^{\infty} |\hat{\psi}_1(a\xi)|^2 \frac{da}{a} = 1$, for a.e. $\xi \in \mathbb{R}$;   (6)

- $\hat{\psi}_2 \in C_c^{\infty}(\mathbb{R})$, supp $\hat{\psi}_2 \subset [-\frac{\sqrt{2}}{4}, \frac{\sqrt{2}}{4}]$, is even, nonnegative, decreasing in $[0, \frac{\sqrt{2}}{4})$, and $\|\psi_2\|_2 = 1$.   (7)

Before presenting the proof of the general characterization result, it is useful to examine the situation where S is a disc. In this case, thanks to the simpler geometry, it is possible to use a more direct argument than in the general case.

2.1 Shearlet Analysis of Circular Edges

Let D_R be the ball in $\mathbb{R}^2$ of radius $R \geq R_0 > 0$, centered at the origin, and let $B_R = \chi_{D_R}$. It was shown in [6] (by refining an incomplete proof in [12]) that the continuous shearlet transform of B_R exactly characterizes the curve ∂D_R.

Theorem 1. *Let* $t \in P = \{t = (t_1, t_2) \in \mathbb{R}^2 : |\frac{t_2}{t_1}| \leq 1\}$.
If $t = t_0 = R(\cos \theta_0, \sin \theta_0)$, *for some* $|\theta_0| \leq \frac{\pi}{4}$, *then*

$$\lim_{a \to 0^+} a^{-\frac{3}{4}} \mathscr{SH}_{\psi} B_R(a, \tan \theta_0, t_0) \neq 0.   \qquad (8)$$

If $t = t_0$ *and* $s \neq \tan \theta_0$, *or if* $t \notin \partial D(0, R)$, *then*

$$\lim_{a \to 0^+} a^{-N} \mathscr{SH}_{\psi} B_R(a, s, t) = 0, \quad \text{for all } N > 0.   \qquad (9)$$

Proof (Sketch). We only sketch the main ideas of this proof, since this result is a special case of Theorem 2, which will be presented below. Also, we will consider the system of horizontal shearlets $\{\psi_{a,s,t}^{(h)}\}$ only, since the analysis for the vertical shearlets is essentially the same. For simplicity of notation, we will drop the upperscript (h).

A direct computation gives:

$$\mathscr{SH}_{\psi} B_R(a, s, t) = \langle \hat{B}_R, \hat{\psi}_{a,s,t} \rangle$$
$$= a^{\frac{3}{4}} \int_{\mathbb{R}} \int_{\mathbb{R}} \hat{\psi}_1(a\xi_1) \, \hat{\psi}_2(a^{-\frac{1}{2}}(\frac{\xi_2}{\xi_1} - s)) e^{2\pi i \langle \xi, t \rangle} \, \hat{B}_R(\xi_1, \xi_2) \, d\xi_1 \, d\xi_2.   \quad (10)$$

For $R = 1$, the Fourier transform $\hat{B}_1(\xi_1, \xi_2)$ is the radial function:

$$\hat{B}_1(\xi_1, \xi_2) = |\xi|^{-1} J_1(2\pi |\xi|),$$

where J_1 is the Bessel function of order 1, whose asymptotic behavior satisfies [17]:

$$J_1(2\pi|\xi|) = \frac{1}{\pi}|\xi|^{-\frac{1}{2}}\cos\left(2\pi|\xi| - \frac{3\pi}{4}\right) + O(|\xi|^{-3/2}) \quad \text{as } |\xi| \to \infty.$$

It is useful to express the integral (10) using polar coordinates. For $\left|\frac{t_2}{t_1}\right| \leq 1$, $|s| \leq \frac{3}{2}$ and $\frac{1}{2}R \leq r \leq 2R$, we write $t = (t_1, t_2)$ as $r(\cos\theta_0, \sin\theta_0)$, where $0 \leq |\theta_0| \leq \frac{\pi}{4}$. Thus, we have:

$$\mathscr{SH}_\psi B_R(a, s, r, \theta_0)$$

$$= a^{\frac{3}{4}} \int_0^\infty \int_0^{2\pi} \hat{\psi}_1(a\rho\cos\theta)\hat{\psi}_2(a^{-\frac{1}{2}}(\tan\theta - s))e^{2\pi i\rho r\cos(\theta - \theta_0)}R^2\,\widehat{B}_1(R\rho)\,d\theta\rho\,d\rho$$

$$= R^2 a^{-\frac{5}{4}} \int_0^\infty \int_0^{2\pi} \hat{\psi}_1(\rho\cos\theta)\hat{\psi}_2(a^{-\frac{1}{2}}(\tan\theta - s))e^{2\pi i\frac{\rho r}{a}\cos(\theta - \theta_0)}\widehat{B}_1(\frac{R\rho}{a})\,d\theta\rho\,d\rho$$

$$= R^2 a^{-\frac{5}{4}} \int_0^\infty \eta(\rho, a, s, r, \theta_0)\,\widehat{B}_1\left(\frac{R\rho}{a}\right)\rho\,d\rho, \tag{11}$$

where

$$\eta(\rho, a, s, r, \theta_0) = \int_0^{2\pi} \hat{\psi}_1(\rho\cos\theta)\,\hat{\psi}_2(a^{-\frac{1}{2}}(\tan\theta - s))\,e^{2\pi i\frac{\rho r}{a}\cos(\theta - \theta_0)}\,d\theta,$$

$$= \eta_1(\rho, a, s, r, \theta_0) - \eta_2(\rho, a, s, r, \theta_0),$$

and

$$\eta_1(\rho, a, s, r, \theta_0) = \int_{-\frac{\pi}{2}}^{\frac{\pi}{2}} \hat{\psi}_1(\rho\cos\theta)\,\hat{\psi}_2(a^{-\frac{1}{2}}(\tan\theta - s))e^{2\pi i\frac{\rho r}{a}\cos(\theta - \theta_0)}\,d\theta; \tag{12}$$

$$\eta_2(\rho, a, s, r, \theta_0) = \int_{-\frac{\pi}{2}}^{\frac{\pi}{2}} \hat{\psi}_1(\rho\cos\theta)\,\hat{\psi}_2(a^{-\frac{1}{2}}(\tan\theta - s))e^{-2\pi i\frac{\rho r}{a}\cos(\theta - \theta_0)}\,d\theta. \tag{13}$$

In the last equality, we have used the fact that $\hat{\psi}_1$ is odd. Hence, using the asymptotic estimate for J_1, for small a we have:

$$\mathscr{SH}_\psi B_R(a, s, r, \theta_0) = a^{\frac{1}{4}}\frac{R^{\frac{1}{2}}}{\pi}(I(a, s, r, \theta_0) + E(a, s, r, \theta_0)),$$

where

$$I(a, s, r, \theta_0) = \int_0^\infty \eta(\rho, a, s, r, \theta_0)\cos\left(\frac{2\pi R\rho}{a} - \frac{3\pi}{4}\right)\rho^{-\frac{1}{2}}\,d\rho,$$

$$E(a, s, r, \theta_0) = \int_0^\infty \eta(\rho, a, s, r, \theta_0)\,O\left(\left(\frac{R\rho}{a}\right)^{-3/2}\right)\rho^{-\frac{1}{2}}\,d\rho.$$

At this point, taking advantage of the assumptions on the support of $\hat{\psi}_1$ and $\hat{\psi}_2$, one can show that, as $a \to 0$, we have that

$$a^{-\beta}E = O(a^{\frac{3}{2}-\beta}) \to 0$$

uniformly for s, r, R, for each $\beta \in (0, 1)$.

It follows that the asymptotic behavior of $\mathscr{SH}_\psi B_R(a, s, r, \theta_0)$, as $a \to 0$, is controlled by the integral $I(a, s, r, \theta_0)$. The rest of the proof is divided into several cases, depending on the values of s and r, which we briefly summarize.

Case 1: $s \neq \tan\theta_0$ (non-normal orientation)

In this case, since the derivative of the phase of the exponential in the integrals (12) and (13) does not vanish, an integration by parts argument shows that $|\eta(\rho, a, s, r, \theta_0)| \leq C_N a^N$, for any $N > 0$. This implies that $I(a, s, r, \theta_0)$ has rapid asymptotic decay, as $a \to 0$.

Case 2: $s = \tan\theta_0$, but $r \neq R$ (away from the boundary)

The original argument in [6] was rather involved. Lemma 2, which is used in the general result, provides a simpler proof for this situation.

Case 3: $s = \tan\theta_0$, $r = R$ (normal orientation)

Setting $s = \tan\theta_0$ and $t = R(\cos\theta_0, \sin\theta_0)$, a direct computation leads to the expression:

$$\lim_{a \to 0} a^{-\frac{3}{4}} |\mathscr{SH}_\psi B_R(a, s, t)| = \lim_{a \to 0} a^{-\frac{3}{4}} |\mathscr{SH}_\psi B_R(a, \tan\theta_0, R, \theta_0)|$$

$$= \left| -\frac{e^{-\frac{3\pi i}{4}}}{2} \int_0^\infty \hat{\psi}_1(\rho\cos\theta_0) \rho^{-\frac{1}{2}} \overline{h(\rho, R)}\, d\rho \right. \quad (14)$$

$$\left. + \frac{e^{\frac{3\pi i}{4}}}{2} \int_0^\infty \hat{\psi}_1(\rho\cos\theta_0) \rho^{-\frac{1}{2}} h(\rho, R)\, d\rho \right|, \quad (15)$$

where

$$h(\rho, R) = \frac{R^{\frac{1}{2}}}{\pi} \int_{-1}^1 \hat{\psi}_2(u\sec^2\theta_0) e^{-\pi i \rho R u^2}\, du.$$

Combining (14) and (15) and using the fact that $\hat{\psi}_2$ is even, it follows that

$$\lim_{a \to 0} a^{-\frac{3}{4}} |\mathscr{SH}_\psi B_R(a, \tan\theta_0, R, \theta_0)|$$

$$= \frac{\sqrt{2R}}{\pi} \left| \int_0^\infty \hat{\psi}_1(\rho\cos\theta_0) \rho^{-\frac{1}{2}} \int_0^1 \hat{\psi}_2(u\sec^2\theta_0) \left(\sin(\pi\rho R u^2) + \cos(\pi\rho R u^2)\right) du\, d\rho \right|.$$

The proof is completed using the fact that $\hat{\psi}_2$ is decreasing and applying the following Lemma, which is also found in [7].

Lemma 1. *Let $\psi_2 \in L^2(\mathbb{R})$ be such that $\|\psi_2\|_2 = 1$, supp $\hat{\psi}_2 \subset [-1,1]$ and $\hat{\psi}_2$ is an even function and is nonnegative and decreasing on $[0,1]$. Then, for each $\rho > 0$, one has*

$$\int_0^1 \hat{\psi}_2(u) \left(\sin(\pi\rho u^2) + \cos(\pi\rho u^2)\right) du > 0.$$

2.2 General 2D Boundaries

As indicated above, the characterization result for the boundaries of 2D regions holds for general bounded planar regions $S \subset \mathbb{R}^2$ whose boundary is piecewise smooth.

More precisely, we assume that the boundary set of S, denoted by ∂S, is a simple curve, of finite length L, which is smooth except possibly for finitely many corner points. To precisely define the corner points, it is useful to introduce a parametrization for ∂S. Namely, let $\alpha(t)$ be the parametrization of ∂S with respect to the arc length parameter t. For any $t_0 \in (0, L)$ and any $j \geq 0$, we assume that $\lim_{t \to t_0^-} \alpha^{(j)}(t) = \alpha^{(j)}(t_0^-)$ and $\lim_{t \to t_0^+} \alpha^{(j)}(t) = \alpha^{(j)}(t_0^+)$ exist. Also, let $\mathbf{n}(t^-), \mathbf{n}(t^+)$ be the outer normal direction(s) of ∂S at $\alpha(t)$ from the left and right, respectively; if these two are equal, we write them as $\mathbf{n}(t)$. Similarly, for the curvature of ∂S, we use the notation $\kappa(t^-), \kappa(t^+)$, and $\kappa(t)$.

We say that $p = \alpha(t_0)$ is a *corner point of ∂S* if either (1) $\alpha'(t_0^-) \neq \pm\alpha'(t_0^+)$ or (2) $\alpha'(t_0^-) = \pm\alpha'(t_0^+)$, but $\kappa(t_0^-) \neq \kappa(t_0^+)$. When (1) holds, we say that p is a *corner point of first type*, while when (2) holds, we say that p is a *corner point of second type*. On the other hand, if $\alpha(t)$ is infinitely many times differentiable at t_0, we say that $\alpha(t_0)$ is a *regular point* of ∂S. Finally, we say that the boundary curve $\alpha(t)$ is *piecewise smooth* if the values $\alpha(t)$ are regular points for all $0 \leq t \leq L$, except for finitely many corner points.

Notice that it is possible to relax the assumptions on the regularity, by assuming that the regular points of ∂S are M times differentiable, for $M \in \mathbb{N}$, rather than infinitely differentiable. All the results presented below can be adapted to the case of piecewise C^M boundary curves, for $M \geq 3$.

Let $p = \alpha(t_0)$ be a regular point and let $s = \tan(\theta_0)$ with $\theta_0 \in (-\frac{\pi}{2}, \frac{\pi}{2})$. Let $\Theta(\theta_0) = (\cos\theta_0, \sin\theta_0)$. We say that s *corresponds to the normal direction of ∂S at p* if $\Theta(\theta_0) = \pm\mathbf{n}(t_0)$. We can proceed similarly when $\alpha(t_0)$ is a corner point. In this case, however, there may be two outer normal directions $\mathbf{n}(t_0^-)$ and $\mathbf{n}(t_0^+)$.

We are now ready to state the following results from [7].

Theorem 2. *Let $B = \chi_S$, where $S \subset \mathbb{R}^2$ satisfies the properties described above.*

(i) If $p \notin \partial S$ then

$$\lim_{a \to 0^+} a^{-N} \mathscr{SH}_\psi B(a, s_0, p) = 0, \quad \text{for all } N > 0. \tag{16}$$

(ii) *If $p \in \partial S$ is a regular point and $s = s_0$ does not correspond to the normal direction of ∂S at p then*

$$\lim_{a \to 0^+} a^{-N} \mathscr{SH}_\psi B(a, s_0, p) = 0, \quad \text{for all } N > 0. \tag{17}$$

(iii) *If $p \in \partial S$ is a regular point and $s = s_0$ corresponds to the normal direction of ∂S at p then*

$$\lim_{a \to 0^+} a^{-\frac{3}{4}} \mathscr{SH}_\psi B(a, s_0, p) \neq 0. \tag{18}$$

In the case where $p \in \partial S$ is a corner point, we have the following result.

Theorem 3. *Let $B = \chi_S$, where $S \subset \mathbb{R}^2$ satisfies the properties described above.*

(i) *If p is a corner point of the first type and $s = s_0$ does not correspond to any of the normal directions of ∂S at p, then*

$$\lim_{a \to 0^+} a^{-\frac{9}{4}} \mathscr{SH}_\psi B(a, s_0, p) < \infty. \tag{19}$$

(ii) *If p is a corner point of the second type and $s = s_0$ does not correspond to any of the normal directions of ∂S at p, then*

$$\lim_{a \to 0^+} a^{-\frac{9}{4}} \mathscr{SH}_\psi B(a, s_0, p) \neq 0. \tag{20}$$

(iii) *If $s = s_0$ corresponds to one of the normal directions of ∂S at p then*

$$\lim_{a \to 0^+} a^{-\frac{3}{4}} \mathscr{SH}_\psi B(a, s_0, p) \neq 0 \tag{21}$$

Theorem 2 shows that, if $p \in \partial S$ is a regular point, the continuous shearlet transform decays rapidly, asymptotically for $a \to 0$, unless $s = s_0$ corresponds to the normal direction of ∂S at p, in which case

$$\mathscr{SH}_\psi B(a, s_0, p) \sim O(a^{\frac{3}{4}}), \text{ as } a \to 0.$$

This result contains the situation where B is a disk, as a special case.

Theorem 3 shows that, at a corner points p, the asymptotic decay of the continuous shearlet transform depends both on the tangent and the curvature at p. If $s = s_0$ corresponds to one of the normal directions of ∂S at p, then the continuous shearlet transform decays as

$$\mathscr{SH}_\psi B(a, s_0, p) \sim O(a^{\frac{3}{4}}), \text{ as } a \to 0.$$

This is the same decay rate as for regular points, when s_0 corresponds to the normal direction (but now there are two normal directions). If p is a corner point of the second type and s does not correspond to any of the normal directions, then

$$\mathscr{SH}_\psi B(a, s_0, p) \sim O(a^{\frac{9}{4}}), \text{ as } a \to 0,$$

which is a faster decay rate than in the normal orientation case. Finally, if p is a corner point of the first type and s_0 does not correspond to any of the normal directions, then, by the theorem, we only know that the asymptotic decay of $|\mathscr{SH}_\psi B(a, s_0, p)|$ is not slower than $O(a^{\frac{9}{4}})$; however, the decay could be faster than $O(a^{\frac{9}{4}})$. For example, as shown in [7], if p is a corner point of a half disk, when s_0 does not correspond to the normal directions, we have that

$$\mathscr{SH}_\psi B(a, s_0, p) \sim O(a^{\frac{9}{4}}), \text{ as } a \to 0.$$

However, when p is corner point of a polygon S and s_0 does not correspond to the normal directions, for any $N \in \mathbb{N}$, there is a constant $C_N > 0$ such that

$$|\mathscr{SH}_\psi B(a, s_0, p)| \leq C_N a^N, \text{ as } a \to 0.$$

2.3 Proofs of Theorems 2 and 3

It is clear that the proof used for the disk in Sect. 2.1 cannot be extended to this case directly since that argument requires an explicit expression of the Fourier transform of the region B. Instead, using the divergence theorem, we can express the Fourier transform of our general region $B \subset \mathbb{R}^2$ as

$$\hat{B}(\xi) = \hat{\chi}_S(\xi) = -\frac{1}{2\pi i |\xi|} \int_{\partial S} e^{-2\pi i \langle \xi, x \rangle} \Theta(\theta) \cdot \mathbf{n}(x) \, d\sigma(x)$$

$$= -\frac{1}{2\pi i \rho} \int_0^L e^{-2\pi i \rho \, \Theta(\theta) \cdot \alpha(t)} \Theta(\theta) \cdot \mathbf{n}(t) \, dt \qquad (22)$$

where $\xi = \rho \Theta(\theta)$, $\Theta(\theta) = (\cos\theta, \sin\theta)$. Notice that this idea for representing the Fourier transform of the characteristic function of a bounded region is used, for example, in [13].

Hence, using (22), we have that

$$\mathscr{SH}_\psi B(a, s, p)$$
$$= \langle B, \psi_{a,s,p} \rangle$$
$$= \int_0^{2\pi} \int_0^\infty \hat{B}(\rho, \theta) \, \overline{\hat{\psi}_{a,s,p}^{(d)}(\rho, \theta)} \, \rho \, d\rho \, d\theta$$
$$= -\frac{1}{2\pi i} \int_0^{2\pi} \int_0^\infty \int_0^L \overline{\hat{\psi}_{a,s,p}^{(d)}(\rho, \theta)} e^{-2\pi i \rho \, \Theta(\theta) \cdot \alpha(t)} \Theta(\theta) \cdot \mathbf{n}(t) \, dt \, d\rho \, d\theta, \qquad (23)$$

where the upper script in $\psi_{a,s,p}^{(d)}$ is either $d = h$, when $|s| \leq 1$, or $d = v$, when $|s| > 1$.

The first useful observation is that the asymptotic decay of the shearlet transform $\mathscr{SH}_\psi B(a, s, p)$, as $a \to 0$, is only determined by the values of the boundary ∂S which are "close" to p. To state this fact, for $\varepsilon > 0$, let $D(\varepsilon, p)$ be the ball in $\mathbb{R}^2$ of radius

ε and center p, and $D^c(\varepsilon, p) = \mathbb{R}^2 \setminus D(\varepsilon, p)$. Hence, using (23), we can write the shearlet transform of B as

$$\mathscr{SH}_\psi B(a, s, p) = I_1(a, s, p) + I_2(a, s, p),$$

where

$$
\begin{aligned}
& I_1(a, s, p) \\
& = -\frac{1}{2\pi i} \int_0^{2\pi} \int_0^\infty \int_{\partial S \cap D(\varepsilon, p)} \overline{\hat{\psi}_{a,s,p}^{(d)}(\rho, \theta)} e^{-2\pi i \rho \, \Theta(\theta) \cdot \alpha(t)} \Theta(\theta) \cdot \mathbf{n}(t) \, dt \, d\rho \, d\theta, \quad (24) \\
& I_2(a, s, p) \\
& = -\frac{1}{2\pi i} \int_0^{2\pi} \int_0^\infty \int_{\partial S \cap D^c(\varepsilon, p)} \overline{\hat{\psi}_{a,s,p}^{(d)}(\rho, \theta)} e^{-2\pi i \rho \, \Theta(\theta) \cdot \alpha(t)} \Theta(\theta) \cdot \mathbf{n}(t) \, dt \, d\rho \, d\theta. \quad (25)
\end{aligned}
$$

Thus, the following Localization Lemma shows that I_2 has rapid asymptotic decay, at fine scales.

Lemma 2 (Localization Lemma). *Let $I_2(a, s, p)$ be given by (25). For any positive integer N, there is a constant $C_N > 0$ such that*

$$|I_2(a, s, p)| \le C_N \, a^{\frac{N}{2}},$$

asymptotically as $a \to 0$, uniformly for all $s \in \mathbb{R}$.

Proof. We will only examine the behavior of $I_2(a, s, p)$ for $|s| \le 1$ (in which case, we use the "horizontal" shearlet transform). The case where $|s| > 1$ is similar. We have:

$$
\begin{aligned}
I_2(a, s, p) &= -\frac{1}{2\pi i} \int_{\partial S \cap D^c(\varepsilon, p)} \int_0^{2\pi} \int_0^\infty \overline{\hat{\psi}_{a,s,p}^{(h)}(\rho, \theta)} e^{-2\pi i \rho \, \Theta(\theta) \cdot \alpha(t)} \Theta(\theta) \cdot \mathbf{n}(t) \, dt \, d\rho \, d\theta \\
&= \frac{-a^{3/4}}{2\pi i} \int_{\partial S \cap D^c(\varepsilon, p)} \int_0^{2\pi} \int_0^\infty \hat{\psi}_1(a\rho \cos\theta) \, \hat{\psi}_2(a^{-1/2}(\tan\theta - s)) \\
&\quad \times e^{2\pi i \rho \, \Theta(\theta) \cdot p} \, d\rho \, d\theta \, e^{-2\pi i \rho \, \Theta(\theta) \cdot \alpha(t)} \Theta(\theta) \cdot \mathbf{n}(t) \, dt \\
&= \frac{-a^{-1/4}}{2\pi i} \int_{\partial S \cap D^c(\varepsilon, p)} \int_0^{2\pi} \int_0^\infty \hat{\psi}_1(\rho \cos\theta) \, \hat{\psi}_2(a^{-1/2}(\tan\theta - s)) \\
&\quad \times e^{2\pi i \frac{\rho}{a} \Theta(\theta) \cdot (p - \alpha(t))} \Theta(\theta) \cdot \mathbf{n}(t) \, d\rho \, d\theta \, dt.
\end{aligned}
$$

By assumption, $\|p - \alpha(t)\| \ge \varepsilon$ for all $\alpha(t) \in \partial S \cap D^c(\varepsilon, p)$. Hence, there is a constant C_p such that $\inf_{x \in \partial S \cap D^c(\varepsilon, p)} |p - x| = C_p$. Let $\mathscr{I} = \{\theta : |\tan\theta - s| \le a^{\frac{1}{2}}\}$, $\mathscr{I}_1 = \{\theta : |\Theta(\theta) \cdot (p - x)| \ge \frac{C_p}{\sqrt{2}}\} \cap \mathscr{I}$, and $\mathscr{I}_2 = \mathscr{I} \setminus \mathscr{I}_1$. Since the vectors $\Theta(\theta), \Theta'(\theta)$ form an orthonormal basis in $\mathbb{R}^2$, it follows that, on the set $\mathscr{I}_2$, we have $|\Theta'(\theta) \cdot (p - x)| \ge \frac{C_p}{\sqrt{2}}$. Hence, we can express the integrals I_2 as a sum of a term where $\theta \in \mathscr{I}_1$ and another term where $\theta \in \mathscr{I}_2$, and integrate by parts as follows.

On $\mathscr{I}_1$, we integrate by parts with respect to the variable ρ; on $\mathscr{I}_2$ we integrate by parts with respect to the variable θ. Doing this repeatedly, it yields that, for any positive integer N, $|I_2(a,s,p)| \leq C_N a^{\frac{N}{2}}$, uniformly in s. This finishes the proof of the lemma. $\quad\square$

Let $\alpha(t)$ be the boundary curve ∂S, with $0 \leq t \leq L$. We may assume that $L > 1$ and $p = (0,0) = \alpha(1)$. When p is a corner point of ∂S, we can write

$$\mathscr{C} = \partial S \cap D(\varepsilon,(0,0)) = \mathscr{C}^- \cup \mathscr{C}^+,$$

where

$$\mathscr{C}^- = \{\alpha(t) : 1 - \varepsilon \leq t \leq 1\}, \quad \mathscr{C}^+ = \{\alpha(t) : 1 \leq t \leq 1+\varepsilon\}. \tag{26}$$

When p is a corner point of ∂S and s corresponds to one of the normal directions of ∂S at p, it will be useful to replace the portion of ∂S near p by the Taylor polynomials of degree 2 on both sides of p. Notice that these two polynomials are not necessarily the same since p is a corner point.

On the other hand, for regular point $p \in \partial S$, rather than using the arclength representation, we let $\mathscr{C}^+ = \{(G^+(u),u),\ 0 \leq u \leq \varepsilon\}$ and $\mathscr{C}^- = \{(G^-(u),u),\ -\varepsilon \leq u \leq 0\}$, where $G^+(u)$ and $G^-(u)$ are smooth functions on $[0,\varepsilon]$ and $[-\varepsilon,0]$, respectively. Without loss of generality, we may assume that $p = (0,0)$ so that $u_0 = 0$ and $G(0) = 0$. Hence, we define the quadratic approximation of S near $p = (0,0)$ by $\partial S_0 = (G_0(u),u)$, where G_0 is the Taylor polynomial of degree 2 of G centered at the origin, given by $G_0(u) = G'(0)u + \frac{1}{2}G''(0)u^2$. Hence, we define $B_0 = \chi_{S_0}$, where S_0 is obtained by replacing the curve ∂S in $B = \chi_S$ with the quadratic curve ∂S_0 near the point $p = (0,0)$. If p is a corner point, then S_0 is obtained by replacing the curve ∂S near p in $B = \chi_S$ with both the left and right quadratic curves ∂S_0 near the point p. For simplicity, we only prove the following result for a regular point p. The argument for a corner point p is similar.

Lemma 3. For any $|s| \leq \frac{3}{2}$, we have

$$\lim_{a \to 0^+} a^{-\frac{3}{4}} \left| \mathscr{SH}_\psi B(a,s,0)) - \mathscr{SH}_\psi B_0(a,s,0) \right| = 0.$$

Proof. Notice that, since we assume $|s| \leq \frac{3}{2}$, we use the system of "horizontal" shearlets only.

Let γ be chosen such that $\frac{3}{8} < \gamma < \frac{1}{2}$ and assume that a is sufficiently small, so that $a^\gamma < \varepsilon$. A direct calculation shows that

$$\left| \mathscr{SH}_\psi B(a,s,0) - \mathscr{SH}_\psi B_0(a,s,0) \right| \leq \int_{\mathbb{R}^2} |\psi_{a,s,0}^{(h)}(x)|\,|\chi_S(x) - \chi_{S_0}(x)|\,dx$$

$$= T_1(a) + T_2(a),$$

where $x = (x_1, x_2) \in \mathbb{R}^2$ and

$$T_1(a) = a^{-\frac{3}{4}} \int_{D(a^\gamma,(0,0))} |\psi^{(h)}(M_{as}^{-1}x)| \, |\chi_S(x) - \chi_{S_0}(x)| \, dx,$$

$$T_2(a) = a^{-\frac{3}{4}} \int_{D^c(a^\gamma,(0,0))} |\psi^{(h)}(M_{as}^{-1}x)| \, |\chi_S(x) - \chi_{S_0}(x)| \, dx.$$

Observe that:

$$T_1(a) \leq C a^{-\frac{3}{4}} \int_{D(a^\gamma,(0,0))} |\chi_S(x) - \chi_{S_0}(x)| \, dx.$$

To estimate the above quantity, it is enough to compute the area between the regions S and S_0. Since G_0 is the Taylor polynomial of G of degree 2, we have

$$T_1(a) \leq C a^{-\frac{3}{4}} \int_{|x| < a^\gamma} |x|^3 dx \leq C a^{4\gamma - \frac{3}{4}}.$$

Since $\gamma > \frac{3}{8}$, the above estimate shows that $T_1(a) = o(a^{\frac{3}{4}})$.

The assumptions on the generator function $\psi^{(h)}$ of the shearlet system $\{\psi_{a,s,t}^{(h)}\}$ imply that, for each $N > 0$, there is a constant $C_N > 0$ such that $|\psi^{(h)}(x)| \leq C_N(1 + |x|^2)^{-N}$. It is easy to see that $(M_{as})^{-1} = A_a B_s$, where $B_s = \begin{pmatrix} 1 & s \\ 0 & 1 \end{pmatrix}$ and $A_a = \begin{pmatrix} a^{-1} & 0 \\ 0 & a^{-\frac{1}{2}} \end{pmatrix}$. Also, it is easy to verify that, for all $|s| \leq \frac{3}{2}$, there is a constant $C_0 > 0$ such that $\|B_s x\|^2 \geq C_0 \|x\|^2$, or $(x_1 + sx_2)^2 + x_2^2 \geq C_0(x_1^2 + x_2^2)$ for all $x \in \mathbb{R}^2$. Thus, for $a < 1$, we can estimate $T_2(a)$ as:

$$T_2(a) \leq C a^{-\frac{3}{4}} \int_{D^c(a^\gamma,(0,0))} |\psi^{(h)}(M_{as}x)| \, dx$$

$$\leq C_N a^{-\frac{3}{4}} \int_{D^c(a^\gamma,(0,0))} \left(1 + (a^{-1}(x_1 + sx_2))^2 + (a^{-\frac{1}{2}}x_2)^2\right)^{-N} dx$$

$$\leq C_N a^{-\frac{3}{4}} \int_{D^c(a^\gamma,(0,0))} \left((a^{-1/2}(x_1 + sx_2))^2 + (a^{-\frac{1}{2}}x_2)^2\right)^{-N} dx$$

$$= C_N a^{N-\frac{3}{4}} \int_{D^c(a^\gamma,(0,0))} (x_1^2 + x_2^2)^{-N} dx$$

$$= C_N a^{N-\frac{3}{4}} \int_{a^\gamma}^\infty r^{1-2N} dr$$

$$= C_N a^{2N(\frac{1}{2}-\gamma)} a^{2\gamma - \frac{3}{4}},$$

where the constant C_0 was absorbed in the constant C_N.

Since $\gamma < \frac{1}{2}$ and N can be chosen arbitrarily large, it follows that $T_2(a) = o(a^{\frac{3}{4}})$.

We can now proceed with the proof of Theorem 2. Here and in the proof of Theorem 3, it will be sufficient to examine the case of the horizontal shearlets only. The case of the vertical ones can be handled in essentially the same way.

Proof of Theorem 2

- *Part (i)* This follows directly from Lemma 2.
- *Part (ii)* Assume that $s = s_0$ does not correspond to any of the normal directions of ∂S at $p = (0,0)$.

We write $s_0 = \tan\theta_0$, where we assume that $|\theta_0| \le \frac{\pi}{4}$. Otherwise, for the case $\frac{\pi}{4} < |\theta_0| \le \frac{\pi}{2}$, one will use the "vertical" shearlet transform, and the argument is very similar to the one we will present below. Hence, we have that

$$I_1(a,s_0,0) = -\frac{a^{-\frac{1}{4}}}{2\pi i} \int_0^\infty \int_0^{2\pi} \hat{\psi}_1(\rho\cos\theta)\,\hat{\psi}_2(a^{-1/2}(\tan\theta - \tan\theta_0))K(a,\rho,\theta)\,d\theta\,d\rho,$$

where

$$K(a,\rho,\theta) = \int_{1-\varepsilon}^{1+\varepsilon} e^{-2\pi i \frac{\rho}{a}\Theta(\theta)\cdot\alpha(t)}\,\Theta(\theta)\cdot\mathbf{n}(t)\,dt.$$

Letting $G(t) \in C_0^\infty(\mathbb{R})$ with $G(t) = 1$ for $|t - 1| \le \frac{\varepsilon}{4}$ and $G(t) = 0$ for $|t - 1| > \frac{3\varepsilon}{4}$, we write

$$I_1(a,s_0,0) = I_{11}(a,s_0,0) + I_{12}(a,s_0,0),$$

where

$$I_{11}(a,s_0,0) = -\frac{a^{-\frac{1}{4}}}{2\pi i} \int_0^\infty \int_0^{2\pi} \hat{\psi}_1(\rho\cos\theta)\,\hat{\psi}_2(a^{-\frac{1}{2}}(\tan\theta - \tan\theta_0))K_1(a,\rho,\theta)\,d\theta\,d\rho,$$

$$I_{12}(a,s_0,0) = -\frac{a^{-\frac{1}{4}}}{2\pi i} \int_0^\infty \int_0^{2\pi} \hat{\psi}_1(\rho\cos\theta)\,\hat{\psi}_2(a^{-\frac{1}{2}}(\tan\theta - \tan\theta_0))K_2(a,\rho,\theta)\,d\theta\,d\rho,$$

and

$$K_1(a,\rho,\theta) = \int_{1-\varepsilon}^{1+\varepsilon} e^{-2\pi i \frac{\rho}{a}\Theta(\theta)\cdot\alpha(t)}\,\Theta(\theta)\cdot\mathbf{n}(t)G(t)\,dt$$

$$K_2(a,\rho,\theta) = \int_{1-\varepsilon}^{1+\varepsilon} e^{-2\pi i \frac{\rho}{a}\Theta(\theta)\cdot\alpha(t)}\,\Theta(\theta)\cdot\mathbf{n}(t)(1 - G(t))\,dt.$$

From the definition of $G(t)$, we have $1 - G(t) = 0$ for $|t - 1| \le \frac{\varepsilon}{4}$. Since the boundary curve $\{\alpha(t),\, 0 \le t \le L\}$ is simple and $p = (0,0) = \alpha(1)$, it follows that there exists a $c_0 > 0$ such that $\|\alpha(t)\| \ge c_0$ for all t with $\frac{\varepsilon}{4} \le |t - 1| \le \varepsilon$. Replacing the set $D^c(\varepsilon, p)$ by the set $\{\alpha(t),\, \frac{\varepsilon}{4} \le |t - 1| \le \varepsilon\}$, one can repeat the argument of Lemma 2 for $I_{12}(a,s_0,0)$ to show that $|I_{12}(a,s_0,0)| \le C_N\,a^N$ for any $N > 0$.

Recall that, when $a \to 0$, we have $\theta \to \theta_0$. Since s_0 does not correspond to the normal direction at p, one can choose ε sufficient small so that $\Theta(\theta)\cdot\alpha'(t) \ne 0$ for $|t - 1| \le \varepsilon$ and for all small a (and hence for θ near θ_0). Also from the assumption on $G(t)$, it follows that $G^{(n)}(1 - \varepsilon) = 0$ and $G^{(n)}(1 + \varepsilon) = 0$ for all $n \ge 0$. Writing

$$e^{-2\pi i \frac{\rho}{a}\Theta(\theta)\cdot\alpha(t)} = \frac{-a}{2\pi i\rho\,\Theta(\theta)\cdot\alpha'(t)}\left(e^{-2\pi i \frac{\rho}{a}\Theta(\theta)\cdot\alpha(t)}\right)',$$

it follows that

$$
\begin{aligned}
K_1(a,\rho,\theta) &= -\frac{a}{2\pi i\rho}\int_{1-\varepsilon}^{1+\varepsilon}\left(e^{-2\pi i\frac{\rho}{a}\Theta(\theta)\cdot\alpha(t)}\right)'\frac{\Theta(\theta)\cdot\mathbf{n}(t)}{\Theta(\theta)\cdot\alpha'(t)}G(t)\,dt \\
&= \frac{ai}{2\pi\rho}\left\{\left(e^{-2\pi i\frac{\rho}{a}\Theta(\theta)\cdot\alpha(t)}\frac{\Theta(\theta)\cdot\mathbf{n}(t)}{\Theta(\theta)\cdot\alpha'(t)}G(t)\right)_{1-\varepsilon}^{1+\varepsilon}+K_3(a,\rho,\theta)\right\} \\
&= \frac{ai}{2\pi\rho}K_3(a,\rho,\theta),
\end{aligned}
$$

where we used the fact that $G(1-\varepsilon)=0$, $G(1+\varepsilon)=0$.

Repeating the above argument for $K_3(a,\rho,\theta)$ and using induction, it follows that for all $N>0$ there exists a $C_N>0$ such that $|K_1(a,\rho,\theta)|\le C_N\,a^N$ and, hence, that $|I_{11}(a,s_0,0)|\le C_N\,a^N$.

- *Part (iii)*   We may assume that $p=(0,0)$ and $s=\tan\theta_0$ for some $|\theta_0|\le\frac{\pi}{4}$. Let S,G,S_0,G_0 be defined as in Lemma 3. Thus, according to Lemma 3, on the curve $\mathscr{C}$, one can use $G_0(u)$ to replace $G^+(u)$ since the approximation error is $o(a^{\frac{3}{4}})$. For simplicity of notation, in the following we will use G to denote G^+.

Using polar coordinates, we can express $I_1(a,0,0)$, evaluated on S_0, as

$$
\begin{aligned}
I_1(a,0,0) &= -\frac{1}{2\pi i a^{\frac{1}{4}}}\int_0^\infty\int_0^{2\pi}\hat{\psi}_1(\rho\cos\theta)\hat{\psi}_2(a^{-\frac{1}{2}}(\tan\theta-\tan\theta_0)) \\
&\quad\times\int_{-\varepsilon}^{\varepsilon}e^{-2\pi i\frac{\rho}{a}(\cos\theta\,G_0(u)+\sin\theta\,u)}(-\cos\theta+\sin\theta\,G_0'(u))\,du\,d\theta\,d\rho.
\end{aligned}
$$

By Lemmas 2 and 3, to complete the proof it is sufficient to show

$$
\lim_{a\to 0^+}a^{-\frac{3}{4}}I_1(a,0,0)\ne 0.
$$

Let $H_\theta(u)=\cos\theta\,G_0(u)+u\sin\theta$ and $A=\frac{1}{2}G''(0)$. Since $s=\tan\theta_0$ corresponds to the normal direction of S_0 at $p=(0,0)$, it follows that $H'_{\theta_0}(0)=0$. This implies that $G'(0)=-\tan\theta_0$, so that $H_\theta(u)=\cos\theta(-u\tan\theta_0+Au^2)+u\sin\theta=Au^2\cos\theta+u(\sin\theta-\cos\theta\tan\theta_0)$. We will consider separately the cases $A\ne 0$ and $A=0$.

Case 1: $A\ne 0$. In this case, we may assume that $A>0$ since the case $A<0$ is similar. We have that

$$
\begin{aligned}
&\int_{-\varepsilon}^{\varepsilon}e^{-2\pi i\frac{\rho}{a}(G_0(u)\cos\theta+u\sin\theta)}(-\cos\theta+G_0'(u)\sin\theta)\,du \\
&= e^{\frac{\rho\pi i}{a}\frac{(\sin\theta-\cos\theta\tan\theta_0)^2}{2A\cos\theta}}\int_{-\varepsilon}^{\varepsilon}e^{-2\pi i\frac{\rho}{a}A(u-u_\theta)^2\cos\theta}(2Au\sin\theta-\cos\theta-\sin\theta\tan\theta_0)\,du \\
&= K_0(\theta,a)+K_1(\theta,a),
\end{aligned}
$$

where $u_\theta = -\frac{\sin\theta - \cos\theta\tan\theta_0}{2A\cos\theta} = -\frac{1}{2A}(\tan\theta - \tan\theta_0)$,

$$K_0(\theta,a) = -(\cos\theta + \sin\theta\tan\theta_0)\,e^{\frac{\rho\pi i}{2A}\frac{(\tan\theta - \tan\theta_0)^2}{a}}\int_{-\varepsilon}^{\varepsilon} e^{-2\pi i\frac{\rho}{a}\cos\theta\,A(u-u_\theta)^2}\,du,$$

$$K_1(\theta,a) = 2A\sin\theta\,e^{\frac{\rho\pi rmi}{2A}\frac{(\tan\theta - \tan\theta_0)^2}{a}}\int_{-\varepsilon}^{\varepsilon} e^{-2\pi i\frac{\rho}{a}\cos\theta\,A(u-u_\theta)^2}u\,du.$$

In the expression of I_1, the interval $[0, 2\pi]$ of the integral in θ can be broken into the subintervals $[-\frac{\pi}{2}, \frac{\pi}{2}]$ and $[\frac{\pi}{2}, \frac{3\pi}{2}]$. On $[\frac{\pi}{2}, \frac{3\pi}{2}]$, we let $\theta' = \theta - \pi$ so that $\theta' \in [-\frac{\pi}{2}, \frac{\pi}{2}]$ and that $\sin\theta = -\sin\theta'$, $\cos\theta = -\cos\theta'$. Using this observation and the fact that $\hat{\psi}_1$ is an odd function, it follows that $I_1(a,0,0) = I_{10}(a,0,0) + I_{11}(a,0,0)$, where for $j = 0, 1$,

$$I_{1j}(a,0,0) = -\frac{1}{2\pi i a^{\frac{1}{4}}}\int_0^\infty \int_{-\frac{\pi}{2}}^{\frac{\pi}{2}} \hat{\psi}_1(\rho\cos\theta)\hat{\psi}_2(a^{-\frac{1}{2}}(\tan\theta))\,K_j(\theta,a)\,d\theta\,d\rho.$$

$$+\frac{1}{2\pi i a^{\frac{1}{4}}}\int_0^\infty \int_{-\frac{\pi}{2}}^{\frac{\pi}{2}} \hat{\psi}_1(\rho\cos\theta)\hat{\psi}_2(a^{-\frac{1}{2}}(\tan\theta))\,K_j(\theta+\pi,a)\,d\theta\,d\rho.$$

For $\theta \in (-\frac{\pi}{2}, \frac{\pi}{2})$, let $t = a^{-\frac{1}{2}}(\tan\theta - \tan\theta_0)$ and $a^{-\frac{1}{2}}u = u'$. Since $a \to 0$ implies $\theta \to \theta_0$, we obtain

$$\lim_{a\to 0^+} a^{-\frac{1}{2}}K_0(\theta,a) = -\sec\theta_0\,e^{\frac{i\pi\rho}{2A}t^2}\int_{-\infty}^{\infty} e^{-2\pi i\rho A\cos\theta_0(u-\frac{t}{2A})^2}\,du$$

$$= -\sec\theta_0\,e^{\frac{i\pi\rho}{2A}t^2}\int_{-\infty}^{\infty} e^{-2\pi i\rho A\cos\theta_0 u^2}\,du.$$

Similarly, we have that

$$\lim_{a\to 0^+} a^{-\frac{1}{2}}K_0(\theta+\pi,a) = \sec\theta_0\,e^{-\frac{i\pi\rho}{2A}t^2}\int_{-\infty}^{\infty} e^{2\pi i\rho A\cos\theta_0 u^2}\,du.$$

Using the calculations of $K_0(\theta,a)$ and $K_0(\theta+\pi,a)$, it follows from the additional factor u inside the integral of K_1 that $K_1(\theta,a) = O(a)$ and $K_1(\theta+\pi,a) = O(a)$. Thus, we obtain

$$\lim_{a\to 0^+} \frac{2\pi i}{a^{3/4}}I_1(a,0,0) = \lim_{a\to 0^+} \frac{2\pi i}{a^{3/4}}I_{10}(a,0,0).$$

Using the fact that $\hat{\psi}_1$ is odd, we have that

$$\lim_{a\to 0^+} 2\pi i a^{-\frac{3}{4}}I_1(a,0,0)$$

$$= \sec\theta_0\int_0^\infty \hat{\psi}_1(\rho)\int_{-1}^1 e^{\frac{\pi i\rho}{2A}t^2}\hat{\psi}_2(t)\,dt\int_{-\infty}^{\infty} e^{-2\pi i\rho A\cos\theta_0 u^2}\,du\,d\rho$$

$$+ \sec\theta_0\int_0^\infty \hat{\psi}_1(\rho)\int_{-1}^1 e^{-\frac{\pi i\rho}{2A}t^2}\hat{\psi}_2(t)\,dt\int_{-\infty}^{\infty} e^{2\pi i\rho A\cos\theta_0 u^2}\,du\,d\rho$$

$$= \sec\theta_0\int_0^\infty \hat{\psi}_1(\rho)\int_{-1}^1 \hat{\psi}_2(t)\,2\Re\left\{e^{\frac{\pi i\rho}{2A}t^2}\int_{-\infty}^{\infty} e^{-2\pi i\rho A\cos\theta_0 u^2}\,du\right\}dt\,d\rho.$$

Recalling the formulas of Fresnel integrals

$$\int_{-\infty}^{\infty} \cos\left(\frac{\pi}{2}x^2\right) dx = \int_{-\infty}^{\infty} \sin\left(\frac{\pi}{2}x^2\right) dx = 1,$$

it follows that

$$\int_{-\infty}^{\infty} \cos(2\pi\rho A \cos\theta_0 x^2) dx = \int_{-\infty}^{\infty} \sin(2\pi\rho A \cos\theta_0 x^2) dx = \frac{1}{2\sqrt{\rho A \cos\theta_0}}.$$

Thus, we conclude that

$$\lim_{a\to 0^+} 2\pi i a^{-\frac{3}{4}} I_1(a,0,0)$$

$$= \frac{(\sec\theta_0)^{\frac{3}{2}}}{\sqrt{A}} \int_0^\infty \frac{\hat{\psi}_1(\rho)}{\sqrt{\rho}} \left(\int_{-1}^1 \cos\left(\frac{\pi\rho}{2A}t^2\right) \hat{\psi}_2(t)\,dt + \int_{-1}^1 \sin\left(\frac{\pi\rho}{2A}t^2\right) \hat{\psi}_2(t)dt \right) d\rho.$$

The last expression is strictly positive by Lemma 1 and by the properties of $\hat{\psi}_1$.

Case 2: $A = 0$. Since, in this case, $G_0(u) = -u\tan\theta_0$ and, hence, $G_0'(u) = -\tan\theta_0$, it follows that

$$\int_{-\varepsilon}^{\varepsilon} e^{-2\pi i \frac{\rho}{a}(G_0(u)\cos\theta + u\sin\theta)} (-\cos\theta + G_0'(u)\sin\theta)\,du$$

$$= -(\cos\theta + \sin\theta\tan\theta_0) \int_{-\varepsilon}^{\varepsilon} e^{-2\pi i \frac{\rho}{a} u \sin\theta}\,du.$$

Using this observation, a direct calculation yields that

$$2\pi i I_1(a,0,0)$$

$$= -a^{-\frac{1}{4}} \int_0^\infty \int_0^{2\pi} \hat{\psi}_1(\rho\cos\theta)\hat{\psi}_2(a^{-\frac{1}{2}}(\tan\theta - \tan\theta_0))$$

$$\times \left(-(\cos\theta + \sin\theta\tan\theta_0) \int_{-\varepsilon}^{\varepsilon} e^{-2\pi i \frac{\rho}{a}(\sin\theta - \cos\theta\tan\theta_0)u}du \right) d\theta\,d\rho$$

$$= a^{-\frac{1}{4}} \int_0^\infty \int_{-\varepsilon}^{\varepsilon} \int_0^{2\pi} \hat{\psi}_1(\rho\cos\theta)\hat{\psi}_2(a^{-\frac{1}{2}}(\tan\theta - \tan\theta_0))$$

$$\times e^{-2\pi i \frac{\rho}{a}\sin\theta u}(\cos\theta + \sin\theta\tan\theta_0)\,d\theta\,du\,d\rho$$

$$= a^{-\frac{1}{4}} \int_0^\infty \int_{-\varepsilon}^{\varepsilon} \int_{-\frac{\pi}{2}}^{\frac{\pi}{2}} \hat{\psi}_1(\rho\cos\theta)\hat{\psi}_2(a^{-\frac{1}{2}}((\tan\theta - \tan\theta_0))$$

$$\times e^{-2\pi i \frac{\rho}{a}(\sin\theta - \cos\theta\tan\theta_0)u}(\cos\theta + \sin\theta\tan\theta_0)\,d\theta\,du\,d\rho$$

$$+ a^{-\frac{1}{4}} \int_0^\infty \int_{-\varepsilon}^{\varepsilon} \int_{-\frac{\pi}{2}}^{\frac{\pi}{2}} \hat{\psi}_1(\rho\cos\theta)\hat{\psi}_2(a^{-\frac{1}{2}}(\tan\theta - \tan\theta_0))$$

$$\times e^{2\pi i \frac{\rho}{a}(\sin\theta - \cos\theta\tan\theta_0)u}(\cos\theta + \sin\theta\tan\theta_0)\,d\theta\,du\,d\rho$$

Using the change of variables $t = a^{-\frac{1}{2}}(\tan\theta - \tan\theta_0)$ and $a^{-\frac{1}{2}}u = u'$, from the last set of equalities we obtain that

$$\lim_{a \to 0^+} 2\pi i\, a^{-\frac{3}{4}} I_1(a,0,0)$$

$$= \int_0^\infty \int_0^\infty \hat{\psi}_1(\rho) \int_{-1}^1 \hat{\psi}_2(t) e^{-2\pi i \cos\theta_0 \rho t u}\, dt\, du\, d\rho$$

$$+ \int_0^\infty \int_0^\infty \hat{\psi}_1(\rho) \int_{-1}^1 \hat{\psi}_2(t) e^{2\pi i \cos\theta_0 \rho t u}\, dt\, du\, d\rho$$

$$= \sec\theta_0\, \hat{\psi}_2(0) \int_0^\infty \frac{\hat{\psi}_1(\rho)}{\rho}\, d\rho > 0.$$

This completes the proof of (iii) and, together with it, the proof of Theorem 2. □

We now prove Theorem 3. Notice that the proofs of parts (i) and (iii) are obtained by modifying the arguments used in (ii) and (iii) of the proof above.

Proof of Theorem 3

- *Part (i)*. The proof follows essentially the same ideas as in the proof of (ii) of Theorem 2.

Assume that $s = s_0$ does not correspond to any of the normal directions of ∂S at $p = (0,0)$. As in the the proof of Theorem 2, we write $s_0 = \tan\theta_0$, where we assume that $|\theta_0| \le \frac{\pi}{4}$.

Breaking the interval $[-\frac{\pi}{2}, \frac{3\pi}{2}]$ into $[-\frac{\pi}{2}, \frac{\pi}{2}]$ and $[\frac{\pi}{2}, \frac{3\pi}{2}]$, and changing the variable $\theta = \theta' + \pi$ for the integral on $[\frac{\pi}{2}, \frac{3\pi}{2}]$, we can write I_1, given by (24), as

$$I_1(a, s_0, (0,0)) = I_{11}(a, s_0, (0,0)) + I_{12}(a, s_0, (0,0)),$$

where, for $j = 1, 2$,

$$I_{1j}(a, s_0, (0,0))$$
$$= -\frac{a^{-\frac{1}{4}}}{2\pi i} \int_0^\infty \int_{-\frac{\pi}{2}}^{\frac{\pi}{2}} \hat{\psi}_1(\rho\cos\theta)\, \hat{\psi}_2(a^{-1/2}(\tan\theta - \tan\theta_0))\, K_j(a, \rho, \theta)\, d\theta\, d\rho,$$

and

$$K_j(a, \rho, \theta) = K_{j1}(a, \rho, \theta) + K_{j2}(a, \rho, \theta),$$

with

$$K_{11}(a, \rho, \theta) = \int_{1-\varepsilon}^1 e^{-2\pi i \frac{\rho}{a}\Theta(\theta)\cdot\alpha(t)}\, \Theta(\theta) \cdot \mathbf{n}(t)\, dt,$$

$$K_{12}(a, \rho, \theta) = \int_1^{1+\varepsilon} e^{-2\pi i \frac{\rho}{a}\Theta(\theta)\cdot\alpha(t)}\, \Theta(\theta) \cdot \mathbf{n}(t)\, dt,$$

$$K_{21}(a, \rho, \theta) = \int_{1-\varepsilon}^1 e^{2\pi i \frac{\rho}{a}\Theta(\theta)\cdot\alpha(t)}\, \Theta(\theta) \cdot \mathbf{n}(t)\, dt,$$

$$K_{22}(a, \rho, \theta) = \int_1^{1+\varepsilon} e^{2\pi i \frac{\rho}{a}\Theta(\theta)\cdot\alpha(t)}\, \Theta(\theta) \cdot \mathbf{n}(t)\, dt.$$

By the support condition on $\hat{\psi}_2$, we have that $\theta \to \theta_0$ as $a \to 0$. Since $s_0 = \tan\theta_0$ does not correspond to any of the normal directions of ∂S at $(0,0)$, it follows that $\Theta(\theta_0) \cdot \alpha'(1) \neq 0$. Hence, for a sufficiently small (in which case θ is near θ_0), there is an $\varepsilon > 0$ sufficiently small, such that $\Theta(\theta) \cdot \alpha'(t) \neq 0$ for all θ near θ_0 and $t \in [1-\varepsilon, 1+\varepsilon]$. Next, writing

$$e^{-2\pi i \frac{\rho}{a}\Theta(\theta)\cdot\alpha(t)} = \frac{-a}{2\pi i \rho\, \Theta(\theta)\cdot\alpha'(t)}\left(e^{-2\pi i \frac{\rho}{a}\Theta(\theta)\cdot\alpha(t)}\right)',$$

and then integrating by parts twice the integral K_{11} with respect to t, we obtain

$$K_{11}(a,\rho,\theta) = -\frac{a}{2\pi i \rho}\int_{1-\varepsilon}^{1}\left(e^{-2\pi i \frac{\rho}{a}\Theta(\theta)\cdot\alpha(t)}\right)'\frac{\Theta(\theta)\cdot\mathbf{n}(t)}{\Theta(\theta)\cdot\alpha'(t)}\,dt$$
$$= K_{111}(a,\rho,\theta) + K_{112}(a,\rho,\theta) + K_{113}(a,\rho,\theta) + O(a^3),$$

where

$$K_{111}(a,\rho,\theta) = -\frac{a}{2\pi i \rho}e^{-2\pi i \frac{\rho}{a}\Theta(\theta)\cdot\alpha(1^-)}\frac{\Theta(\theta)\cdot\mathbf{n}(1^-)}{\Theta(\theta)\cdot\alpha'(1^-)}$$

$$K_{112}(a,\rho,\theta) = \frac{a}{2\pi i \rho}e^{-2\pi i \frac{\rho}{a}\Theta(\theta)\cdot\alpha(1-\varepsilon)}\frac{\Theta(\theta)\cdot\mathbf{n}(1-\varepsilon)}{\Theta(\theta)\cdot\alpha'(1-\varepsilon)}$$

$$K_{113}(a,\rho,\theta) = \frac{a^2}{(2\pi i \rho)^2}\left(e^{-2\pi i \frac{\rho}{a}\Theta(\theta)\cdot\alpha(t)}\frac{1}{\Theta(\theta)\cdot\alpha'(t)}\left(\frac{\Theta(\theta)\cdot\mathbf{n}(t)}{\Theta(\theta)\cdot\alpha'(t)}\right)'\right)\bigg|_{1-\varepsilon}^{1}.$$

Similarly, one can write

$$K_{12}(a,\rho,\theta) = K_{121}(a,\rho,\theta) + K_{122}(a,\rho,\theta) + K_{123}(a,\rho,\theta) + O(a^3).$$

Accordingly, we write

$$I_{11}(a,s_0,p) = I_{111}(a,s_0,p) + I_{112}(a,s_0,p) + I_{113}(a,s_0,p) + O(a^3),$$

where, for $l = 1,2,3$,

$$I_{11l}(a,s_0,p) = -\frac{a^{-\frac{1}{4}}}{2\pi i}\int_0^\infty\int_0^{2\pi}\hat{\psi}_1(\rho\cos\theta)\,\hat{\psi}_2(a^{-1/2}(\tan\theta - \tan\theta_0))$$
$$\times\,(K_{11l}(a,\rho,\theta) + K_{12l}(a,\rho,\theta))\,d\theta\,d\rho.$$

Similarly, for the integral I_{12} we write

$$I_{12}(a,s_0,p) = I_{121}(a,s_0,p) + I_{122}(a,s_0,p) + I_{123}(a,s_0,p) + O(a^3),$$

where, for $l = 1,2,3$,

$$I_{12l}(a,s_0,p) = -\frac{a^{-\frac{1}{4}}}{2\pi i}\int_0^\infty\int_0^{2\pi}\hat{\psi}_1(\rho\cos\theta)\,\hat{\psi}_2(a^{-1/2}(\tan\theta - \tan\theta_0))$$
$$\times\,(K_{21l}(a,\rho,\theta) + K_{22l}(a,\rho,\theta))\,d\theta\,d\rho,$$

and the terms K_{211}, K_{221} are constructed as the corresponding terms K_{111}, K_{121}. A direct computation shows that

$$K_{111}(a,\rho,\theta) + K_{121}(a,\rho,\theta) + K_{211}(a,\rho,\theta) + K_{221}(a,\rho,\theta) = 0,$$

and this implies that[1]

$$I_{111}(a,s_0,p) + I_{121}(a,s_0,p) = 0.$$

Since ∂S is simple, it follows that $\alpha(1-\varepsilon) \neq (0,0)$ and $\alpha(1+\varepsilon) \neq (0,0)$. Therefore, by the argument used in the proof of Lemma 2, it follows that, for any $N > 0$,

$$|I_{112}(a,s_0,p)| \leq C_N a^N, \quad \text{as } a \to 0.$$

Similarly, one has

$$|I_{122}(a,s_0,p)| \leq C_N a^N, \quad \text{as } a \to 0.$$

It only remains to analyze the terms I_{113}, I_{123}. To do that, notice that each one of the elements $K_{113}, K_{123}, K_{213}, K_{223}$, is made out of two terms, one at $t = 1 \pm \varepsilon$ and one at $t = 1$. As for the integrals $I_{112}(a,s_0,p)$ and $I_{122}(a,s_0,p)$, also in this case the K terms evaluated at $t = 1 \pm \varepsilon$ have fast asymptotic decay as $a \to 0$, and can be included in negligible part $O(a^3)$. Thus, in order to determine the asymptotic decay rate for $I_{113}(a,s_0,p) + I_{123}(a,s_0,p)$, one only needs to analyze the corresponding K terms corresponding to $t = 1$. To do that, let $\kappa(t)$ be the curvature of ∂S at $\alpha(t)$. By the Frenet's formula [3], we have that

$$\alpha''(t) = \kappa(t)\mathbf{n}(t), \quad \mathbf{n}'(t) = -\kappa(t)\alpha'(t).$$

Hence, using these equalities and the fact that the pair $\{\alpha'(t), \mathbf{n}(t)\}$ is an orthonormal basis in $\mathbb{R}^2$, we have

$$\left(\frac{\Theta(\theta) \cdot \mathbf{n}(t)}{\Theta(\theta) \cdot \alpha'(t)} \right)'$$
$$= \frac{(\Theta(\theta) \cdot \mathbf{n}'(t))(\Theta(\theta) \cdot \alpha'(t)) - (\Theta(\theta) \cdot \alpha''(t))(\Theta(\theta) \cdot \mathbf{n}(t))}{(\Theta(\theta) \cdot \alpha'(t))^2}$$
$$= -\frac{\kappa(t)\left((\Theta(\theta) \cdot \alpha'(t))(\Theta(\theta) \cdot \alpha'(t)) + (\Theta(\theta) \cdot \mathbf{n}(t))(\Theta(\theta) \cdot \mathbf{n}(t))\right)}{(\Theta(\theta) \cdot \alpha'(t))^2}$$
$$= -\frac{\kappa(t)|\Theta(\theta)|^2}{(\Theta(\theta) \cdot \alpha'(t))^2} = -\frac{\kappa(t)}{(\Theta(\theta) \cdot \alpha'(t))^2}.$$

It follows from the above observations that

$$\lim_{a \to 0^+} \left(\frac{\rho}{a}\right)^2 (K_{113}^-(a,s_0,p) + K_{123}^+(a,s_0,p))$$
$$= \frac{1}{(2\pi i)^2} \left(\frac{\kappa(1^+)}{(\Theta(\theta_0) \cdot \alpha'(1^+))^3} - \frac{\kappa(1^-)}{(\Theta(\theta_0) \cdot \alpha'(1^-))^3} \right). \tag{27}$$

---

[1] Notice: the assumption that $\hat{\psi}_1$ is odd makes this cancellation possible. By contrast, the generating function of the curvelet system, which is defined in polar coordinates, does not share this property.

Similarly, one has

$$\lim_{a \to 0^+} \left(\frac{p}{a}\right)^2 (K_{213}^-(a, s_0, p) + K_{223}^+(a, s_0, p))$$

$$= \frac{1}{(2\pi i)^2} \left(\frac{\kappa(1^+)}{(\Theta(\theta_0) \cdot \alpha'(1^+))^3} - \frac{\kappa(1^-)}{(\Theta(\theta_0) \cdot \alpha'(1^-))^3}\right). \tag{28}$$

Finally, by making the change of variables $u = a^{-\frac{1}{2}}(\tan\theta - \tan\theta_0)$ in I_{113} and I_{123}, and applying (27) and (28), we obtain

$$\lim_{a \to 0^+} a^{-\frac{9}{4}} (I_{113}(a, s_0, p) + I_{123}(a, s_0, p)) = -A,$$

where

$$A = \frac{\cos^2\theta_0}{2\pi^2} \left(\frac{\kappa(1^+)}{(\Theta(\theta_0) \cdot \alpha'(1^+))^3} - \frac{\kappa(1^-)}{(\Theta(\theta_0) \cdot \alpha'(1^-))^3}\right)$$

$$\times \int_0^\infty \hat{\psi}_1(\rho\cos\theta_0) \, d\rho \int_{-1}^1 \hat{\psi}_2(u) \, du < \infty. \tag{29}$$

This completes the proof of part (i).

- *Part (ii).* It is sufficient to show that, if p is a corner point of the second type, $A \neq 0$, where A is given by (29).

  In fact, we have that

  $$\alpha'(1^+) = \alpha'(1^-) \quad \text{or} \quad \alpha'(1^+) = -\alpha'(1^-).$$

  If $\alpha'(1^+) = \alpha'(1^-)$, from $\kappa(1^+) \neq \kappa(1^-)$, we have

  $$A = \frac{\cos^2\theta_0}{2\pi^2} \frac{\kappa(1^+) - \kappa(1^-)}{(\Theta(\theta_0) \cdot \alpha'(1))^3} \int_0^\infty \hat{\psi}_1(\rho\cos\theta_0) \, d\rho \int_{-1}^1 \hat{\psi}_2(u) \, du \neq 0.$$

  If, on the other hand, $\alpha'(1^+) = -\alpha'(1^-)$, then it follows that $\mathbf{n}(1^+) = -\mathbf{n}(1^-)$. Since $\alpha''(1^+) = -\kappa(1^+)\mathbf{n}(1^-)$ and $\alpha''(1^-) = \kappa(1^-)\mathbf{n}(1^-)$, and since $\kappa(1^+) \neq \kappa(1^-)$, we see that it is not possible to have $\kappa(1^+) = \kappa(1^-) = 0$. Since we know $\kappa(1^+) \geq 0$, and $\kappa(1^-) \geq 0$, it follows that $\kappa(1^+) + \kappa(1^-) > 0$. Thus, also in this case, we have

  $$A = \frac{\cos^2\theta_0}{2\pi^2} \frac{\kappa(1^+) + \kappa(1^-)}{(\Theta(\theta_0) \cdot \alpha'(1^+))^3} \int_0^\infty \hat{\psi}_1(\rho\cos\theta_0) \, d\rho \int_{-1}^1 \hat{\psi}_2(u) \, du \neq 0.$$

  This completes the proof of part (ii).

- *Part (iii).* By the assumptions on the corner points, it follows that if s corresponds to $\mathscr{C}^-$ at p, then it cannot correspond to $\mathscr{C}^+$ at p. From part (i), we see that it is enough to consider $\mathscr{C}^-$ or $\mathscr{C}^+$. Thus, we may assume that s corresponds to the outer normal direction of $\mathscr{C}^+$ at $p = (0,0)$. Since the argument for this case is very similar to the one for (iii) of Theorems 2, to save the notations here we assume $\theta_0 = 0$ so that $s = 0$.

Let S, G, S_0, G_0 be defined as in Lemma 3. Thus, according to Lemma 3, on $\mathscr{C}^+$, we one can use $G_0(u)$ to replace $G^+(u)$ since the approximation error is $o(a^{\frac{3}{4}})$. For simplicity, in the following we will use G to denote G^+.

Using polar coordinates, we can express $I_1(a,0,0)$, evaluated on S_0, as

$$I_1(a,0,0) = -\frac{1}{2\pi i a^{\frac{1}{4}}} \int_0^\infty \int_0^{2\pi} \hat{\psi}_1(\rho \cos\theta) \hat{\psi}_2(a^{-\frac{1}{2}} \tan\theta) \int_0^\varepsilon e^{-2\pi i \frac{\rho}{a}(\cos\theta G_0(u) + \sin\theta\, u)}$$
$$\times (-\cos\theta + \sin\theta\, G_0'(u))\, du\, d\theta\, d\rho.$$

By Lemmas 2 and 3, to complete the proof it is sufficient to show

$$\lim_{a\to 0^+} a^{-\frac{3}{4}} I_1(a,0,0) \neq 0.$$

Since $\theta_0 = 0$, we have $\mathbf{n}(p) = (1,0)$. It follows that $G_0(0) = 0$, $G_0'(0) = 0$ so that $G_0(u) = \frac{1}{2} G''(0) u^2$. Let $A = \frac{1}{2} G''(0)$. We will consider separately the case $A \neq 0$ and $A = 0$.

Case 1: $A \neq 0$. In this case, we may assume that $A > 0$, since the case $A < 0$ is similar. We have that

$$\int_0^\varepsilon e^{-2\pi i \frac{\rho}{a}(G_0(u)\cos\theta + u\sin\theta)}(-\cos\theta + G_0'(u)\sin\theta)\, du$$
$$= e^{\frac{\rho\pi i}{a} \frac{\sin^2\theta}{2\cos\theta A}} \int_0^\varepsilon e^{-2\pi i \frac{\rho}{a}\cos\theta A(u-u_\theta)^2}(-\cos\theta + 2Au\sin\theta)\, du$$
$$= K_0(\theta,a) + K_1(\theta,a),$$

where $u_\theta = -\frac{\sin\theta}{2A\cos\theta}$ and

$$K_0(\theta,a) = -\cos\theta e^{\frac{\rho\pi i}{a} \frac{\sin^2\theta}{2A\cos\theta}} \int_0^\varepsilon e^{-2\pi i \frac{\rho}{a}\cos\theta A(u-u_\theta)^2}\, du,$$
$$K_1(\theta,a) = 2A\sin\theta e^{\frac{\rho\pi i}{a} \frac{\sin^2\theta}{2A\cos\theta}} \int_0^\varepsilon e^{-2\pi i \frac{\rho}{a}\cos\theta A(u-u_\theta)^2}\, u\, du.$$

In the expression of I_1, the interval $[0, 2\pi]$ of the integral in θ can be broken into the subintervals $[-\frac{\pi}{2}, \frac{\pi}{2}]$ and $[\frac{\pi}{2}, \frac{3\pi}{2}]$. On $[\frac{\pi}{2}, \frac{3\pi}{2}]$, we let $\theta' = \theta - \pi$ so that $\theta' \in [-\frac{\pi}{2}, \frac{\pi}{2}]$ and that $\sin\theta = -\sin\theta'$, $\cos\theta = -\cos\theta'$. Using this observation and the fact that $\hat{\psi}_1$ is an odd function, it follows that $I_1(a,0,0) = I_{10}(a,0,0) + I_{11}(a,0,0)$, where for $j = 0,1$,

$$I_{1j}(a,0,0) = -\frac{1}{2\pi i a^{\frac{1}{4}}} \int_0^\infty \int_{-\frac{\pi}{2}}^{\frac{\pi}{2}} \hat{\psi}_1(\rho\cos\theta) \hat{\psi}_2(a^{-\frac{1}{2}}(\tan\theta)) K_j(\theta,a)\, d\theta\, d\rho.$$
$$+\frac{1}{2\pi i a^{\frac{1}{4}}} \int_0^\infty \int_{-\frac{\pi}{2}}^{\frac{\pi}{2}} \hat{\psi}_1(\rho\cos\theta) \hat{\psi}_2(a^{-\frac{1}{2}}(\tan\theta)) K_j(\theta + \pi, a)\, d\theta\, d\rho.$$

For $\theta \in (-\frac{\pi}{2},\frac{\pi}{2})$, let $t = a^{-\frac{1}{2}}\tan\theta$ (and $a^{-\frac{1}{2}}u = u'$). Recall that as $a \to 0$, θ approaches $\theta_0 = 0$ so that $\cos\theta \to 1$. We have

$$\lim_{a\to0^+} a^{-\frac{1}{2}} K_0(\theta,a) = -e^{\frac{i\pi\rho}{2A}t^2}\int_0^\infty e^{-2\pi i\rho A(u-\frac{t}{2A})^2}\,du$$

$$= -e^{\frac{i\pi\rho}{2A}t^2}\int_0^\infty e^{-2\pi i\rho Au^2}\,du - e^{\frac{i\pi\rho}{2A}t^2}\int_{-\frac{t}{2A}}^0 e^{-2\pi i\rho Au^2}\,du.$$

Similarly, we have

$$\lim_{a\to0^+} a^{-\frac{1}{2}} K_0(\theta+\pi,a) = -e^{-\frac{i\pi\rho}{2A}t^2}\int_0^\infty e^{2\pi i\rho Au^2}\,du + e^{-\frac{i\pi\rho}{2A}t^2}\int_{-\frac{t}{2A}}^0 e^{2\pi i\rho Au^2}\,du.$$

Since $\sin\theta = O(a^{\frac{1}{2}})$ (due to the conditions on the support of $\hat\psi$), based on the calculation of $K_0(\theta,a)$ and $K_0(\theta+\pi,a)$, it follows that $K_1(\theta,a) = O(a)$, $K_1(\theta+\pi,a) = O(a)$. Thus,

$$\lim_{a\to0^+}\frac{2\pi i}{a^{3/4}} I_1(a,0,0) = \lim_{a\to0^+}\frac{2\pi i}{a^{3/4}} I_{10}(a,0,0).$$

Finally, using the facts that $\hat\psi_2$ is even, that the function $\int_{-\frac{t}{2A}}^0 e^{2\pi i\rho Au^2}\,du$ is an odd function of t, and using the formulas of Fresnel integral as in the proof of Theorem 2, we conclude that

$$\lim_{a\to0^+} 2\pi i\, a^{-\frac{3}{4}} I_1(a,0,0) = \int_0^\infty \hat\psi_1(\rho)\int_{-1}^1 e^{\frac{\pi i\rho}{2A}t^2}\hat\psi_2(t)\,dt\int_0^\infty e^{-2\pi i\rho Au^2}\,du\,d\rho$$

$$+ \int_0^\infty \hat\psi_1(\rho)\int_{-1}^1 e^{-\frac{\pi i\rho}{2A}t^2}\hat\psi_2(t)\,dt\int_0^\infty e^{2\pi i\rho Au^2}\,du\,d\rho$$

$$= \frac{1}{\sqrt{A}}\int_0^\infty\frac{\hat\psi_1(\rho)}{\sqrt\rho}\left(\int_{-1}^1\cos\left(\frac{\pi i\rho}{2A}t^2\right)\hat\psi_2(t)dt + \int_{-1}^1\sin\left(\frac{\pi i\rho}{2A}t^2\right)\hat\psi_2(t)dt\right)d\rho.$$

The last expression is strictly positive by Lemma 1, and by the properties of $\hat\psi_1$.

Case 2: $A = 0$. Since, in this case, $G_0(u) = 0$, it follows that

$$\int_0^\varepsilon e^{-2\pi i\frac{\rho}{a}(G_0(u)\cos\theta+u\sin\theta)}(-\cos\theta + G_0'(u)\sin\theta)\,du = -\cos\theta\int_0^\varepsilon e^{-2\pi i\frac{\rho}{a}u\sin\theta}\,du.$$

It follows that

$$2\pi i I_1(a,0,0)$$

$$= -a^{-\frac{1}{4}}\int_0^\infty\int_0^{2\pi}\hat\psi_1(\rho\cos\theta)\hat\psi_2(a^{-\frac{1}{2}}\tan\theta)\left(-\cos\theta\int_0^\varepsilon e^{-2\pi i\frac{\rho}{a}\sin\theta\,u}du\right)d\theta\,d\rho$$

$$= a^{-\frac{1}{4}}\int_0^\infty\int_0^\varepsilon\int_0^{2\pi}\hat\psi_1(\rho\cos\theta)\hat\psi_2(a^{-\frac{1}{2}}\tan\theta)e^{-2\pi i\frac{\rho}{a}\sin\theta u}\cos\theta\,d\theta\,du\,d\rho$$

$$= a^{-\frac{1}{4}}\int_0^\infty\int_0^\varepsilon\int_{-\frac{\pi}{2}}^{\frac{\pi}{2}}\hat\psi_1(\rho\cos\theta)\hat\psi_2(a^{-\frac{1}{2}}(\tan\theta))e^{-2\pi i\frac{\rho}{a}\sin\theta u}\cos\theta\,d\theta\,du\,d\rho$$

$$+ a^{-\frac{1}{4}}\int_0^\infty\int_0^\varepsilon\int_{-\frac{\pi}{2}}^{\frac{\pi}{2}}\hat\psi_1(\rho\cos\theta)\hat\psi_2(a^{-\frac{1}{2}}(\tan\theta))e^{2\pi i\frac{\rho}{a}\sin\theta u}\cos\theta\,d\theta\,du\,d\rho$$

Using the change of variables $t = a^{-\frac{1}{2}} \tan \theta$ (and $a^{-\frac{1}{2}} u = u'$) we obtain

$$\lim_{a \to 0^+} 2\pi i a^{-\frac{3}{4}} I_1(a,0,0)$$

$$= \int_0^\infty \int_0^\infty \hat{\psi}_1(\rho) \int_{-1}^1 \hat{\psi}_2(t) e^{-2\pi i \rho t u} dt \, du \, d\rho + \int_0^\infty \int_0^\infty \hat{\psi}_1(\rho) \int_{-1}^1 \hat{\psi}_2(t) e^{2\pi i \rho t u} dt \, du \, d\rho$$

$$= (\hat{\psi}_2(0))^2 \int_0^\infty \frac{\hat{\psi}_1(\rho)}{\rho} d\rho > 0.$$

This completes the proof of (iii) and, together with it, the proof of Theorem 3. $\square$

2.4 Extensions and Generalizations

The results presented above are limited to the analysis of the continuous shearlet transform of characteristic functions of sets. It is clear that, to provide a more realistic model for images containing edges, it would be useful to consider a more general class of compactly supported functions, which are not necessarily constant or piecewise constant.

Unfortunately, the analysis of this situation is significantly more complicated and cannot be derived directly using the techniques developed above. This is due to the fact that one main technical tool which was used to deduce Theorems 2 and 3 is the divergence theorem (cf. (22)), which allows us to conveniently express the Fourier transform of $B = \chi_S$, the characteristic function of a set S. If χ_S is replaced by $g\chi_S$, this produces a convolution in the Fourier domain, and this has a dramatic impact on all the arguments used above, even in the simplified situation where g can be expanded using a Taylor polynomial.

Despite the fact that a general result for the characterization of singularities of piecewise smooth functions is not known at the moment, it is still possible to derive some useful observations which take advantage of the directional sensitivity of the continuous shearlet transform. Following the approach in [6], let Ω be a bounded open subset of $\mathbb{R}^2$ and assume a smooth partition

$$\Omega = \bigcup_{n=1}^L \Omega_n \cup \Gamma,$$

where:

(i) For each $n = 1, \ldots, L$, Ω_n is a connected open domain
(ii) Each boundary $\partial_\Omega \Omega_n$ is generated by a C^3 curve γ_n and each of the boundary curves γ_n can be parametrized as $(\rho(\theta)\cos\theta, \rho(\theta)\sin\theta)$ where $\rho(\theta) : [0, 2\pi) \to [0, 1]$ is a radius function
(iii) $\Gamma = \bigcup_{n=1}^L \partial_\Omega \Omega_n$, where $\partial_\Omega X$ denotes the relative topological boundary in Ω of $X \subset \Omega$

Hence, we define the space $E^{1,3}(\Omega)$ as the collection of functions which are compactly supported in Ω and have the form

$$f(x) = \sum_{n=1}^{L} f_n(x) \chi_{\Omega_n}(x) \text{ for } x \in \Omega \backslash \Gamma$$

where, for each $n = 1, \ldots, L$, $f_n \in C_0^1(\Omega)$ with $\sum_{|\alpha| \leq 1} \|D^\alpha f_n\|_\infty \leq C$ for some $C > 0$, and the sets Ω_n are pairwise disjoint in measure. The functions $E^{1,3}(\Omega)$ are a variant of the cartoon-like images, where the set Γ describes the boundaries of different objects. Similar image models are commonly used, for example, in the variational approach to image processing [2, Chap. 3]. Notice that each term $u_n(x) = f_n(x) \chi_{\Omega_n}(x)$ models the relatively homogeneous interior of a single object and that the definition above does not specify the function value along the boundary set Γ.

For each x in a C^3 component of Γ, we define the *jump of f at x*, denoted by $[f]_x$, to be

$$[f]_x = \lim_{\varepsilon \to 0^+} f(x + \varepsilon v_x) - f(x - \varepsilon v_x)$$

where v_x is an unit normal vector along Γ at x. Also, for $x \in \mathbb{R}^2$, $L > 0$, we denote the cube of center x and side-length $2L$ by $Q(x, L)$; that is, $Q(x, L) = [-L, L]^2 + x$. For $k = (k_1, k_2) \in \mathbb{Z}^2$, let $M \in \mathbb{N}$ be sufficiently large so that each of the boundary curves γ_n may be parametrized as either $(E(t_2), t_2)$ or $(t_1, E(t_1))$ in $Q(\frac{k}{M}, \frac{1}{M})$ if $Q(\frac{k}{M}, \frac{1}{M}) \cap \Gamma \neq \emptyset$. We have the following result from [6].

Theorem 4. *Let $f \in E^{1,3}(\Omega)$ and suppose that the boundary curve γ_n, for some n, is parametrized as $(E(t_2), t_2)$ in $Q(\frac{k}{M}, \frac{1}{M})$ for some $k \in \mathbb{Z}^2$, and that $t = (E(t_2), t_2) \in Q(\frac{k}{M}, \frac{1}{2M})$ for some t_2. If $s = -E'(t_2)$, there exist positive constants C_1 and C_2 such that*

$$C_1 |[f]_t| \leq \lim_{a \to 0^+} a^{-\frac{3}{4}} |\mathscr{SH}_\psi f(a, s, t)| \leq C_2 |[f]_t|. \tag{30}$$

If $s \neq -E'(t_2)$, then

$$\lim_{a \to 0^+} a^{-\frac{3}{4}} |\mathscr{SH}_\psi f(a, s, t)| = 0. \tag{31}$$

This shows that, on the discontinuity curve, if s corresponds to the normal orientation, then the continuous shearlet transform of f decays as $O(a^{-\frac{3}{4}})$, provided $|[f]_t| \neq 0$. However, if s does not correspond to the normal orientation, we can only claim that $\mathscr{SH}_\psi f(a, s, t)$ decays faster than $O(a^{-\frac{3}{4}})$.

3 Extension to Higher Dimensions

The characterization of boundary regions using the continuous shearlet transform extends to the 3D setting [8, 9]. However, due to the more complicated geometry, the arguments for the 2D case do not carry over directly. More precisely, the main difficulties arise in dealing with irregular points on the boundary of solid regions, for which there are still some open problems as will be described below.

3.1 3D Continuous Shearlet Transform

The construction of shearlet systems in 3D follows essentially the same ideas as the 2D construction. Also, in this case, it is convenient to use separate shearlet systems defined in different subregions of the frequency space. This leads to the definition of three pyramid-based systems, associated with the pyramidal regions:

$$\mathcal{P}_1 = \{(\xi_1,\xi_2,\xi_3) \in \mathbb{R}^3 : |\xi_1| \geq 2, \, |\tfrac{\xi_2}{\xi_1}| \leq 1 \text{ and } |\tfrac{\xi_3}{\xi_1}| \leq 1\},$$

$$\mathcal{P}_2 = \{(\xi_1,\xi_2,\xi_3) \in \mathbb{R}^3 : |\xi_1| \geq 2, \, |\tfrac{\xi_2}{\xi_1}| > 1 \text{ and } |\tfrac{\xi_3}{\xi_1}| \leq 1\},$$

$$\mathcal{P}_3 = \{(\xi_1,\xi_2,\xi_3) \in \mathbb{R}^3 : |\xi_1| \geq 2, \, |\tfrac{\xi_2}{\xi_1}| \leq 1 \text{ and } |\tfrac{\xi_3}{\xi_1}| > 1\}.$$

For $\xi = (\xi_1,\xi_2,\xi_3) \in \mathbb{R}^3$, $\xi_1 \neq 0$, let $\psi^{(d)}$, $d = 1,2,3$ be defined by

$$\hat{\psi}^{(1)}(\xi) = \hat{\psi}^{(1)}(\xi_1,\xi_2,\xi_3) = \hat{\psi}_1(\xi_1)\,\hat{\psi}_2(\tfrac{\xi_2}{\xi_1}),\,\hat{\psi}_2(\tfrac{\xi_3}{\xi_1}),$$

$$\hat{\psi}^{(2)}(\xi) = \hat{\psi}^{(2)}(\xi_1,\xi_2,\xi_3) = \hat{\psi}_1(\xi_2)\,\hat{\psi}_2(\tfrac{\xi_1}{\xi_2}),\,\hat{\psi}_2(\tfrac{\xi_3}{\xi_2}),$$

$$\hat{\psi}^{(3)}(\xi) = \hat{\psi}^{(3)}(\xi_1,\xi_2,\xi_3) = \hat{\psi}_1(\xi_3)\,\hat{\psi}_2(\tfrac{\xi_2}{\xi_3}),\,\hat{\psi}_2(\tfrac{\xi_1}{\xi_3}),$$

where ψ_1, ψ_2 satisfy the same assumptions as in the 2D case. Hence, for $d = 1,2,3$, the *3D pyramid-based continuous shearlet systems for* $L^2(\mathcal{P}_d)^\vee$ are the systems

$$\{\psi^{(d)}_{a,s_1,s_2,t} : 0 \leq a \leq \tfrac{1}{4}, -\tfrac{3}{2} \leq s_1 \leq \tfrac{3}{2}, -\tfrac{3}{2} \leq s_2 \leq \tfrac{3}{2}, t \in \mathbb{R}^3\} \tag{32}$$

where $\psi^{(d)}_{a,s_1,s_2,t}(x) = |\det M^{(d)}_{as_1 s_2}|^{-\frac{1}{2}}\psi^{(d)}((M^{(d)}_{as_1 s_2})^{-1}(x-t))$, and

$$M^{(1)}_{as_1 s_2} = \begin{pmatrix} a & -a^{1/2}s_1 & -a^{1/2}s_2 \\ 0 & a^{1/2} & 0 \\ 0 & 0 & a^{1/2} \end{pmatrix}, \quad M^{(2)}_{as_1 s_2} = \begin{pmatrix} a^{1/2} & 0 & 0 \\ -a^{1/2}s_1 & a & -a^{1/2}s_2 \\ 0 & 0 & a^{1/2} \end{pmatrix},$$

$$M^{(3)}_{as_1 s_2} = \begin{pmatrix} a^{1/2} & 0 & 0 \\ 0 & a^{1/2} & 0 \\ -a^{1/2}s_1 & -a^{1/2}s_2 & a \end{pmatrix}.$$

Notice that the elements of the shearlet systems $\psi^{(d)}_{a,s_1,s_2,t}$ are well-localized waveforms associated with various scales, controlled by a, various orientations, controlled by the two shear variables s_1, s_2 and various locations, controlled by t. Similar to the 2D case, in each pyramidal region the shearing variables are only allowed to vary over a compact set.

For $f \in L^2(\mathbb{R}^3)$, we define the *3D (fine-scale) pyramid-based continuous shearlet transform* $f \to \mathcal{SH}_\psi f(a,s_1,s_2,t)$, for $a > 0$, $s_1, s_2 \in \mathbb{R}$, $t \in \mathbb{R}^3$ by

$$\mathcal{SH}_\psi f(a,s_1,s_2,t) = \begin{cases} \langle f, \psi^{(1)}_{a,s_1,s_2,t} \rangle & \text{if } |s_1|,|s_2| \leq 1, \\ \langle f, \psi^{(2)}_{a,\frac{1}{s_1},\frac{s_2}{s_1},t} \rangle & \text{if } |s_1| > 1, |s_2| \leq |s_1| \\ \langle f, \psi^{(3)}_{a,\frac{s_1}{s_2},\frac{1}{s_2},t} \rangle & \text{if } |s_2| > 1, |s_2| > |s_1|. \end{cases}$$

That is, depending on the values of the shearing variables, the 3D continuous shear-let transform corresponds to one specific pyramid-based shearlet system. As above, we are only interested in the continuous shearlet transform at "fine scales," as a approaches 0, since this is all we need for the analysis of the singularities of f.

3.2 Characterization of 3D Boundaries

The 3D continuous shearlet transform shares the same properties as its 2D counter-part in terms of the ability to characterize the geometry of the set of singularities of a function or distribution f. In particular, it is possible to derive a characterization for the boundary set of some rather general solid regions.

To present these results, let us define the type of surfaces which will be con-sidered. Let $B = \chi_\Omega$, where Ω is a subset of $\mathbb{R}^3$ whose boundary $\partial\Omega$ is a two-dimensional manifold. We say that $\partial\Omega$ is *piecewise smooth* if:

(i) $\partial\Omega$ is a C^∞ manifold except possibly for finitely many separating C^3 curves on $\partial\Omega$.
(ii) at each point on a separating curve, $\partial\Omega$ has exactly two outer normal vectors, which are not on the same line.

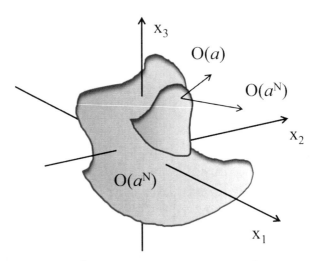

Fig. 3 General region $\Omega \subset \mathbb{R}^3$ with piecewise smooth boundary $\partial\Omega$. The continuous shear-let transform of $B = \chi_\Omega$ has rapid asymptotic decay except when the location variable t is on $\partial\Omega$ and the shearing variables (s_1, s_2) correspond to the normal direction at t; in this case $\mathscr{SH}_\psi B(a, s_1, s_2, t) \sim a$, as $a \to 0$

Let the outer normal vector of $\partial\Omega$ be $\mathbf{n}_p = \pm(\cos\theta_0\sin\phi_0, \sin\theta_0\sin\phi_0, \cos\phi_0)$ for some $\theta_0 \in [0, 2\pi]$, $\phi_0 \in [0, \pi]$. We say that $s = (s_1, s_2)$ corresponds to the normal direction $\mathbf{n}_p$ if $s_1 = a^{-\frac{1}{2}}\tan\theta_0$, $s_2 = a^{-\frac{1}{2}}\cot\phi_0\sec\theta_0$. Notice that this definition excludes, in particular, surfaces containing cusps, such as the vertex of a cone, since at the moment no argument is known for dealing with the type of points.

The following theorem shows that the behavior of the 3D continuous shearlet transform is consistent with the one found in dimension 2. Namely, for a bounded region in $\mathbb{R}^3$ whose boundary is a piecewise smooth two-dimensional manifold, the continuous shearlet transform of B, denoted by $\mathscr{SH}_\psi B(a, s_1, s_2, t)$, has rapid asymptotic decay as $a \to 0$ for all locations $t \in \mathbb{R}^3$, except when t is on the boundary of Ω and the orientation variables s_1, s_2 correspond to the normal direction of the boundary surface at t, or when t is on a separating curve and the orientation variables s_1, s_2 correspond to the normal direction of the boundary surface at t (see Fig. 3). Thus, as in the 2D case, the continuous shearlet transform provides a description of the geometry of $\partial\Omega$ through the asymptotic decay of $\mathscr{SH}_\psi B(a, s_1, s_2, t)$, at fine scales.

Theorem 5. *Let Ω be a bounded region in $\mathbb{R}^3$ and denote its boundary by $\partial\Omega$. Assume that $\partial\Omega$ is a piecewise smooth two-dimensional manifold. Let γ_j, $j = 1, 2, \ldots, m$ be the separating curves of $\partial\Omega$. Then we have*

(i) If $t \notin \partial\Omega$ then

$$\lim_{a\to 0^+} a^{-N}\,\mathscr{SH}_\psi B(a, s_1, s_2, t) = 0, \quad \textit{for all } N > 0.$$

(ii) If $t \in \partial\Omega \setminus \bigcup_{j=1}^m \gamma_j$ and (s_1, s_2) does not correspond to the normal direction of $\partial\Omega$ at t, then

$$\lim_{a\to 0^+} a^{-N}\,\mathscr{SH}_\psi B(a, s_1, s_2, t) = 0, \quad \textit{for all } N > 0.$$

(iii) If $t \in \partial\Omega \setminus \bigcup_{j=1}^m \gamma_j$ and $s = (s_1, s_2)$ corresponds to the normal direction of $\partial\Omega$ at t or $t \in \bigcup_{j=1}^m \gamma_j$ and $s = (s_1, s_2)$ corresponds to one of the two normal directions of $\partial\Omega$ at t, then

$$\lim_{a\to 0^+} a^{-1}\,\mathscr{SH}_\psi B(a, s_1, s_2, t) \neq 0.$$

(iv) If $t \in \gamma_j$ and (s_1, s_2) does not correspond to the normal directions of $\partial\Omega$ at t, then there is a constant C (possibly $C = 0$) such that

$$\lim_{a\to 0^+} a^{-3/2}\,\mathscr{SH}_\psi B(a, s_1, s_2, t) = C.$$

Hence, similar to the 2D case, the continuous shearlet transform decays rapidly away from the boundary and on the boundary, for nonnormal orientations. The decay rate is only $O(a^1)$ at the boundary, for normal orientation. However, the situation on the separating curves of the surface is less sharp, in the sense that, for normal orientations, the decay rate is $O(a^1)$, but, for nonnormal orientations, we can only say that the decay rate is of the order of or faster than $O(a^{3/2})$ (C could be zero). However there are examples where the decay rate on a separating curve, for nonnormal orientation, is exactly $O(a^{3/2})(C \neq 0)$. Thus, the rate $a^{\frac{3}{2}}$ in (iv) cannot be improved. In the following, we only make a few observations about the proof. We refer the reader to [9] for a complete proof of this result.

As in the 2D case, the starting point is the divergence theorem, which allows us to write the Fourier transform of B as

$$\hat{B}(\xi) = \hat{\chi}_\Omega(\xi) = -\frac{1}{2\pi i |\xi|^2} \int_{\partial\Omega} e^{-2\pi i \langle \xi, x \rangle} \xi \cdot \mathbf{n}(x) \, d\sigma(x),$$

where $\mathbf{n}$ is the outer normal vector to $\partial\Omega$ at x. Next, using spherical coordinates, we have that

$$\mathscr{SH}_\psi B(a, s_1, s_2, t) = \langle B, \psi_{a,s_1,s_2,t} \rangle = I_1(a, s_1, s_2, t) + I_2(a, s_1, s_2, t),$$

where

$$I_1(a, s_1, s_2, t) = \int_0^{2\pi} \int_0^\pi \int_0^\infty T_1(\rho, \theta, \phi) \, \overline{\hat{\psi}_{a,s_1,s_2,t}}(\rho, \theta, \phi) \rho^2 \sin\phi \, d\rho \, d\phi \, d\theta$$

$$I_2(a, s_1, s_2, t) = \int_0^{2\pi} \int_0^\pi \int_0^\infty T_2(\rho, \theta, \phi) \, \overline{\hat{\psi}_{a,s_1,s_2,t}}(\rho, \theta, \phi) \rho^2 \sin\phi \, d\rho \, d\phi \, d\theta$$

and

$$T_1(\rho, \theta, \phi) = -\frac{1}{2\pi i \rho} \int_{P_\varepsilon(t)} e^{-2\pi i \rho \Theta(\theta, \phi) \cdot x} \Theta(\theta, \phi) \cdot \mathbf{n}(x) \, d\sigma(x)$$

$$T_2(\rho, \theta, \phi) = -\frac{1}{2\pi i \rho} \int_{\partial\Omega \backslash P_\varepsilon(t)} e^{-2\pi i \rho \Theta(\theta, \phi) \cdot x} \Theta(\theta, \phi) \cdot \mathbf{n}(x) \, d\sigma(x),$$

where $P_\varepsilon(t) = \partial\Omega \cap \beta_\varepsilon(t)$, and $\beta_\varepsilon(t)$ is a ball of radius ε and center t. Notice that I_2 is associated with the term T_2 which is evaluated away from the location t of the continuous shearlet transform. Hence, a localization result similar to Lemma 2 shows that I_2 is rapidly decreasing as $a \to 0$. The rest of the proof, when t is located on a regular point of $\partial\Omega$, is similar to Theorem 2. By contrast, the situation where t is on a separating curve requires a different method than the one contained in the proof of Theorem 3.

Acknowledgments

The authors acknowledge support from NSF grant DMS 1008900/1008907; D.L. also acknowledges support from NSF grant DMS (Career) 1005799.

References

1. E. Candès and D. Donoho, *Continuous curvelet transform: I. Resolution of the wavefront set*, Appl. Comput. Harmon. Anal. **19** (2005), 162–197.
2. T. Chan and J. Shen, Image Processing and Analysis, SIAM, Philadelphia, 2005.
3. M. Do Carmo, *Differential geometry of Curves and Surfaces*, Prentice Hall, 1976.
4. P. Grohs, *Continuous shearlet frames and resolution of the wavefront set*, Monatsh. Math. **164** (2011), 393–426.
5. G. Easley, K. Guo, and D. Labate, *Analysis of Singularities and Edge Detection using the Shearlet Transform*, Proceedings of SAMPTA '09, Marseille 2009.
6. K. Guo, D. Labate and W. Lim, *Edge Analysis and identification using the Continuous Shearlet Transform*, Appl. Comput. Harmon. Anal. **27** (2009), 24–46.
7. K. Guo, and D. Labate, *Characterization and analysis of edges using the continuous shearlet transform*, SIAM Journal on Imaging Sciences **2** (2009), 959–986.
8. K. Guo, and D. Labate, *Analysis and Detection of Surface Discontinuities using the 3D Continuous Shearlet Transform*, Appl. Comput. Harmon. Anal. **30** (2011), 231–242.
9. K. Guo, and D. Labate, *Characterization of Piecewise-Smooth Surfaces Using the 3D Continuous Shearlet Transform*, to appear in J. Fourier Anal. Appl. (2012)
10. S. Jaffard, Y. Meyer, *Wavelet methods for pointwise regularity and local oscillations of functions*, Memoirs of the AMS, **123** n.587 (1996).
11. S. Jaffard, *Pointwise smoothness, two-microlocalization and wavelet coefficients*, Publications Matematiques **35** (1991), 155–168
12. G. Kutyniok and D. Labate, *Resolution of the wavefront set using continuous shearlets*, Trans. Amer. Math. Soc. **361** (2009), 2719–2754.
13. C. Herz, *Fourier transforms related to convex sets*, Ann. of Math. **75** (1962), 81–92.
14. M. Holschneider, *Wavelets. Analysis tool*, Oxford University Press, Oxford, 1995.
15. S. Mallat, *A Wavelet Tour of Signal Processing*, Academic Press, San Diego, 1998.
16. Y. Meyer, *Wavelets and Operators*, Cambridge Stud. Adv. Math. vol. 37, Cambridge Univ. Press, Cambridge, UK, 1992.
17. E. M. Stein, *Harmonic Analysis: real-variable methods, orthogonality, and oscillatory integrals,* Princeton University Press, Princeton, NJ, 1993.
18. S. Yi, D. Labate, G. R. Easley, and H. Krim, *A Shearlet approach to Edge Analysis and Detection,* IEEE Trans. Image Process **18**(5) (2009), 929–941.

Multivariate Shearlet Transform, Shearlet Coorbit Spaces and Their Structural Properties

Stephan Dahlke, Gabriele Steidl, and Gerd Teschke

Abstract This chapter is devoted to the generalization of the continuous shearlet transform to higher dimensions as well as to the construction of associated smoothness spaces and to the analysis of their structural properties, respectively. To construct canonical scales of smoothness spaces, so-called shearlet coorbit spaces, and associated atomic decompositions and Banach frames we prove that the general coorbit space theory of Feichtinger and Gröchenig is applicable for the proposed shearlet setting. For the two-dimensional case we show that for large classes of weights, variants of Sobolev embeddings exist. Furthermore, we prove that for natural subclasses of shearlet coorbit spaces which in a certain sense correspond to "cone-adapted shearlets" there exist embeddings into homogeneous Besov spaces. Moreover, the traces of the same subclasses onto the coordinate axis can again be identified with homogeneous Besov spaces. These results are based on the characterization of Besov spaces by atomic decompositions and rely on the fact that shearlets with compact support can serve as analyzing vectors for shearlet coorbit spaces. Finally, we demonstrate that the proposed multivariate shearlet transform can be used to characterize certain singularities.

Key words: Atomic decompositions, Banach frames, Coorbit theory, Embeddings, Multivariate shearlet transform, Singularity analysis, Smoothness spaces, Traces

S. Dahlke
Philipps-Universität Marburg, FB12 Mathematik und Informatik, Hans-Meerwein Straße, Lahnberge, 35032 Marburg, Germany, e-mail: dahlke@mathematik.uni-marburg.de

G. Steidl
Universität Kaiserslautern, Fachbereich Mathematik, Gottlieb-Daimler-Straße 48/516, 67663 Kaiserslautern, Germany
steidl@mathematik.uni-kl.de

G. Teschke
Hochschule Neubrandenburg - University of Applied Sciences, Institute for Computational Mathematics in Science and Technology, Bodaer Straße 2, 17033 Neubrandenburg, Germany
teschke@hs-nb.de

G. Kutyniok and D. Labate (eds.), *Shearlets: Multiscale Analysis for Multivariate Data*, Applied and Numerical Harmonic Analysis, DOI 10.1007/978-0-8176-8316-0_4, © Springer Science+Business Media, LLC 2012

1 Introduction

In the context of *directional* signal analysis and information retrieval several approaches have been suggested such as ridgelets [3], curvelets [4], contourlets [17], shearlets [29], and many others. Among all these approaches, the shearlet transform stands out because it is related to group theory, i.e., this transform can be derived from a square integrable representation $\pi : \mathscr{S} \to \mathscr{U}(L_2(\mathbb{R}^2))$ of a certain group $\mathscr{S}$, the so-called shearlet group, see [10]. An admissible function with respect to this group is called a *shearlet*. Therefore, in the context of the shearlet transform, all the powerful tools of group representation theory can be exploited.

For analyzing data in $\mathbb{R}^d$, $d \geq 3$, we have to generalize the shearlet transform to higher dimensions. The first step toward a higher dimensional shearlet transform is the identification of a suitable shear matrix. Given an d-dimensional vector space V and a k-dimensional subspace W of V, a reasonable model reads as follows: the shear should fix the space W and translate all vectors parallel to W. That is, for $V = W \oplus W'$ and $v = w + w'$, the shear operation S can be described as $S(v) = w + (w' + s(w'))$, where s is a linear mapping from W' to W. Then, with respect to an appropriate basis of V, the shear operation S corresponds to a block matrix of the form

$$S = \begin{pmatrix} I_k & s^{\mathrm{T}} \\ 0 & I_{d-k} \end{pmatrix}, \qquad s \in \mathbb{R}^{d-k,k}.$$

Then, we are faced with the problem how to choose the block s. Since we want to end up with a square integrable group representation, one has to be careful. Usually, the number of parameters has to fit together with the space dimension, otherwise the resulting group would be either too large or too small. Since we have d degrees of freedom related with the translates and one degree of freedom related with the dilation, $d - 1$ degrees of freedom for the shear component would be optimal. Therefore, one natural choice would be $s \in \mathbb{R}^{d-1,1}$, i.e., $k = 1$. Indeed, we show that with this choice the associated multivariate shearlet transform can be interpreted as a square integrable group representation of a $(2d)$-parameter group, the full shearlet group. It is a remarkable fact that this choice is in some sense a canonical one, other $(d-1)$-parameter choices might lead to nice group structures, but the representation will usually not be square integrable. Another approach, which we do not discuss in this chapter, involves shear matrices of Toeplitz type. We refer the interested reader to [9, 14].

With a square integrable group representation at hand, there is a very natural link to another useful concept, namely the coorbit space theory introduced by Feichtinger and Gröchenig in a series of papers [18, 19, 20, 21, 24]. By means of the coorbit space theory, it is possible to derive in a very natural way scales of smoothness spaces associated with the group representation. In this setting, the smoothness of functions is measured by the decay of the associated voice transform. Moreover, by a tricky discretization of the representation, it is possible to obtain (Banach) frames for these smoothness spaces. Fortunately, it turns out that for our multivariate continuous shearlet transform, all the necessary conditions for the application of the coorbit space theory can be established, so that we end up with new canonical smoothness spaces, the multivariate shearlet coorbit spaces, together with their atomic decompositions and Banach frames for these spaces.

Once these new smoothness spaces are established some natural questions arise. How do these spaces really look like? Are there "nice" sets of functions that are dense in these spaces? What are the relations to classical smoothness spaces such as Besov spaces? Do there exist embeddings into Besov spaces? And do there exist generalized versions of Sobolev embedding theorems for shearlet coorbit spaces? Moreover, can the associated trace spaces be identified? We shall provide some first answers to these questions. We concentrate on the two-dimensional case where we show that for natural subclasses of shearlet coorbit spaces which correspond to "cone-adapted shearlets," there exist embeddings into homogeneous Besov spaces and that for the same subclasses, the traces onto the coordinate axis can again be identified with homogeneous Besov spaces. The general d-dimensional scenario requires more sophisticated techniques than presented here and is the content of future work, see [8].

Finally, an interesting issue of the two-dimensional continuous shearlet transform is the fact that it can be used to analyze singularities. Indeed, as outlined in [32], see also [5] for curvelets, it turns out that the decay of the continuous shearlet transform exactly describes the location and orientation of certain singularities. By our approach these characterizations carry over to higher dimensions.

2 Multivariate Continuous Shearlet Transform

In this section, we introduce the shearlet transform on $L_2(\mathbb{R}^d)$. This requires the generalization of the parabolic dilation matrix and of the shear matrix. We will start with a rather general definition of shearlet groups in Sect. 2.1 and then restrict ourself to those groups having square integrable representations in Sect. 2.2.

In the following, let I_d denote the (d,d)-identity matrix and 0_d, respectively, 1_d the vectors with d entries 0, respectively 1.

2.1 Unitary Representations of the Shearlet Group

We define *dilation matrices* depending on one parameter $a \in \mathbb{R}^* := \mathbb{R} \setminus \{0\}$ by

$$A_a := \mathrm{diag}\,(a_1(a),\ldots,a_d(a)),$$

where $a_1(a) := a$ and $a_j(a) = a^{\alpha_j}$ with $\alpha_j \in (0,1)$, $j = 2,\ldots,d$. In order to have directional selectivity, the dilation factors at the diagonal of A_a should be chosen in an anisotropic way, i.e., $|a_k(a)|$, $k = 2, \ldots, d$, should increase less than linearly in a as $a \to \infty$. Our favorite choice will be

$$A_a := \begin{pmatrix} a & 0_{d-1}^{\mathsf{T}} \\ 0_{d-1} & \mathrm{sgn}(a)|a|^{\frac{1}{d}} I_{d-1} \end{pmatrix}. \tag{1}$$

In Sect. 6, we will see that this choice leads to an increase of the shearlet transform at hyperplane singularities as $|a| \to 0$. Consequently, this enables us to detect special directional information. For fixed $k \in \{1, \ldots, d\}$, we define our *shear matrices* by

$$S_s = \begin{pmatrix} I_k & s^{\mathsf{T}} \\ 0_{d-k,k} & I_{d-k} \end{pmatrix}, \qquad s \in \mathbb{R}^{d-k,k}. \tag{2}$$

The shear matrices form a subgroup of $GL_d(\mathbb{R})$.

Remark 1. Shear matrices on $\mathbb{R}^d$ were also considered in [28], see also [34]. We want to show the relation of those matrices to our setting (2). The authors in [28] call $S \in \mathbb{R}^{d,d}$ a *general shear matrix* if

$$(I_d - S)^2 = 0_{d,d}. \tag{3}$$

Of course, our matrices in (2) fulfill this condition. Condition (3) is equivalent to the fact that S decomposes as

$$S = P^{-1} \, \mathrm{diag}(J_1, \ldots, J_r, 1_{d-2r}) P, \quad J_j := \begin{pmatrix} 1 & 1 \\ 0 & 1 \end{pmatrix}, \quad r \leq d/2.$$

With $P := (p_1, \ldots, p_d)$ and $P^{-1} = (q_1, \ldots, q_d)^{\mathsf{T}}$ this can be written as

$$S = I_d + \sum_{j=1}^{r} q_{2j-1} p_{2j}^{\mathsf{T}}, \quad \text{with} \quad p_{2j}^{\mathsf{T}} q_{2i-1} = 0, \ i, j = 1, \ldots, r.$$

Matrices of the type $S_{qp} := I_d + q \, p^{\mathsf{T}}$ with $p^{\mathsf{T}} q = 0$ are called *elementary shear matrices*. The general shear matrices do not form a group. In particular, the product of two elementary shear matrices $S_{q_1 p_1}$ and $S_{q_2 p_2}$ is again a shear matrix if and only if the matrices commute which is the case if and only if $p_1^{\mathsf{T}} q_2 = p_2^{\mathsf{T}} q_1 = 0$. Then, $S_{q_1 p_1} S_{q_2 p_2} = I_n + \sum_{j=1}^{2} q_j p_j^{\mathsf{T}}$ holds true. Hence, we see that any general shear matrix is the product of elementary shear matrices. In [28] any subgroup of $GL_d(\mathbb{R})$ generated by *finitely* many pairwise commuting elementary matrices is called a *shear group*. A shear group is maximal if it is not a proper subgroup of any other shear group. It is not difficult to show that maximal shear groups are those of the form

$$G := \left\{ I_d + \left(\sum_{i=1}^{k} c_i q_i \right) \left(\sum_{j=1}^{d-k} d_j p_j^{\mathsf{T}} \right) : c_i, d_j \in \mathbb{R} \right\}, \quad p_j^{\mathsf{T}} q_i = 0,$$

with linearly independent vectors q_i, $i = 1,\ldots,k$, respectively, p_j, $j = 1,\ldots,k$. Let $\{\tilde{q}_i : i = 1,\ldots,k\}$ be the dual basis of $\{q_i : i = 1,\ldots,k\}$ in the linear space V spanned by these vectors and let $\{\tilde{p}_j : j = 1,\ldots,d-k\}$ be the dual basis of $\{p_j : j = 1,\ldots,d-k\}$ in $V^\perp$. Set $P := (q_1,\ldots,q_k,\tilde{p}_1,\ldots,\tilde{p}_{d-k})$ so that $P^{-1} = (\tilde{q}_1,\ldots,\tilde{q}_k,p_1,\ldots,p_{d-k})^{\mathrm{T}}$. Then, we see that for all $S \in G$

$$P^{-1}SP = \begin{pmatrix} I_k & cd^{\mathrm{T}} \\ 0_{d-k,k} & I_{d-k} \end{pmatrix}, \quad c = (c_1,\ldots,c_k)^{\mathrm{T}}, d = (d_1,\ldots,d_{d-k})^{\mathrm{T}}.$$

In other words, up to a basis transform, the maximal shear groups G coincide with our block matrix groups in (2).

Note that admissible subgroups of the semidirect product of the Heisenberg group and the symplectic group were examined in [6]. Some important progress in the construction of multivariate directional systems has been achieved for the curvelet case in [2] and for surfacelets in [35].

For our shearlet transform we have to combine dilation matrices and shear matrices. Let $A_{a,1} := \mathrm{diag}\,(a_1,\ldots,a_k)$ and $A_{a,2} := \mathrm{diag}\,(a_{k+1},\ldots,a_d)$. We will use the relations

$$S_s^{-1} = \begin{pmatrix} I_k & -s^{\mathrm{T}} \\ 0_{d-k,k} & I_{d-k} \end{pmatrix} \quad \text{and} \quad S_s A_a S_{s'} A_{a'} = S_{s+A_{a,2}^{-1}s'A_{a,1}} A_{aa'}. \tag{4}$$

For the special setting in (1), the last relation simplifies to

$$S_s A_a S_{s'} A_{a'} = S_{s+|a|^{1-\frac{1}{d}}s'} A_{aa'}.$$

Lemma 1. *The set $\mathbb{R}^* \times \mathbb{R}^{d-k,k} \times \mathbb{R}^d$ endowed with the operation*

$$(a,s,t) \circ (a',s',t') = (aa', s + A_{a,2}^{-1}s'A_{a,1}, t + S_s A_a t')$$

is a locally compact group $\mathbb{S}$. The left and right Haar measures on $\mathbb{S}$ are given by

$$d\mu_l(a,s,t) = \frac{|\det A_{a,2}|^{k-1}}{|a||\det A_{a,1}|^{d-k+1}}\,da\,ds\,dt \quad \text{and} \quad d\mu_r(a,s,t) = \frac{1}{|a|}\,da\,ds\,dt.$$

Proof. By the left relation in (4) it follows that $e := (1,0_{d-k,k},0_d)$ is the neutral element in $\mathbb{S}$ and that the inverse of $(a,s,t) \in \mathbb{R}^* \times \mathbb{R}^{d-1} \times \mathbb{R}^d$ is given by

$$(a,s,t)^{-1} = (a^{-1}, -A_{a,2}sA_{a,1}^{-1}, -A_a^{-1}S_s^{-1}t).$$

By straightforward computation it can be checked that the multiplication is associative.

Further, we have for a function F on $\mathbb{S}$ that

$$\int_{\mathbb{S}} F\big((a',s',t') \circ (a,s,t)\big)\,d\mu_l(a,s,t)$$

$$= \int_{\mathbb{R}} \int_{\mathbb{R}^{k(d-k)}} \int_{\mathbb{R}^d} F(a'a, s' + A_{a',2}^{-1}sA_{a',1}, t' + S_{s'}A_{a'}t)\,d\mu_l(a,s,t).$$

By substituting $\tilde{t} := t' + S_{s'}A_{a'}t$, i.e., $d\tilde{t} = |\det A_{a'}| dt$ and $\tilde{s} := s' + A_{a',2}^{-1} s A_{a',1}$, i.e., $d\tilde{s} = |\det A_{a',1}|^{d-k}/|\det A_{a',2}|^k ds$ and $\tilde{a} := a'a$ this can be rewritten as

$$\int_{\mathbb{R}^d} \int_{\mathbb{R}^{k(d-k)}} \int_{\mathbb{R}} F(\tilde{a},\tilde{s},\tilde{t}) \frac{1}{|\det A_{a'}|} \frac{|\det A_{a',2}|^k}{|\det A_{a',1}|^{d-k}} \frac{1}{|a'|} \frac{|a'||\det A_{a',1}|^{d-k+1}}{|\det A_{a',2}|^{k-1}} d\mu_l(\tilde{a},\tilde{s},\tilde{t})$$

so that $d\mu_l$ is indeed the left Haar measure on $\mathbb{S}$. Similarly, we can verify that $d\mu_r$ is the right Haar measure on $\mathbb{S}$. □

In the following, we use only the left Haar measure and the abbreviation $d\mu = d\mu_l$. For $f \in L_2(\mathbb{R}^d)$ we define

$$\pi(a,s,t)f(x) = f_{a,s,t}(x) := |\det A_a|^{-\frac{1}{2}} f(A_a^{-1} S_s^{-1}(x-t)). \tag{5}$$

It is easy to check that $\pi : \mathbb{S} \to \mathcal{U}(L_2(\mathbb{R}^d))$ is a mapping from $\mathbb{S}$ into the group $\mathcal{U}(L_2(\mathbb{R}^d))$ of unitary operators on $L_2(\mathbb{R}^d)$. Recall that a *unitary representation* of a locally compact group G with the left Haar measure μ on a Hilbert space $\mathcal{H}$ is a homomorphism π from G into the group of unitary operators $\mathcal{U}(\mathcal{H})$ on $\mathcal{H}$ which is continuous with respect to the strong operator topology.

Lemma 2. *The mapping π defined by (5) is a unitary representation of $\mathbb{S}$.*

Proof. We verify that π is a homomorphism. Let $\psi \in L_2(\mathbb{R}^d)$, $x \in \mathbb{R}^d$, and (a,s,t), $(a',s',t') \in \mathbb{S}$. Using (4) we obtain

$$\begin{aligned}
\pi(a,s,t)(\pi(a',s',t')\psi)(x) &= |\det A_a|^{-\frac{1}{2}} \left(\pi(a',s',t')\psi\right)(A_a^{-1}S_s^{-1}(x-t)) \\
&= |\det A_{aa'}|^{-\frac{1}{2}} \psi(A_{a'}^{-1}S_{s'}^{-1}(A_a^{-1}S_s^{-1}(x-t)-t')) \\
&= |\det A_{aa'}|^{-\frac{1}{2}} \psi(A_{a'}^{-1}S_{s'}^{-1}A_a^{-1}S_s^{-1}(x-(t+S_sA_at'))) \\
&= |\det A_{aa'}|^{-\frac{1}{2}} \psi(A_{aa'}^{-1}S_{s+A_{a,2}^{-1}s'A_{a,1}}^{-1}(x-(t+S_sA_at'))) \\
&= \pi((a,s,t)\circ(a',s',t'))\psi(x).
\end{aligned}$$

□

2.2 Square Integrable Representations of the Shearlet Group

A nontrivial function $\psi \in L_2(\mathbb{R}^d)$ is called *admissible*, if

$$\int_{\mathbb{S}} |\langle \psi, \pi(a,s,t)\psi \rangle|^2 d\mu(a,s,t) < \infty.$$

If π is irreducible and there exits at least one admissible function $\psi \in L_2(\mathbb{R}^d)$, then π is called *square integrable* . By the following remark we will only consider a special setting in the rest of this chapter.

Remark 2. Assume that our shear matrix has the form (2) with $s^T = (s_{ij})_{i,j=1}^{k,d-k} \in \mathbb{R}^{k,d-k}$. Let s contain N different entries (variables). We assume that $N \geq d - 1$ since we have one dilation parameter and otherwise the group becomes too small. Then, we obtain instead of (12)

$$\int_S |\langle f, \psi_{a,s,t} \rangle|^2 \, d\mu(a,s,t)$$

$$= \int_{\mathbb{R}} \int_{\mathbb{R}^d} \int_{\mathbb{R}^N} |\hat{f}(\omega)|^2 |\det A_a| |\hat{\psi}(A_a \left(\frac{\tilde{\omega}_1}{\tilde{\omega}_2 + s\tilde{\omega}_1} \right))|^2 \, d\mu(a,s,t) \qquad (6)$$

where $\tilde{\omega}_1 := (\omega_1, \ldots, \omega_k)^T$, $\tilde{\omega}_2 := (\omega_{k+1}, \ldots, \omega_d)^T$ and the *Fourier transform* $\mathscr{F} f_{a,s,t} = \hat{f}_{a,s,t}$ of $f_{a,s,t}$ is given by

$$\hat{f}_{a,s,t}(\omega) = \int_{\mathbb{R}^d} f_{a,s,t}(x) e^{-2\pi i \langle x, \omega \rangle} \, dx$$

$$= |\det A_a|^{\frac{1}{2}} e^{-2\pi i \langle t, \omega \rangle} \hat{f}(A_a^T S_s^T \omega)$$

$$= |\det A_a|^{\frac{1}{2}} e^{-2\pi i \langle t, \omega \rangle} \hat{f} \left(A_a \left(\frac{\tilde{\omega}_1}{\tilde{\omega}_2 + s\tilde{\omega}_1} \right) \right). \qquad (7)$$

Now we can use the following substitution procedure:

$$\xi_{k+1} := (\omega_{k+1} + s_{11}\omega_1 + \cdots + s_{1k}\omega_k), \qquad (8)$$

i.e., $d\xi_{k+1} = |\omega_1| ds_{11}$ and with corresponding modifications if some of the s_{1j}, $j > 1$ are the same as s_{11}. Then, we replace s_{11} in the other rows of $\tilde{\omega}_2 + s\tilde{\omega}_1$ where it appears by (8). Next we continue to substitute the second row if it contains an integration variable from s ($\neq s_{11}$). Continuing this substitution process up to the final row we have at the end replaced the lower $d - k$ values in $\hat{\psi}$ by $d - r$, $r \leq k$ variables $\xi_1 = \xi_{j_1}, \ldots, \xi_{j_{d-r}}$ and some functions depending only on $a, \omega, \xi_{j_1}, \ldots, \xi_{j_{d-r}}$. Consequently, the integrand depends only on these variables. However, we have to integrate over $a, \omega, \xi_{j_1}, \ldots, \xi_{j_{d-r}}$ and over the remaining $N - (d - r)$ variables from s. But then the integral in (6) becomes infinity unless $N = d - r$. Since $d - 1 \leq N$ this implies $r = k = 1$, i.e., our choice of S_s with (9).

By Remark 2 we will deal only with shear matrices (2) with $k = 1$, i.e., with

$$S = \begin{pmatrix} 1 & s^T \\ 0_{d-1} & I_{d-1} \end{pmatrix}, \qquad s \in \mathbb{R}^{d-1}. \qquad (9)$$

and with dilation matrices of the form (1). Then, we have that

$$d\mu(a,s,t) = \frac{1}{|a|^{d+1}} \, da \, ds \, dt.$$

Then, the following result shows that the unitary representation π defined in (5) is square integrable.

Theorem 1. *A function $\psi \in L_2(\mathbb{R}^d)$ is admissible if and only if it fulfills the* admissibility condition

$$C_\psi := \int_{\mathbb{R}^d} \frac{|\hat{\psi}(\omega)|^2}{|\omega_1|^d}\, d\omega < \infty. \tag{10}$$

If ψ is admissible, then, for any $f \in L_2(\mathbb{R}^d)$, the following equality holds true:

$$\int_{\mathbb{S}} |\langle f, \psi_{a,s,t}\rangle|^2\, d\mu(a,s,t) = C_\psi \|f\|_{L_2(\mathbb{R}^d)}^2. \tag{11}$$

In particular, the unitary representation π is irreducible and hence square integrable.

Proof. Employing the Plancherel theorem and (7), we obtain

$$\begin{aligned}
\int_{\mathbb{S}} |\langle f, \psi_{a,s,t}\rangle|^2\, d\mu(a,s,t) &= \int_{\mathbb{S}} |f * \psi_{a,s,0}^*(t)|^2\, dt\, ds\, \frac{da}{|a|^{d+1}} \\
&= \int_{\mathbb{R}} \int_{\mathbb{R}^{d-1}} \int_{\mathbb{R}^d} |\hat{f}(\omega)|^2 |\hat{\psi}^*_{a,s,0}(\omega)|^2\, d\omega\, ds\, \frac{da}{|a|^{d+1}} \\
&= \int_{\mathbb{R}} \int_{\mathbb{R}^{d-1}} \int_{\mathbb{R}^d} |\hat{f}(\omega)|^2 |\det A_a| |\hat{\psi}(A_a^{\mathsf{T}} S_s^{\mathsf{T}} \omega)|^2\, d\omega\, ds\, \frac{da}{|a|^{d+1}} \quad (12) \\
&= \int_{\mathbb{R}} \int_{\mathbb{R}^d} \int_{\mathbb{R}^{d-1}} |\hat{f}(\omega)|^2 \frac{|\det A_{a,2}|}{|a|^d} |\hat{\psi}\left(\begin{matrix} a\omega_1 \\ A_{a,2}(\tilde{\omega}+\omega_1 s) \end{matrix}\right)|^2\, ds\, d\omega\, da,
\end{aligned}$$

where $\psi_{a,s,0}^*(x) = \overline{\psi_{a,s,0}(-x)}$. Substituting $\tilde{\xi} := A_{a,2}(\tilde{\omega}+\omega_1 s)$, i.e., $|\det A_{a,2}| |\omega_1|^{d-1}\, ds = d\tilde{\xi}$, we obtain

$$\int_{\mathbb{S}} |\langle f, \psi_{a,s,t}\rangle|^2\, d\mu = |a|^{-d} \int_{\mathbb{R}} \int_{\mathbb{R}^d} \int_{\mathbb{R}^{d-1}} |\hat{f}(\omega)|^2 |\omega_1|^{-(d-1)} |\hat{\psi}\left(\begin{matrix} a\omega_1 \\ \tilde{\xi}\end{matrix}\right)|^2\, d\tilde{\xi}\, d\omega\, da.$$

Next, we substitute $\xi_1 := a\omega_1$, i.e., $\omega_1\, da = d\xi_1$ which results in

$$\begin{aligned}
\int_{\mathbb{S}} |\langle f, \psi_{a,s,t}\rangle|^2\, d\mu &= \int_{\mathbb{R}} \int_{\mathbb{R}^d} \int_{\mathbb{R}^{d-1}} |\hat{f}(\omega)|^2 \frac{|\omega_1|^d}{|\xi_1|^d |\omega_1|^d} |\hat{\psi}\left(\begin{matrix}\xi_1 \\ \tilde{\xi}\end{matrix}\right)|^2\, d\tilde{\xi}\, d\omega\, d\xi_1 \\
&= C_\psi \|f\|_{L_2(\mathbb{R}^d)}^2.
\end{aligned}$$

Setting $f := \psi$, we see that ψ is admissible if and only if C_ψ is finite.

The irreducibility of π follows from (11) in the same way as in [11]. □

2.3 Continuous Shearlet Transform

A function $\psi \in L_2(\mathbb{R}^d)$ fulfilling the admissibility condition (10) is called a *continuous shearlet*, the transform $\mathscr{SH}_\psi : L_2(\mathbb{R}^d) \to L_2(\mathbb{S})$,

$$\mathscr{SH}_\psi f(a,s,t) := \langle f, \psi_{a,s,t} \rangle = (f * \psi_{a,s,0}^*)(t),$$

continuous shearlet transform and $\mathbb{S}$ defined in Lemma 1 with (9) a *shearlet group*.

Remark 3. An example of a continuous shearlet can be constructed as follows: Let ψ_1 be an admissable wavelet with $\hat{\psi}_1 \in C^\infty(\mathbb{R})$ and supp $\hat{\psi}_1 \subseteq [-2, -\frac{1}{2}] \cup [\frac{1}{2}, 2]$, and let ψ_2 be such that $\hat{\psi}_2 \in C^\infty(\mathbb{R}^{d-1})$ and supp $\hat{\psi}_2 \subseteq [-1, 1]^{d-1}$. Then, the function $\psi \in L^2(\mathbb{R}^d)$ defined by

$$\hat{\psi}(\omega) = \hat{\psi}(\omega_1, \tilde{\omega}) = \hat{\psi}_1(\omega_1) \, \hat{\psi}_2 \left(\frac{1}{\omega_1} \tilde{\omega} \right)$$

is a continuous shearlet. The support of $\hat{\psi}$ is depicted for $\omega_1 \geq 0$ in Fig. 1.

Remark 4. In [34] the authors consider admissible subgroups G of $GL_d(\mathbb{R})$, i.e., those subgroups for which the semidirect product with the translation group gives rise to a square integrable representation $\pi(g,t)f(x) = |\det g|^{-\frac{1}{2}} f(g^{-1}(x-t))$. Let $\triangle$ denotes the modular function on G, i.e., $d\mu_l(g) = \triangle(g) \, d\mu_r(g)$ and write $\triangle \equiv |\det|$ meaning that $\triangle(g) = |\det g|$ for all $g \in G$. Then, [34] contains the following result:

1. If G is admissible, then $\triangle \not\equiv |\det|$ and $G_x^0 := \{g \in G : gx = x\}$ are compact for a.e. $x \in \mathbb{R}^d$.
2. If $\triangle \not\equiv |\det|$ and for a.e. $x \in \mathbb{R}^d$ there exits $\varepsilon(x) > 0$ such that $G_x^\varepsilon : \{g \in G : |gx - x| \leq \varepsilon(x)\}$ is compact, then G is admissible.

Unfortunately, the above conditions "just fail" to be a characterization of admissibility by the "ε-gap" in the compactness condition. In our case we have that $\triangle \not\equiv |\det|$ since $|a|^{-d} \neq |a| |a|^{\alpha_2 + \cdots + \alpha_d}$ for $|a| \neq 1$. Further, $G_x^0 = (1, 0_{d-1})$ and $G_x^\varepsilon = \{(a,s) : |a| \in [1 - \varepsilon_1, 1 + \varepsilon_1], s_j \in [-\varepsilon_j, \varepsilon_j], j = 2, \ldots, d\}$ for some small ε_j, so that the necessary condition (1) and the sufficient condition (2) are fulfilled.

3 General Concept of Coorbit Space Theory

In this section, we want to briefly recall the basic facts concerning the coorbit theory as developed by Feichtinger and Gröchenig in a series of papers [18, 19, 20, 21]. This theory is based on square-integrable group representations and has the following important advantages:

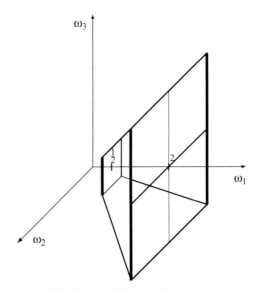

Fig. 1 Support of the shearlet $\hat{\psi}$ in Remark 3 for $\omega_1 \geq 0$

- The theory is universal in the following sense: Given a Hilbert space $\mathcal{H}$, a square-integrable representation of a group G and a nonempty set of so-called analyzing functions, the whole abstract machinery can be applied.
- The approach provides us with natural families of smoothness spaces, the coorbit spaces. They are defined as the collection of all elements in the Hilbert space $\mathcal{H}$ for which the voice transform associated with the group representation has a certain decay. In many cases, e.g., for the affine group and the Weyl-Heisenberg group, these coorbit spaces coincide with classical smoothness spaces such as Besov and modulation spaces, respectively.
- The Feichtinger–Gröchenig theory not only give rise to Hilbert frames in $\mathcal{H}$ but also to frames in scales of the associated coorbit spaces. Moreover, not only Hilbert spaces but also Banach spaces can be handled.
- The discretization process that produces the frame does not take place in $\mathcal{H}$ (which might look ugly and complicated), but on the topological group at hand (which is usually a more handy object), and is transported to $\mathcal{H}$ by the group representation.

First of all, in Sect. 3.1, we explain how the coorbit spaces can be established. Then, in Sect. 3.2, we discuss the discretization problem, i.e., we outline the basic steps to construct Banach frames for these spaces. The facts are mainly taken from [24].

3.1 General Coorbit Spaces

Fix an irreducible, unitary, continuous representation π of a σ-compact group G in a Hilbert space $\mathcal{H}$. Let w be real-valued, continuous, sub-multiplicative weight on $\mathbb{S}$, i.e., $w(gh) \leq w(g)w(h)$ for all $g, h \in \mathbb{S}$. Furthermore, we will always assume that the weight function w satisfies all the coorbit-theory conditions as stated in [24, Sect. 2.2]. To define our coorbit spaces we need the set

$$\mathscr{A}_w := \{\psi \in L_2(\mathbb{R}^d) : V_\psi(\psi) = \langle \psi, \pi(\cdot)\psi \rangle \in L_{1,w}\}$$

of *analyzing vectors*. In particular, we assume that our weight is symmetric with respect to the modular function, i.e., $w(g) = w(g^{-1})\triangle(g^{-1})$. Starting with an ordinary weight function w, its symmetric version can be obtained by $w^\#(g) := w(g) + w(g^{-1})\triangle(g^{-1})$. It was proved in Lemma 2.4 of [18] that $\mathscr{A}_w = \mathscr{A}_{w^\#}$.

For an analyzing vector ψ we can consider the space

$$\mathscr{H}_{1,w} := \{f \in L_2(\mathbb{R}^d) : V_\psi(f) = \langle f, \pi(\cdot)\psi \rangle \in L_{1,w}(G)\},$$

with norm $\|f\|_{\mathscr{H}_{1,w}} := \|V_\psi(f)\|_{L_{1,w}(G)}$ and its anti-dual $\mathscr{H}_{1,w}^\sim$, the space of all continuous conjugate-linear functionals on $\mathscr{H}_{1,w}$. The spaces $\mathscr{H}_{1,w}$ and $\mathscr{H}_{1,w}^\sim$ are π-invariant Banach spaces with continuous embeddings $\mathscr{H}_{1,w} \hookrightarrow \mathscr{H} \hookrightarrow \mathscr{H}_{1,w}^\sim$. Then, the inner product on $L_2(\mathbb{R}^d) \times L_2(\mathbb{R}^d)$ extends to a sesquilinear form on $\mathscr{H}_{1,w}^\sim \times \mathscr{H}_{1,w}$, therefore for $\psi \in \mathscr{H}_{1,w}$ and $f \in \mathscr{H}_{1,w}^\sim$ the *extended representation coefficients*

$$V_\psi(f)(g) := \langle f, \pi(g)\psi \rangle_{\mathscr{H}_{1,w}^\sim \times \mathscr{H}_{1,w}}$$

are well-defined.

Let m be a w-moderate weight on G which means that $m(xyz) \leq w(x)m(y)w(z)$ for all $x, y, z \in G$ and moreover, for $1 \leq p \leq \infty$, let

$$L_{p,m}(G) := \{F \text{ measurable} : Fm \in L_p(G)\}.$$

We can now define Banach spaces which are called *coorbit spaces* by

$$\mathscr{H}_{p,m} := \{f \in \mathscr{H}_{1,w}^\sim : V_\psi(f) \in L_{p,m}(G)\}, \quad \|f\|_{\mathscr{H}_{p,m}} := \|V_\psi f\|_{L_{p,m}(G)}.$$

Note that the definition of $\mathscr{H}_{p,m}$ is independent of the analyzing vector ψ and of the weight w in the sense that $\tilde{w}$ with $w(g) \leq C\tilde{w}(g)$ for all $g \in G$ and $\mathscr{A}_{\tilde{w}} \neq \{0\}$ give rise to the same space, see [18, Theorem 4.2]. In applications, one may start with some sub-multiplicative weight m and use the symmetric weight $w := m^\#$ for the definition of $\mathscr{A}_w$. Obviously, we have that m is w-moderate.

3.2 Atomic Decompositions and Banach Frames

The Feichtinger–Gröchenig theory also provides us with a machinery to construct atomic decompositions and Banach frames for the coorbit spaces introduced above. To this end, the subset $\mathscr{B}_w$ of $\mathscr{A}_w$ has to be nonempty:

$$\mathscr{B}_w := \{\psi \in L_2(\mathbb{R}^d) : V_\psi(\psi) \in \mathscr{W}(C_0, L_{1,w})\},$$

where $\mathscr{W}(C_0, L_{1,w})$ is the *Wiener-Amalgam space*

$$\mathscr{W}(C_0, L_{1,w}) := \{F : \|(L_x\chi_\mathscr{Q})F\|_\infty \in L_{1,w}\}, \quad \|(L_x\chi_\mathscr{Q})F\|_\infty = \sup_{y \in x\mathscr{Q}} |F(y)|$$

$L_x f(y) := f(x^{-1}y)$ is the *left translation* and $\mathscr{Q}$ is a relatively compact neighborhood of the identity element in G, see [24]. Note that in general $\mathscr{B}_w$ is defined with respect to the right version $\mathscr{W}^R(C_0, L_{1,w}) := \{F : \|(R_x\chi_\mathscr{Q})F\|_\infty = \sup_{y \in \mathscr{Q}x^{-1}} |F(y)| \in L_{1,w}\}$ of the Wiener-Amalgam space, where $R_x f(y) := f(yx)$ denotes the *right translation*. Since $V_\psi(\psi)(g) = V_\psi(\psi)(g^{-1})$ and assuming that $\mathscr{Q} = \mathscr{Q}^{-1}$ both definitions of $\mathscr{B}_w$ coincide. It follows that $\mathscr{B}_w \subset \mathscr{H}_{1,w}$.

Moreover, a (countable) family $X = (g_\lambda)_{\lambda \in \Lambda}$ in G is said to be *U-dense* if $\cup_{\lambda \in \Lambda} g_\lambda U = G$, and *separated* if for some compact neighborhood Q of e we have $g_i Q \cap g_j Q = \emptyset, i \neq j$, and *relatively separated* if X is a finite union of separated sets.

Then, the following decomposition theorem, which was proved in a general setting in [18, 19, 20], says that discretizing the representation by means of an U-dense set produces an atomic decomposition for $\mathscr{H}_{p,m}$. Furthermore, given such an atomic decomposition, the theorem provides conditions under which a function f is completely determined by its moments $\langle f, \pi(g_\lambda)\psi \rangle$ and how f can be reconstructed from these moments.

Theorem 2. *Let* $1 \leq p \leq \infty$ *and* $\psi \in \mathscr{B}_w$, $\psi \neq 0$. *Then, there exists a (sufficiently small) neighborhood U of e so that for any U-dense and relatively separated set $X = (g_\lambda)_{\lambda \in \Lambda}$ the set $\{\pi(g_\lambda)\psi : \lambda \in \Lambda\}$ provides an atomic decomposition and a Banach frame for $\mathscr{H}_{p,m}$:*

Atomic decompositions: *If* $f \in \mathscr{H}_{p,m}$, *then*

$$f = \sum_{\lambda \in \Lambda} c_\lambda(f)\pi(g_\lambda)\psi$$

where the sequence of coefficients depends linearly on f and satisfies

$$\|(c_\lambda(f))_{\lambda \in \Lambda}\|_{\ell_{p,m}} \leq C\|f\|_{\mathscr{H}_{p,m}}$$

with a constant C depending only on ψ and with $\ell_{p,m}$ being defined by

$$\ell_{p,m} := \{c = (c_\lambda)_{\lambda \in \Lambda} : \|c\|_{\ell_{p,m}} := \|cm\|_{\ell_p} < \infty\},$$

where $m = (m(g_\lambda))_{\lambda \in \Lambda}$. Conversely, if $(c_\lambda(f))_{\lambda \in \Lambda} \in \ell_{p,m}$, then $f = \sum_{\lambda \in \Lambda} c_\lambda \pi(g_\lambda) \psi$ is in $\mathcal{H}_{p,m}$ and

$$\|f\|_{\mathcal{H}_{p,m}} \leq C' \|(c_\lambda(f))_{\lambda \in \Lambda}\|_{\ell_{p,m}}.$$

Banach frames: *The set $\{\pi(g_\lambda)\psi : \lambda \in \Lambda\}$ is a Banach frame for $\mathcal{H}_{p,m}$ which means that*

(i) $f \in \mathcal{H}_{p,m}$ *if and only if* $(\langle f, \pi(g_\lambda)\psi \rangle_{\mathcal{H}_{1,m}^\sim \times \mathcal{H}_{1,m}})_{\lambda \in \Lambda} \in \ell_{p,m}$;

(ii) *There exists two constants* $0 < D \leq D' < \infty$ *such that*

$$D\|f\|_{\mathcal{H}_{p,m}} \leq \|(\langle f, \pi(g_\lambda)\psi \rangle_{\mathcal{H}_{1,m}^\sim \times \mathcal{H}_{1,m}})_{\lambda \in \Lambda}\|_{\ell_{p,m}} \leq D'\|f\|_{\mathcal{H}_{p,m}};$$

(iii) *There exists a bounded, linear reconstruction operator $\mathcal{R}$ from $\ell_{p,m}$ to $\mathcal{H}_{p,m}$ such that* $\mathcal{R}\left((\langle f, \pi(g_\lambda)\psi \rangle_{\mathcal{H}_{1,m}^\sim \times \mathcal{H}_{1,m}})_{\lambda \in \Lambda}\right) = f$.

4 Multivariate Shearlet Coorbit Theory

In this section we want to establish the coorbit theory based on the square integrable representation (5) of the shearlet group $\mathbb{S}$ defined with (1) and (9). We mainly follow the lines of [11] and [12].

4.1 Shearlet Coorbit Spaces

We consider weight functions $w(a,s,t) = w(a,s)$ that are locally integrable with respect to a and s, i.e., $w \in L_1^{loc}(\mathbb{R}^d)$ and fulfill the requirements made at the beginning of Sect. 3.1.

In order to construct the coorbit spaces related to the shearlet group we have to ensure that there exists a function $\psi \in L_2(\mathbb{R}^d)$ such that

$$\mathcal{SH}_\psi(\psi) = \langle \psi, \pi(a,s,t)\psi \rangle \in L_{1,w}(\mathbb{S}). \tag{13}$$

Concerning the integrability of group representations we also mention [26]. To this end, we need a preliminary lemma on the support of ψ which is shown in [12].

Lemma 3. *Let $a_1 > a_0 \geq \alpha > 0$ and $b = (b_1, \ldots, b_{d-1})^{\mathsf{T}}$ be a vector with positive components. Suppose that supp $\hat{\psi} \subseteq ([-a_1, -a_0] \cup [a_0, a_1]) \times Q_b$, where $Q_b := [-b_1, b_1] \times \cdots \times [-b_{d-1}, b_{d-1}]$. Then, $\hat{\psi}\hat{\psi}_{a,s,0} \not\equiv 0$ implies that $a \in \left[-\frac{a_1}{a_0}, -\frac{a_0}{a_1}\right] \cup \left[\frac{a_0}{a_1}, \frac{a_1}{a_0}\right]$ and $s \in Q_c$, where $c := \frac{1+(a_1/a_0)^{1/d}}{a_0} b$.*

Now we can prove the required property (13) of $\mathcal{SH}_\psi(\psi)$.

Theorem 3. *Let ψ be a Schwartz function such that supp $\hat{\psi} \subseteq ([-a_1, -a_0] \cup [a_0, a_1]) \times Q_b$. Then, we have that $\mathcal{SH}_\psi(\psi) \in L_{1,w}(\mathbb{S})$, i.e.,*

$$\|\langle \psi, \pi(\cdot)\psi \rangle\|_{L_{1,w}(\mathbb{S})} = \int_{\mathbb{S}} |\mathcal{SH}_\psi(\psi)(a,s,t)| w(a,s,t) \, d\mu(a,s,t) < \infty.$$

Proof. Straightforward computation gives

$$
\begin{aligned}
\|\langle \psi, \pi(\cdot)\psi\rangle\|_{L_{1,w}(\mathbb{S})} &= \int_{\mathbb{R}}\int_{\mathbb{R}^{d-1}}\int_{\mathbb{R}^d}|\langle \psi, \psi_{a,s,t}\rangle|\, w(a,s)\,\mathrm{d}t\mathrm{d}s\frac{\mathrm{d}a}{|a|^{d+1}} \\
&= \int_{\mathbb{R}}\int_{\mathbb{R}^{d-1}}\int_{\mathbb{R}^d}|\psi * \psi_{a,s,0}^*(t)|\, w(a,s)\,\mathrm{d}t\mathrm{d}s\frac{\mathrm{d}a}{|a|^{d+1}} \\
&= \int_{\mathbb{R}}\int_{\mathbb{R}^{d-1}}\int_{\mathbb{R}^d}|\mathscr{F}^{-1}\mathscr{F}\left(\psi * \psi_{a,s,0}^*\right)(t)|\,\mathrm{d}t\, w(a,s)\,\mathrm{d}s\frac{\mathrm{d}a}{|a|^{d+1}} \\
&= \int_{\mathbb{R}}\int_{\mathbb{R}^{d-1}}\|\mathscr{F}\left(\psi * \psi_{a,s,0}^*\right)\|_{\mathscr{F}^{-1}L_1}\, w(a,s)\,\mathrm{d}s\frac{\mathrm{d}a}{|a|^{d+1}} \\
&= \int_{\mathbb{R}}\int_{\mathbb{R}^{d-1}}\|\hat{\psi}\,\bar{\hat{\psi}}_{a,s,0}\|_{\mathscr{F}^{-1}L_1}\, w(a,s)\,\mathrm{d}s\frac{\mathrm{d}a}{|a|^{d+1}},
\end{aligned}
$$

where $\|f\|_{\mathscr{F}^{-1}L_1(\mathbb{R}^d)} := \int_{\mathbb{R}^d}|\mathscr{F}^{-1}f(x)|\,\mathrm{d}x$. By Lemma 3 this can be rewritten as

$$
\|\langle \psi, \pi(\cdot)\psi\rangle\|_{L_{1,w}(\mathbb{S})} = \left(\int_{-a_1/a_0}^{-a_0/a_1} + \int_{a_0/a_1}^{a_1/a_0}\right)\int_{Q_c}\|\hat{\psi}\,\hat{\psi}_{a,s,0}^*\|_{\mathscr{F}^{-1}L_1(\mathbb{R}^d)}\, w(a,s)\,\mathrm{d}s\frac{\mathrm{d}a}{|a|^{d+1}},
$$

which is obviously finite. □

For ψ satisfying (13) we can, therefore, consider the space

$$
\mathscr{H}_{1,w} = \{f \in L_2(\mathbb{R}^d) : \mathscr{SH}_\psi(f) \in L_{1,w}(\mathbb{S})\}
$$

and its anti-dual $\mathscr{H}_{1,w}^{\sim}$. By the reasoning of Sect. 3.1, the *extended representation coefficients*

$$
\mathscr{SH}_\psi(f)(a,s,t) = \langle f, \pi(a,s,t)\psi\rangle_{\mathscr{H}_{1,w}^{\sim}\times\mathscr{H}_{1,w}}
$$

are well-defined and, for $1 \le p \le \infty$, we can define Banach spaces which we call *shearlet coorbit spaces* by

$$
\mathscr{SC}_{p,m} := \{f \in \mathscr{H}_{1,w}^{\sim} : \mathscr{SH}_\psi(f) \in L_{p,m}(\mathbb{S})\}
$$

with norms $\|f\|_{\mathscr{SC}_{p,m}} := \|\mathscr{SH}_\psi f\|_{L_{p,m}(\mathbb{S})}$.

4.2 Shearlet Atomic Decompositions and Shearlet Banach Frames

In a first step, we have to determine, for a compact neighborhood U of $e \in \mathbb{S}$ with non-void interior, U-dense sets, see Sect. 3.2.

Lemma 4. *Let U be a neighborhood of the identity in $\mathbb{S}$, and let $\alpha > 1$ and $\beta, \tau > 0$ be defined such that*

$$
[\alpha^{\frac{1}{d}-1}, \alpha^{\frac{1}{d}}) \times [-\tfrac{\beta}{2}, \tfrac{\beta}{2})^{d-1} \times [-\tfrac{\tau}{2}, \tfrac{\tau}{2})^d \subseteq U.
$$

Then, the sequence

$$\{(\varepsilon\alpha^j, \beta\alpha^{j(1-\frac{1}{d})}k, S_{\beta\alpha^{j(1-\frac{1}{d})}k}A_{\alpha^j}\tau m) : j \in \mathbb{Z}, k \in \mathbb{Z}^{d-1}, m \in \mathbb{Z}^d, \varepsilon \in \{-1,1\}\} \quad (14)$$

is U-dense and relatively separated.

For a proof we refer to [12]. Then, as shown in [12], the following decomposition theorem says that discretizing the representation by means of this U-dense set produces an atomic decomposition for $\mathscr{S}\mathscr{C}_{p,m}$.

Theorem 4. *Let* $1 \leq p \leq \infty$ *and* $\psi \in \mathscr{B}_w$, $\psi \neq 0$. *Then there exists a (sufficiently small) neighborhood* U *of* e *so that for any* U-*dense and relatively separated set* $X = ((a,s,t)_\lambda)_{\lambda\in\Lambda}$ *the set* $\{\pi(g_\lambda)\psi : \lambda \in \Lambda\}$ *provides an atomic decomposition and a Banach frame for* $\mathscr{S}\mathscr{C}_{p,m}$:

Atomic decompositions: *If* $f \in \mathscr{S}\mathscr{C}_{p,m}$, *then*

$$f = \sum_{\lambda\in\Lambda} c_\lambda(f)\pi((a,s,t)_\lambda)\psi \quad (15)$$

where the sequence of coefficients depends linearly on f *and satisfies*

$$\|(c_\lambda(f))_{\lambda\in\Lambda}\|_{\ell_{p,m}} \leq C\|f\|_{\mathscr{S}\mathscr{C}_{p,m}}.$$

Conversely, if $(c_\lambda(f))_{\lambda\in\Lambda} \in \ell_{p,m}$, *then* $f = \sum_{\lambda\in\Lambda} c_\lambda\pi((a,s,t)_\lambda)\psi$ *is in* $\mathscr{S}\mathscr{C}_{p,m}$ *and*

$$\|f\|_{\mathscr{S}\mathscr{C}_{p,m}} \leq C'\|(c_\lambda(f))_{\lambda\in\Lambda}\|_{\ell_{p,m}}.$$

Banach frames: *The set* $\{\pi(g_\lambda)\psi : \lambda \in \Lambda\}$ *is a Banach frame for* $\mathscr{S}\mathscr{C}_{p,m}$ *which means that*

(i) $f \in \mathscr{S}\mathscr{C}_{p,m}$ *if and only if* $(\langle f, \pi((a,s,t)_\lambda)\psi\rangle_{\mathscr{H}_{1,w}^{\sim}\times\mathscr{H}_{1,w}})_{\lambda\in\Lambda} \in \ell_{p,m}$;

(ii) *There exists two constants* $0 < D \leq D' < \infty$ *such that*

$$D\|f\|_{\mathscr{S}\mathscr{C}_{p,m}} \leq \|(\langle f, \pi((a,s,t)_\lambda)\psi\rangle_{\mathscr{H}_{1,w}^{\sim}\times\mathscr{H}_{1,w}})_{\lambda\in\Lambda}\|_{\ell_{p,m}} \leq D'\|f\|_{\mathscr{S}\mathscr{C}_{p,m}};$$

(iii) *There exists a bounded, linear reconstruction operator* $\mathscr{R}$ *from* $\ell_{p,m}$ *to* $\mathscr{S}\mathscr{C}_{p,m}$ *such that* $\mathscr{R}\left((\langle f, \pi((a,s,t)_\lambda)\psi\rangle_{\mathscr{H}_{1,w}^{\sim}\times\mathscr{H}_{1,w}})_{\lambda\in\Lambda}\right) = f$.

4.3 Nonlinear Approximation

In Sect. 4.2, we established atomic decompositions of functions from the shearlet coorbit spaces $\mathscr{S}\mathscr{C}_{p,m}$ by means of special discretized shearlet systems $(\psi_\lambda)_{\lambda\in\Lambda}$, $\Lambda \subset \mathbb{S}$. From the computational point of view, this naturally leads us to the question of the quality of approximation schemes in $\mathscr{SC}_{p,m}$ using only a finite number of elements from $(\psi_\lambda)_{\lambda \in \Lambda}$.

In this section, we will focus on the nonlinear approximation scheme of *best N-term approximation*, i.e., of approximating functions f of $\mathscr{SC}_{p,m}$ in an "optimal" way by a linear combination of precisely N elements from $(\psi_\lambda)_{\lambda \in \Lambda}$. In order to study the quality of best N-term approximation we will prove estimates for the asymptotic behavior of the approximation error.

Let us now delve more into the specific setting we are considering here. Let U be a properly chosen small neighborhood of e in $\mathbb{S}$. Further, let $\Lambda \subset \mathbb{S}$ be a relatively separated, U-dense sequence, which exists by Lemma 4. Then, the associated shearlet system

$$\{\psi_\lambda = \psi_{a,s,t} : \lambda = (a,s,t) \in \Lambda\} \tag{16}$$

can be employed for atomic decompositions of elements from $\mathscr{SC}_{p,m}$, where $1 \leq p < \infty$. Indeed, by Theorem 4, for any $f \in \mathscr{SC}_{p,m}$, we have

$$f = \sum_{\lambda \in \Lambda} c_\lambda \psi_\lambda$$

with $(c_\lambda)_{\lambda \in \Lambda}$ depending linearly on f, and

$$C_1 \|f\|_{\mathscr{SC}_{p,m}} \leq \|(c_\lambda)_{\lambda \in \Lambda}\|_{\ell_{p,m}} \leq C_2 \|f\|_{\mathscr{SC}_{p,m}}$$

with constants C_1, C_2 being independent of f. We intend to approximate functions f from the shearlet coorbit spaces $\mathscr{SC}_{p,m}$ by elements from the nonlinear manifolds Σ_n, $n \in \mathbb{N}$, which consist of all functions $S \in \mathscr{SC}_{p,m}$ whose expansions with respect to the shearlet system $(\psi_\lambda)_{\lambda \in \Lambda}$ from (16) have at most n nonzero coefficients, i.e.,

$$\Sigma_n := \left\{ S \in \mathscr{SC}_{p,m} : S = \sum_{\lambda \in \Gamma} d_\lambda \psi_\lambda, \Gamma \subseteq \Lambda, \#\Gamma \leq n \right\}.$$

Then, we are interested in the asymptotic behavior of the error

$$E_n(f)_{\mathscr{SC}_{p,m}} := \inf_{S \in \Sigma_n} \|f - S\|_{\mathscr{SC}_{p,m}}.$$

Usually, the order of approximation which can be achieved depends on the regularity of the approximated function as measured in some associated smoothness space. For instance, for nonlinear wavelet approximation, the order of convergence is determined by the regularity as measured in a specific scale of Besov spaces, see [15]. For nonlinear approximation based on Gabor frames, it has been shown in [27] that the "right" smoothness spaces are given by a specific scale of modulation spaces. An extension of these relations to systems arising from the Weyl-Heisenberg group and α-modulation spaces has been studied in [7].

In our case it turns out that a result from [27], i.e., an estimate in one direction, carries over. The basic ingredient in the proof of the theorem is the following lemma which has been shown in [27], see also [16].

Lemma 5. *Let $0 < p < q \leq \infty$. Then, there exist a constant $D_p > 0$ independent of q such that, for all decreasing sequences of positive numbers $a = (a_i)_{i=1}^{\infty}$, we have*

$$2^{-1/p}\|a\|_{\ell_p} \leq \left(\sum_{n=1}^{\infty} \frac{1}{n} (n^{1/p-1/q} E_{n,q}(a))^p \right)^{1/p} \leq D_p \|a\|_{\ell_p},$$

where $E_{n,q}(a) := \left(\sum_{i=n}^{\infty} a_i^q \right)^{1/q}$.

The following theorem, which provides an upper estimate for the asymptotic behavior of $E_n(f)_{\mathscr{SC}_{p,m}}$ was proved in [11].

Theorem 5. *Let $(\psi_\lambda)_{\lambda \in \Lambda}$ be a discrete shearlet system as in (16), and let $1 \leq p < q < \infty$. Then, there exist a constant $C = C(p,q) < \infty$ such that, for all $f \in \mathscr{SC}_{p,m}$, we have*

$$\left(\sum_{n=1}^{\infty} \frac{1}{n} \left(n^{1/p-1/q} E_n(f)_{\mathscr{SC}_{q,m}} \right)^p \right)^{1/p} \leq C\|f\|_{\mathscr{SC}_{p,m}}.$$

5 Structure of Shearlet Coorbit Spaces

In this section we provide some first structural properties of the shearlet coorbit spaces. We use the notation $f \lesssim g$ for the relation $f \leq Cg$ with some generic constant $C \geq 0$, and the notation "$\sim$" stands for equivalence up to constants which are independent of the involved parameters.

The subsequent analysis is limited to the *two-dimensional* case (more general concepts are provided in [8]), and we show that

- For large classes of weights, variants of Sobolev embeddings exist;
- For natural subclasses which in a certain sense correspond to the "cone adapted shearlets" [32], there exists embeddings into (homogeneous) Besov spaces;
- For the same subclass, the traces onto the coordinate axis can again be identified with homogeneous Besov spaces.

The two-dimensional approach heavily relies on the atomic decomposition techniques. We have seen that the coorbit space theory naturally gives rise to Banach frames, and therefore, by using the associated norm equivalences, all the tasks outlined above can be studied by means of weighted sequence norms. In particular, based on the general analysis in [30], quite recently this technique has been applied to derive new embedding and trace results for Besov spaces [36].

To make this approach really powerful, it is very convenient and sometimes even necessary to work with *compactly supported* building blocks. In the shearlet case, this is a nontrivial problem, since usually the analyzing shearlets are band-limited functions, see Theorem 3. For the specific case of "cone-adapted shearlets", quite recently a first solution has been provided in [31]. We refer to the overview article [33] for a detailed discussion. As the "cone adapted shearlets" do not really fit into the group theoretical setting, we provide a new construction of families of compactly supported shearlets. We show that indeed a compactly supported function with sufficient smoothness and enough vanishing moments can serve as an analyzing vector for shearlet coorbit spaces, i.e., we show that $\mathscr{A}_w$ contains shearlets with compact support. To this end, we need the following auxiliary lemma which is a modification of Lemma 11.1.1 in [25].

Lemma 6. *For $r > 1$ and $\alpha > 0$, the following estimate holds true*

$$I(x) := \int_{\mathbb{R}} (1+|t|)^{-r}(1+\alpha|x-t|)^{-r}\,dt \leq C\left(\frac{1}{\alpha}(1+|x|)^{-r} + (1+\alpha|x|)^{-r}\right).$$

Proof. Let

$$\mathscr{N}_x := \left\{t \in \mathbb{R} : |t-x| \leq \frac{|x|}{2}\right\}, \quad \mathscr{N}_x^c := \left\{t \in \mathbb{R} : |t-x| > \frac{|x|}{2}\right\}.$$

Then, we obtain for $t \in \mathscr{N}_x$ by $|x| - |t| \leq |t-x| \leq |x|/2$ that $|t| \geq |x|/2$ and consequently

$$(1+|t|)^{-r} \leq \left(1 + \frac{|x|}{2}\right)^{-r} \leq 2^r(1+|x|)^{-r}.$$

Now the above integral can be estimated as follows:

$$I(x) = \int_{\mathscr{N}_x}(1+|t|)^{-r}(1+\alpha|x-t|)^{-r}dt + \int_{\mathscr{N}_x^c}(1+|t|)^{-r}(1+\alpha|x-t|)^{-r}dt$$

$$\leq 2^r(1+|x|)^{-r}\int_{\mathscr{N}_x}(1+\alpha|x-t|)^{-r}dt + \left(1 + \alpha\frac{|x|}{2}\right)^{-r}\int_{\mathscr{N}_x^c}(1+|t|)^{-r}dt$$

$$\leq 2^r\frac{1}{\alpha}(1+|x|)^{-r}\int_{\mathbb{R}}(1+|u|)^{-r}du + 2^r(1+\alpha|x|)^{-r}\int_{\mathbb{R}}(1+|t|)^{-r}dt.$$

This implies the assertion. □

Theorem 6. *For some $D > 0$, let $Q_D := [-D,D] \times [-D,D]$ and let $\psi(x) \in L_2(\mathbb{R}^2)$ fulfill supp $\psi \in Q_D$. Suppose that the weight function satisfies $w(a,s,t) = w(a) \leq |a|^{-\rho_1} + |a|^{\rho_2}$ for $\rho_1, \rho_2 \geq 0$ and that*

$$|\hat{\psi}(\omega_1, \omega_2)| \lesssim \frac{|\omega_1|^n}{(1+|\omega_1|)^r}\frac{1}{(1+|\omega_2|)^r}$$

with $n \geq \max(\frac{1}{4}+\rho_2, \frac{9}{4}+\rho_1)$ and $r > n + \max(\frac{7}{4}+\rho_2, \frac{9}{4}+\rho_1)$. Then, we have that $\mathscr{SH}_\psi(\psi) \in L_{1,w}(\mathbb{S})$, i.e.,

$$I := \int_{\mathbb{S}}|\mathscr{SH}_\psi(\psi)(g)|\,w(g)\,d\mu(g) < \infty.$$

Proof. First, we have by the support property of ψ that $\mathscr{SH}_\psi(\psi) = \langle \psi, \psi_{a,s,t}\rangle \neq 0$ requires $(x_1, x_2) \in Q_D$ and

$$-D \le \frac{\operatorname{sgn} a}{\sqrt{|a|}} (x_2 - t_2) \le D,$$

$$-D \le \frac{1}{a}(x_1 - t_1 - s(x_2 - t_2)) \le D.$$

Hence, $\langle \psi, \psi_{a,s,t} \rangle \ne 0$ implies that

$$-D(1 + \sqrt{|a|}) \le t_2 \le D(1 + \sqrt{|a|}),$$

$$-D\left(1 + |a| + |s|(2 + \sqrt{|a|})\right) \le t_1 \le D\left(1 + |a| + |s|(2 + \sqrt{|a|})\right).$$

Using this relation we obtain that

$$I \le \int_{\mathbb{R}^*} \int_{\mathbb{R}} 4D^2 (1 + \sqrt{|a|}) \left(1 + |a| + |s|(2 + \sqrt{|a|})\right) |\langle \psi, \psi_{a,s,t} \rangle| \, ds \, w(a) \frac{da}{|a|^3}.$$

Next, Plancherel's equality together with (7) and the decay assumptions on $\hat{\psi}$ yield

$$I \le C \int_{\mathbb{R}^*} \int_{\mathbb{R}} (1 + \sqrt{|a|}) \left(1 + |a| + |s|(2 + \sqrt{|a|})\right) |\langle \hat{\psi}, \hat{\psi}_{a,s,t} \rangle| \, ds \, w(a) \frac{da}{|a|^3}$$

$$\le C \int_{\mathbb{R}^*} \int_{\mathbb{R}} \left(\underbrace{1 + |a|^{\frac{1}{2}} + |a| + |a|^{\frac{3}{2}}}_{=: p_3(|a|^{\frac{1}{2}})} + |s|(\underbrace{2 + 3|a|^{\frac{1}{2}} + a}_{=: p_2(|a|^{\frac{1}{2}})}) \right) J(a,s) \, ds \, w(a) \frac{da}{|a|^3},$$

where $p_k \in \Pi_k$ are polynomials of degree $\le k$, $|\mathscr{SH}_\psi \psi(a,s,t)| \le J(a,s)$ and

$$J(a,s) := |a|^{\frac{3}{4}}$$
$$\times \int_{\mathbb{R}} \int_{\mathbb{R}} \frac{|\omega_1|^n}{(1 + |\omega_1|)^r} \frac{1}{(1 + |\omega_2|)^r} \frac{|a\omega_1|^n}{(1 + |a\omega_1|)^r} \frac{1}{(1 + \sqrt{|a|}\, |s\omega_1 + \omega_2|)^r} \, d\omega_2 d\omega_1$$
$$= \int_{\mathbb{R}} \frac{|\omega_1|^n}{(1 + |\omega_1|)^r} \frac{|a\omega_1|^n}{(1 + |a\omega_1|)^r} \int_{\mathbb{R}} \frac{1}{(1 + |\omega_2|)^r} \frac{1}{(1 + \sqrt{|a|}\, |s\omega_1 + \omega_2|)^r} \, d\omega_2 d\omega_1.$$

The inner integral can be estimated by Lemma 6 which results in

$$J(a,s) \le C |a|^{n + \frac{3}{4}}$$
$$\times \int_{\mathbb{R}} \frac{|\omega_1|^n}{(1 + |\omega_1|)^r} \frac{|\omega_1|^n}{(1 + |a\omega_1|)^r} \left(\frac{1}{\sqrt{|a|}\,(1 + |s\omega_1|)^r} + \frac{1}{(1 + \sqrt{|a|}\,|s\omega_1|)^r} \right) d\omega_1.$$

Now we obtain

$$I \le C \left(\int_{\mathbb{R}^*} \int_{\mathbb{R}} \int_{\mathbb{R}} |a|^{n - \frac{11}{4}} (p_3 + |s|p_2) \frac{|\omega_1|^{2n}}{(1 + |\omega_1|)^r (1 + |a\omega_1|)^r} \frac{w(a)}{(1 + |s\omega_1|)^r} \, ds \, d\omega_1 \, da \right.$$
$$\left. + \int_{\mathbb{R}^*} \int_{\mathbb{R}} \int_{\mathbb{R}} |a|^{n - \frac{9}{4}} (p_3 + |s|p_2) \frac{|\omega_1|^{2n}}{(1 + |\omega_1|)^r (1 + |a\omega_1|)^r} \frac{w(a)}{(1 + \sqrt{|a|}\,|s\omega_1|)^r} \, ds \, d\omega_1 \, da \right).$$

Since the integrand is even in ω_1, s, and a this can be further simplified as

$$
I \leq C \left(\int_0^\infty a^{n-\frac{11}{4}} p_3(\sqrt{a}) \int_0^\infty \frac{\omega_1^{2n}}{(1+\omega_1)^r(1+a\omega_1)^r} \int_0^\infty \frac{w(a)}{(1+s\omega_1)^r} \, ds \, d\omega_1 \, da \right.
$$

$$
+ \int_0^\infty a^{n-\frac{11}{4}} p_2(\sqrt{a}) \int_0^\infty \frac{\omega_1^{2n}}{(1+\omega_1)^r(1+a\omega_1)^r} \int_0^\infty \frac{w(a)s}{(1+s\omega_1)^r} \, ds \, d\omega_1 \, da
$$

$$
+ \int_0^\infty a^{n-\frac{9}{4}} p_3(\sqrt{a}) \int_0^\infty \frac{\omega_1^{2n}}{(1+\omega_1)^r(1+a\omega_1)^r} \int_0^\infty \frac{w(a)}{(1+\sqrt{a}s\omega_1)^r} \, ds \, d\omega_1 \, da
$$

$$
\left. + \int_0^\infty a^{n-\frac{9}{4}} p_2(\sqrt{a}) \int_0^\infty \frac{\omega_1^{2n}}{(1+\omega_1)^r(1+a\omega_1)^r} \int_0^\infty \frac{w(a)s}{(1+\sqrt{a}s\omega_1)^r} \, ds \, d\omega_1 \, da \right).
$$

Substituting $t := s\omega_1$ with $dt = \omega_1 \, ds$ in the first two integrals and $t := \sqrt{a}s\omega_1$ with $dt = \sqrt{a}\omega_1 \, ds$ in the last two integrals, we obtain for $r > 2$ that

$$
I \leq C \left(\int_0^\infty \frac{\omega_1^{2n-1}}{(1+\omega_1)^r} \int_0^\infty a^{n-\frac{11}{4}} p_3(\sqrt{a}) \frac{w(a)}{(1+a\omega_1)^r} \, da \, d\omega_1 \right.
$$

$$
+ \int_0^\infty \frac{\omega_1^{2n-2}}{(1+\omega_1)^r} \int_0^\infty a^{n-\frac{11}{4}} p_2(\sqrt{a}) \frac{w(a)}{(1+a\omega_1)^r} \, da \, d\omega_1
$$

$$
+ \int_0^\infty \frac{\omega_1^{2n-1}}{(1+\omega_1)^r} \int_0^\infty a^{n-\frac{11}{4}} p_3(\sqrt{a}) \frac{w(a)}{(1+a\omega_1)^r} \, da \, d\omega_1
$$

$$
\left. + \int_0^\infty \frac{\omega_1^{2n-2}}{(1+\omega_1)^r} \int_0^\infty a^{n-\frac{13}{4}} p_2(\sqrt{a}) \frac{w(a)}{(1+a\omega_1)^r} \, da \, d\omega_1 \right).
$$

Substituting $b := a\omega_1$ with $db = \omega_1 da$ and bounding w accordingly we conclude further that

$$
I \leq C \left(\int_0^\infty \frac{\omega_1^{n+\frac{3}{4}+\rho_1}}{(1+\omega_1)^r} \int_0^\infty p_3\left(\sqrt{\frac{b}{\omega_1}}\right) \frac{b^{n-\frac{11}{4}-\rho_1}}{(1+b)^r} \, db \, d\omega_1 \right.
$$

$$
+ \int_0^\infty \frac{\omega_1^{n-\frac{1}{4}+\rho_1}}{(1+\omega_1)^r} \int_0^\infty p_2\left(\sqrt{\frac{b}{\omega_1}}\right) \frac{b^{n-\frac{11}{4}-\rho_1}}{(1+b)^r} \, db \, d\omega_1
$$

$$
+ \int_0^\infty \frac{\omega_1^{n+\frac{1}{4}+\rho_1}}{(1+\omega_1)^r} \int_0^\infty p_2\left(\sqrt{\frac{b}{\omega_1}}\right) \frac{b^{n-\frac{13}{4}-\rho_1}}{(1+b)^r} \, db \, d\omega_1
$$

$$
+ \int_0^\infty \frac{\omega_1^{n+\frac{3}{4}-\rho_2}}{(1+\omega_1)^r} \int_0^\infty p_3\left(\sqrt{\frac{b}{\omega_1}}\right) \frac{b^{n-\frac{11}{4}+\rho_2}}{(1+b)^r} \, db \, d\omega_1
$$

$$
+ \int_0^\infty \frac{\omega_1^{n-\frac{1}{4}-\rho_2}}{(1+\omega_1)^r} \int_0^\infty p_2\left(\sqrt{\frac{b}{\omega_1}}\right) \frac{b^{n-\frac{11}{4}+\rho_2}}{(1+b)^r} \, db \, d\omega_1
$$

$$
\left. + \int_0^\infty \frac{\omega_1^{n+\frac{1}{4}-\rho_2}}{(1+\omega_1)^r} \int_0^\infty p_2\left(\sqrt{\frac{b}{\omega_1}}\right) \frac{b^{n-\frac{13}{4}+\rho_2}}{(1+b)^r} \, db \, d\omega_1 \right).
$$

Since $p_k \in \Pi_k$, $k = 2, 3$ we see that the integrals are finite if $n \geq \max(\frac{1}{4} + \rho_2, \frac{9}{4} + \rho_1)$ and $r > n + \max(\frac{7}{4} + \rho_2, \frac{9}{4} + \rho_1)$. This finishes the proof. $\square$

By the help of the following corollary which was proved in [13] we additionally establish $\psi \in \mathcal{B}_w$ and therewith the existence of atomic decompositions and Banach frames for $\mathscr{SC}_{p,m}$.

Corollary 1. *Let $\psi(x) \in L_2(\mathbb{R}^2)$ fulfill supp $\psi \in Q_D$. Suppose that the weight function satisfies $w(a, s, t) = w(a) \leq |a|^{-\rho_1} + |a|^{\rho_2}$ for $\rho_1, \rho_2 \geq 0$ and that*

$$|\hat{\psi}(\omega_1, \omega_2)| \lesssim \frac{|\omega_1|^n}{(1 + |\omega_1|)^r} \frac{1}{(1 + |\omega_2|)^r}$$

for sufficiently large n and r. Then, we have that $\psi \in \mathcal{B}_w$.

5.1 Atomic Decomposition of Besov Spaces

Let us recall the characterization of homogeneous Besov spaces $B_{p,q}^{\sigma}$ from [23], see also [30, 37]. For inhomogeneous Besov spaces we refer to [36]. For $\alpha > 1$, $D > 1$, and $K \in \mathbb{N}_0$, a K times differentiable function a on $\mathbb{R}^d$ is called a *K-atom* if the following two conditions are fulfilled:

1. supp $a \subset DQ_{j,m}(\mathbb{R}^d)$ for some $m \in \mathbb{R}^d$, where $Q_{j,m}(\mathbb{R}^d)$ denotes the cube in $\mathbb{R}^d$ centered at $\alpha^{-j}m$ with sides parallel to the coordinate axes and side length $2\alpha^{-j}$.
2. $|D^\gamma a(x)| \leq \alpha^{|\gamma|j}$ for $|\gamma| \leq K$.

Now the homogeneous Besov spaces can be characterized as follows.

Theorem 7. *Let $D > 1$, $\sigma > 0$, and $K \in \mathbb{N}_0$ with $K \geq 1 + \lfloor \sigma \rfloor$ be fixed. Let $1 \leq p \leq \infty$. Then, $f \in B_{p,q}^{\sigma}$ if and only if it can be represented[1] as*

$$f(x) = \sum_{j \in \mathbb{Z}} \sum_{l \in \mathbb{Z}^d} \lambda(j, l) a_{j,l}(x), \qquad (17)$$

where the $a_{j,l}$ are K-atoms with supp $a_{j,l} \subset DQ_{j,l}(\mathbb{R}^d)$ and

$$\|f\|_{B_{p,q}^{\sigma}} \sim \inf \left(\sum_{j \in \mathbb{Z}} \alpha^{j(\sigma - \frac{d}{p})q} \left(\sum_{l \in \mathbb{Z}^d} |\lambda(j, l)|^p \right)^{\frac{q}{p}} \right)^{\frac{1}{q}}$$

where the infimum is taken over all admissible representations (17).

In this section, we are mainly interested in weights

$$m(a, s, t) = m(a) := |a|^{-r}, r \geq 0$$

---

[1] In the sense of distributions, a-posteriori implying norm convergence for $p < \infty$.

and use the abbreviation

$$\mathscr{SC}_{p,r} := \mathscr{SC}_{p,m}.$$

For simplicity, we further assume in the following that we can use $\beta = \tau = 1$ in the U-dense, relatively separated set (14) and restrict ourselves to the case $\varepsilon = 1$. In other words, we assume that $f \in \mathscr{SC}_{p,r}$ can be written as

$$f(x) = \sum_{j \in \mathbb{Z}} \sum_{k \in \mathbb{Z}} \sum_{l \in \mathbb{Z}^2} c(j,k,l) \pi(\alpha^{-j}, \beta\alpha^{-j/2}k, S_{\alpha^{-j/2}k} A_{\alpha^{-j}l}) \psi(x)$$

$$= \sum_{j \in \mathbb{Z}} \sum_{k \in \mathbb{Z}} \sum_{l \in \mathbb{Z}^2} c(j,k,l) \alpha^{\frac{3}{4}j} \psi(\alpha^j x_1 - \alpha^{j/2}kx_2 - l_1, \alpha^{j/2}x_2 - l_2). \quad (18)$$

To derive reasonable trace and embedding theorems, it is necessary to introduce the following subspaces of $\mathscr{SC}_{p,r}$. For fixed $\psi \in B_w$ we denote by $\mathscr{SCC}_{p,r}$ the closed subspace of $\mathscr{SC}_{p,r}$ consisting of those functions which are representable as in (18) but with integers $|k| \le \alpha^{j/2}$. As we shall see in the sequel for each of these ψ the resulting spaces $\mathscr{SCC}_{p,r}$ embed in the same scale of Besov spaces, and the same holds true for the trace theorems.

5.2 A Density Result

In most of the classical smoothness spaces like Sobolev and Besov spaces dense subsets of "nice" functions can be identified. Typically, the set of Schwartz functions S serves as such a dense subset. We refer to [1] and any book of Hans Triebel for further information. By the following theorem the same is true for our shearlet coorbit spaces.

Theorem 8. *Let*

$$S_0 := \left\{ f \in S : |\hat{f}(\omega)| \le \frac{\omega_1^{2\alpha}}{(1 + \|\omega\|^2)^{2\alpha}} \ \forall \, \alpha > 0 \right\}$$

and $m(a,s,t) = m(a,s) := |a|^r (\frac{1}{|a|} + |a| + |s|)^n$ for some $r \in \mathbb{R}, n \ge 0$. Then the set of Schwartz functions forms a dense subset of the shearlet coorbit space $\mathscr{SC}_{p,m}$.

Proof. As in [11, Theorem 4.7] it can be shown that S_0 is at least contained in $\mathscr{SC}_{p,m}$. (Note that in [11] the weight $(\frac{1}{|a|} + |a|)^r (\frac{1}{|a|} + |a| + |s|)^n$, $r,n > 0$ which is not smaller than 1 was considered.) It remains to show the density. To this end, we observe from Theorem 3 that certain band-limited Schwartz functions can be used as analyzing shearlets. Now let us recall that the atomic decomposition in (15) has to be understood as a limit of *finite* linear combinations with respect to the shearlet coorbit norm. However, every finite linear combination of Schwartz functions is again a Schwartz function, hence (15) implies that we have found for any $f \in \mathscr{SC}_{p,m}$ a sequence of Schwartz functions which converges to f. $\quad \square$

5.3 *Traces on the Real Axes*

In this section which is based on [13], we investigate the traces of functions lying in certain subspaces of $\mathscr{SC}_{p,r}$ with respect to the horizontal and vertical axes, respectively. With larger technical effort it is also possible to prove trace theorems with respect to more general lines.

Theorem 9. *Let* $\mathrm{Tr}_h f$ *denote the restriction of* f *to the (horizontal)* x_1-*axis, i.e.,* $(\mathrm{Tr}_h f)(x_1) := f(x_1, 0)$. *Then* $\mathrm{Tr}_h(\mathscr{SC}\mathscr{C}_{p,r}) \subset B^{\sigma_1}_{p,p}(\mathbb{R}) + B^{\sigma_2}_{p,p}(\mathbb{R})$, *where*

$$B^{\sigma_1}_{p,p}(\mathbb{R}) + B^{\sigma_2}_{p,p}(\mathbb{R}) := \{h \mid h = h_1 + h_2, h_1 \in B^{\sigma_1}_{p,p}(\mathbb{R}), h_2 \in B^{\sigma_2}_{p,p}(\mathbb{R})\}$$

and the parameters σ_1 *and* σ_2 *satisfy the conditions*

$$\sigma_1 = r - \frac{5}{4} + \frac{3}{2p}, \quad \sigma_2 = r - \frac{3}{4} + \frac{1}{p}.$$

Note that $\sigma_1 \le \sigma_2$ for $p \ge 2$.

Proof. Using (18) we split f into $f = f_1 + f_2$, where

$$f_1(x_1, x_2) := \sum_{j \ge 0} \sum_{|k| \le \alpha^{j/2}} \sum_{l \in \mathbb{Z}^2} c(j,k,l) \alpha^{\frac{3}{4}j} \psi(\alpha^j x_1 - \alpha^{j/2} k x_2 - l_1, \alpha^{j/2} x_2 - l_2), \quad (19)$$

$$f_2(x_1, x_2) := \sum_{j < 0} \sum_{l \in \mathbb{Z}^2} c(j,0,l) \alpha^{\frac{3}{4}j} \psi(\alpha^j x_1 - l_1, \alpha^{j/2} x_2 - l_2). \quad (20)$$

By Corollary 1 we can choose ψ compactly supported in $[-D, D] \times [-D, D]$ for some $D > 1$. Moreover, we can assume that $|D_1^\gamma \psi| \le 1$ for $0 \le \gamma \le K := \max\{K_1, K_2\}$, where $K_1 := 1 + \lfloor \sigma_1 \rfloor$, $K_2 := 1 + \lfloor \sigma_2 \rfloor$ and where $D_1 \psi$ denotes the derivative with respect to the first component of ψ. Now $\mathrm{Tr}_h f$ can be written as

$$\mathrm{Tr}_h f(x_1) = f(x_1, 0) = \sum_{j \in \mathbb{Z}} \sum_{|k| \le \alpha^{j/2}} \sum_{l \in \mathbb{Z}^2} c(j,k,l) \alpha^{\frac{3}{4}j} \psi(\alpha^j x_1 - l_1, -l_2)$$

$$= \sum_{j \in \mathbb{Z}} \sum_{l_1 \in \mathbb{Z}} \sum_{|k| \le \alpha^{j/2}} \sum_{|l_2| \le D} c(j,k,l_1,l_2) \alpha^{\frac{3}{4}j} \psi(\alpha^j x_1 - l_1, -l_2)$$

$$= \sum_{j \ge 0} \sum_{l_1 \in \mathbb{Z}} \lambda(j,l_1) a_{j,l_1}(x_1) + \sum_{j < 0} \sum_{l_1 \in \mathbb{Z}} \lambda(j,l_1) a_{j,l_1}(x_1)$$

$$= \mathrm{Tr}_h f_1(x_1) + \mathrm{Tr}_h f_2(x_1),$$

where for $j \ge 0$,

$$a_{j,l_1}(x_1) := \begin{cases} \lambda(j,l_1)^{-1} \alpha^{\frac{3}{4}j} \sum_{|k| \le \alpha^{j/2}} \sum_{|l_2| \le D} c(j,k,l_1,l_2) \psi(\alpha^j x_1 - l_1, -l_2) & \text{if } \lambda(j,l_1) \ne 0, \\ 0 & \text{otherwise,} \end{cases}$$

$$\lambda(j,l_1) := \alpha^{\frac{3}{4}j} \sum_{|k| \le \alpha^{j/2}} \sum_{|l_2| \le D} |c(j,k,l_1,l_2)|,$$

and for $j < 0$

$$
a_{j,l_1}(x_1) := \begin{cases} \lambda(j,l_1)^{-1} \alpha^{\frac{3}{4}j} \sum\limits_{|l_2| \le D} c(j,0,l_1,l_2) \psi(\alpha^j x_1 - l_1, -l_2) & \text{if } \lambda(j,l_1) \ne 0, \\ 0 & \text{otherwise,} \end{cases}
$$

$$
\lambda(j,l_1) := \alpha^{\frac{3}{4}j} \sum\limits_{|l_2| \le D} |c(j,0,l_1,l_2)|.
$$

We have that supp $\psi(\alpha^j x_1 - l_1, -l_2) \subset DQ_{j,l_1}(\mathbb{R})$ which is also true for all a_{j,l_1} and by construction we know further that $|D^\gamma a_{j,l_1}| \le \alpha^{j\gamma}, 0 \le \gamma \le K$. Thus, the a_{j,l_1} are K_1-atoms on $\mathbb{R}$. Next, we consider

$$
\|\mathrm{Tr}_h f_1\|_{B^{\sigma_1}_{p,p}} \lesssim \left(\sum_{j \ge 0} \alpha^{j(\sigma_1 - \frac{1}{p})p} \sum_{l_1 \in \mathbb{Z}} |\lambda(j,l_1)|^p \right)^{\frac{1}{p}}
$$

$$
= \left(\sum_{j \ge 0} \alpha^{jp(\sigma_1 + \frac{3}{4} - \frac{1}{p})} \sum_{l_1 \in \mathbb{Z}} \Big(\sum_{|k| \le \alpha^{j/2}} \sum_{|l_2| \le D} |c(j,k,l_1,l_2)| \Big)^p \right)^{\frac{1}{p}}.
$$

Since $\left(\sum_{i=1}^N |z_i| \right)^p \le N^{p-1} \sum_{i=1}^N |z_i|^p$ and the set $\{k \in \mathbb{Z} : |k| \le \alpha^{j/2}\}$ contains $C\alpha^{j/2}$ elements we can estimate

$$
\|\mathrm{Tr}_h f_1\|_{B^{\sigma_1}_{p,p}} \lesssim \left(\sum_{j \ge 0} \alpha^{jp(\sigma_1 + \frac{5}{4} - \frac{3}{2p})} \sum_{|k| \le \alpha^{j/2}} \sum_{l \in \mathbb{R}^2} |c(j,k,l)|^p \right)^{\frac{1}{p}}
$$

$$
\lesssim \left(\sum_{j \in \mathbb{Z}} \alpha^{jpr} \sum_{k \in \mathbb{Z}} \sum_{l \in \mathbb{R}^2} |c(j,k,l)|^p \right)^{\frac{1}{p}} \lesssim \|f\|_{\mathscr{SCC}_{p,r}}
$$

with $r = \sigma_1 + \frac{5}{4} - \frac{3}{2p}$. In the same way we obtain that

$$
\|\mathrm{Tr}_h f_2\|_{B^{\sigma_2}_{p,p}} \lesssim \left(\sum_{j < 0} \alpha^{jp(\sigma_2 + \frac{3}{4} - \frac{1}{p})} \sum_{l \in \mathbb{R}^2} |c(j,0,l)|^p \right)^{\frac{1}{p}}
$$

$$
\lesssim \left(\sum_{j \in \mathbb{Z}} \alpha^{jpr} \sum_{k \in \mathbb{Z}} \sum_{l \in \mathbb{R}^2} |c(j,k,l)|^p \right)^{\frac{1}{p}} \lesssim \|f\|_{\mathscr{SCC}_{p,r}}
$$

with $r = \sigma_2 + \frac{3}{4} - \frac{1}{p}$. This completes the proof. $\quad \square$

By the following corollary the restriction to $\mathscr{SCC}_{p,r}$ is not necessary for $p = 1$.

Corollary 2. *For $p = 1$, the embedding $\mathrm{Tr}_h(\mathscr{SC}_{1,r}) \subset B^\sigma_{1,1}(\mathbb{R})$ with $\sigma = r - \frac{3}{4} + \frac{1}{p}$ holds true.*

Proof. Following the lines of the previous proof, where the summation with respect to k is over $\mathbb{Z}$, we obtain

$$
\|\mathrm{Tr}_h f\|_{B^\sigma_{1,1}} \lesssim \sum_{j \in \mathbb{Z}} \alpha^{j((\sigma + \frac{3}{4})p - 1)} \sum_{l_1 \in \mathbb{Z}} \sum_{k \in \mathbb{Z}} \sum_{|l_2| \le D} |c(j,k,l_1,l_2)| \le C\|f\|_{\mathscr{SC}_{1,r}}
$$

with $r = \sigma + \frac{3}{4} - \frac{1}{p}$ and we are done. $\quad \square$

Let us turn to traces on the vertical axis.

Theorem 10. *Let* $\mathrm{Tr}_v f$ *denote the restriction of* f *to the (vertical)* x_2-*axis, i.e.,* $(\mathrm{Tr}_v f)(x_2) := f(0, x_2)$. *Then, the embedding* $\mathrm{Tr}_v(\mathscr{SCC}_{p,r}) \subset B^{\sigma_1}_{p,p}(\mathbb{R}) + B^{\sigma_2}_{p,p}(\mathbb{R})$, *holds true, where* σ_1 *is the largest number such that*

$$\sigma_1 + \lfloor \sigma_1 \rfloor \leq 2r - \frac{9}{2} + \frac{3}{p} \quad \text{and} \quad \sigma_2 = 2r - \frac{3}{2} + \frac{1}{p}.$$

Proof. As in (19) and (20) we split f into $f = f_1 + f_2$, where we can choose ψ compactly supported in $[-D, D] \times [-D, D]$ for some $D > 1$ and normalized such that the derivatives of order $0 \leq \gamma \leq K$ with $K := \max\{K_1, K_2\}$, where $K_1 := 1 + \lfloor \sigma_1 \rfloor$, $K_2 := 1 + \lfloor \sigma_2 \rfloor$ are not larger than 1. By the support assumption on ψ we have that

$$\alpha^{-j/2}(l_2 - D) \leq x_2 \leq \alpha^{-j/2}(l_2 + D),$$
$$-kl_2 - D(1 + |k|) \leq l_1 \leq -kl_2 + D(1 + |k|).$$

Let $I_{k,l_2} := \{r \in \mathbb{Z} : |r + kl_2| \leq D(1 + |k|)\}$. Now we obtain that

$$\mathrm{Tr}_v f(x_2) = f(0, x_2) = \sum_{j \in \mathbb{Z}} \sum_{|k| \leq \alpha^{j/2}} \sum_{l \in \mathbb{Z}^2} c(j, k, l) \alpha^{\frac{3}{4}j} \psi(-\alpha^{j/2} k x_2 - l_1, \alpha^{j/2} x_2 - l_2).$$

This can be rewritten as

$$f(0, x_2) = \sum_{j \geq 0} \sum_{l_2 \in \mathbb{Z}} \lambda(j, l_2) a_{j,l_2}(x_2) + \sum_{j < 0} \sum_{l_2 \in \mathbb{Z}} \lambda(j, l_2) a_{j,l_2}(x_2)$$
$$= \mathrm{Tr}_v f_1(x_2) + \mathrm{Tr}_v f_2(x_2),$$

where for $j \geq 0$,

$$a_{j,l_2}(x_2) := \lambda(j, l_2)^{-1} \alpha^{\frac{3 + 2K_1}{4}j} \sum_{|k| \leq \alpha^{j/2}} \sum_{l_1 \in I_{k,l_2}} c(j, k, l_1, l_2) \alpha^{-K_1 j/2} \psi(-\alpha^{j/2} k x_2 - l_1, \alpha^{j/2} x_2 - l_2)$$

if $\lambda(j, l_2) \neq 0$ and $a_{j,l_2}(x_2) = 0$ otherwise and

$$\lambda(j, l_2) := \alpha^{\frac{3 + 2K_1}{4}j} \sum_{|k| \leq \alpha^{j/2}} \sum_{l_1 \in I_{k,l_2}} |c(j, k, l_1, l_2)|$$

and for $j < 0$,

$$a_{j,l_2}(x_2) := \lambda(j, l_2)^{-1} \alpha^{\frac{3}{4}j} \sum_{|l_1| \leq D} c(j, 0, l_1, l_2) \psi(-l_1, \alpha^{j/2} x_2 - l_2)$$

if $\lambda(j, l_2) \neq 0$ and $a_{j,l_2}(x_2) = 0$ otherwise and

$$\lambda(j, l_2) := \alpha^{\frac{3}{4}j} \sum_{|l_1| \leq D} |c(j, 0, l_1, l_2)|.$$

We have that supp $\psi(-\alpha^{j/2}kx_2 - l_1, \alpha^{j/2}x_2 - l_2) \subset DQ_{j,l_2}(\mathbb{R})$, where the cube is considered with respect to $\sqrt{\alpha}$ now. This is also true for a_{j,l_2}. For $j \geq 0$ we conclude by $|k| \leq \alpha^{j/2}$ that $\alpha^{-Kj/2}|D^\gamma \psi(-\alpha^{j/2}kx_2 - l_1, \alpha^{j/2}x_2 - l_2)| \leq \alpha^{\frac{j}{2}\gamma}$ and consequently $|D^\gamma a_{j,l_2}| \leq \alpha^{\frac{j}{2}\gamma}$, $\gamma \leq K_1$. For $j < 0$ we also have that $|D^\gamma a_{j,l_2}| \leq \alpha^{\frac{j}{2}\gamma}$. Thus, a_{j,l_2} are K_1-atoms. We get

$$\|\mathrm{Tr}_\nu f_1\|_{B^{\sigma_1}_{p,p}} \lesssim \left(\sum_{j\in\mathbb{Z}} \alpha^{\frac{j}{2}(\sigma_1 - \frac{1}{p})p} \sum_{l_2\in\mathbb{Z}} |\lambda(j,l_2)|^p \right)^{\frac{1}{p}}$$

$$\leq \left(\sum_{j\geq 0} \alpha^{\frac{j}{2}(\sigma_1 - \frac{1}{p})p} \alpha^{\frac{j}{2}(\frac{3+2K_1}{2})p} \alpha^{\frac{j}{2}(2-\frac{2}{p})p} \sum_{|k|\leq\alpha^{j/2}} \sum_{l\in\mathbb{R}^2} |c(j,k,l)|^p \right)^{\frac{1}{p}}$$

$$\leq \left(\sum_{j\in\mathbb{Z}} \alpha^{\frac{j}{2}(\sigma_1 + \frac{7}{2} + K_1 - \frac{3}{p})p} \sum_{|k|\leq\alpha^{j/2}} \sum_{l\in\mathbb{R}^2} |c(j,k,l)|^p \right)^{\frac{1}{p}}$$

$$\leq \left(\sum_{j\in\mathbb{Z}} \alpha^{\frac{j}{2}(\sigma_1 + \frac{7}{2} + 1 + \lfloor\sigma_1\rfloor - \frac{3}{p})p} \sum_{|k|\leq\alpha^{j/2}} \sum_{l\in\mathbb{R}^2} |c(j,k,l)|^p \right)^{\frac{1}{p}}$$

$$\leq \left(\sum_{j\in\mathbb{Z}} \alpha^{jpr} \sum_{k\in\mathbb{Z}} \sum_{l\in\mathbb{R}^2} |c(j,k,l)|^p \right)^{\frac{1}{p}}$$

$$\lesssim \|f\|_{\mathscr{SC}_{p,r}}$$

with $r \geq \frac{1}{2}(\sigma_1 + \lfloor\sigma_1\rfloor + \frac{9}{2} - \frac{3}{p})$. Analogously, we can compute

$$\|\mathrm{Tr}_\nu f_2\|_{B^{\sigma_2}_{p,p}} \lesssim \left(\sum_{j\in\mathbb{Z}} \alpha^{\frac{j}{2}(\sigma_2 - \frac{1}{p})p} \sum_{l_2\in\mathbb{Z}} |\lambda(j,l_2)|^p \right)^{\frac{1}{p}}$$

$$\leq \left(\sum_{j<0} \alpha^{\frac{j}{2}(\sigma_2 - \frac{1}{p} + \frac{3}{2})p} \sum_{l\in\mathbb{R}^2} |c(j,0,l)|^p \right)^{\frac{1}{p}}$$

$$\leq \left(\sum_{j\in\mathbb{Z}} \alpha^{jpr} \sum_{k\in\mathbb{Z}} \sum_{l\in\mathbb{R}^2} |c(j,k,l)|^p \right)^{\frac{1}{p}}$$

$$\lesssim \|f\|_{\mathscr{SC}_{p,r}}$$

with $r = \frac{1}{2}(\sigma_2 + \frac{3}{2} - \frac{1}{p})$ and we are done. $\square$

Remark 5. An alternative way to obtain trace results would be first to apply the Besov embedding discussed in the next section and afterward the classical trace theorem for homogeneous Besov spaces. Let us briefly discuss the relation between these different approaches. For simplicity we restrict ourselves to the positives scales and traces to the x_2-axis. Usually, an application of trace theorems in Besov spaces leads to a loss of smoothness of order $1/p$, that is $\mathrm{Tr}(B^s_{pp}(\mathbb{R}^d)) = B^{s-1/p}_{pp}(\mathbb{R}^{d-1})$, see [23]. Let the coorbit space smoothness index r be fixed. Depending on the concrete values of r and p, the direct and the indirect approach can yield the same result. However, in specific cases it turns out that the direct approach is superior as we gain some smoothness: Let $2r - \frac{9}{2} + \frac{3}{p} = 2\kappa + \alpha$ with $\kappa \in \mathbb{Z}$ and $\alpha \in [0,2)$. Then, we have for $\alpha \in [0,1)$ by Theorem 10 that $\sigma_1 = \kappa + \alpha$. On the other hand, in case $\alpha + \frac{1}{p} \in [1,2)$ an application of Theorem 11 yields $\mathscr{SCC}_{p,r} \subset B_{p,p}^{\tilde{\sigma}_1}$, where $\tilde{\sigma}_1 = \kappa + 1 - \varepsilon$ for arbitrary small $\varepsilon > 0$. Consequently, applying the trace theorem for Besov spaces yields smoothness $\tilde{\sigma}_1 - 1/p = \kappa + 1 - \varepsilon - 1/p < \kappa + \alpha = \sigma_1$.

5.4 Embedding Results

In this section, we prove embedding results of certain subspaces of shearlet coorbit spaces into (sums of) homogeneous Besov spaces. But first we provide a result concerning the embedding within shearlet coorbit spaces. In [18, Sect. 5.7] some embedding theorems for general $L_{p,m}$ coorbit spaces were given. In particular, the authors mentioned that for a fixed weight m, these spaces are monotonically increasing with p. The following corollary is a special results in this direction.

Corollary 3. For $1 \leq p_1 \leq p_2 \leq \infty$ the embedding $\mathscr{SC}_{p_1,r} \subset \mathscr{SC}_{p_2,r}$ holds true. Introducing the "smoothness spaces" $\mathscr{G}_p^r := \mathscr{SC}_{p,r+d(\frac{1}{2}-\frac{1}{p})}$, this implies the continuous embedding

$$\mathscr{G}_{p_1}^{r_1} \subset \mathscr{G}_{p_2}^{r_2}, \quad \text{if} \quad r_1 - \frac{d}{p_1} = r_2 - \frac{d}{p_2}.$$

For convenience we add the simple proof.

Proof. By Theorem 4 we obtain that

$$\|f\|_{\mathscr{SC}_{p_2,r}} \lesssim \|(c_\varepsilon(j,k,l))\|_{\ell_{p_2,r}} \lesssim \left(\sum_{j \in \mathbb{Z}} \alpha^{jrp_2} \sum_{\substack{k,l \\ \varepsilon \in \{-1,1\}}} |c_\varepsilon(j,k,l)|^{p_2} \right)^{\frac{1}{p_2}},$$

where $c_\varepsilon(j,k,l)$ is the coefficient belonging in the representation (15) with respect to (14) to the function $\pi(\varepsilon \alpha^{-j}, \sigma \alpha^{-j/2}k, S_{\sigma \alpha^{-j/2}k} A_{\alpha^{-j}} \tau l) \psi$. Since $\ell_{p_1} \subset \ell_{p_2}$ for $p_1 \leq p_2$ we get finally that

$$\|f\|_{\mathscr{SC}_{p_2,r}} \lesssim \left(\sum_{j \in \mathbb{Z}} \alpha^{jrp_2} \left(\sum_{\substack{k,l \\ \varepsilon \in \{-1,1\}}} |c_\varepsilon(j,k,l)|^{p_1} \right)^{\frac{p_2}{p_1}} \right)^{\frac{1}{p_2}}$$

$$\lesssim \left(\sum_{j \in \mathbb{Z}} \alpha^{jrp_1} \sum_{\substack{k,l \\ \varepsilon \in \{-1,1\}}} |c_\varepsilon(j,k,l)|^{p_1} \right)^{\frac{1}{p_1}} \lesssim \|f\|_{\mathscr{SC}_{p_1,r}}.$$

$\square$

Next we state our final result.

Theorem 11. The embedding $\mathscr{SCC}_{p,r} \subset B_{p,p}^{\sigma_1}(\mathbb{R}^2) + B_{p,p}^{\sigma_2}(\mathbb{R}^2)$, holds true, where σ_1 is the largest number such that

$$\sigma_1 + \lfloor \sigma_1 \rfloor \leq 2r - \frac{9}{2} + \frac{4}{p}, \quad \text{and} \quad \sigma_2 - \frac{\lfloor \sigma_2 \rfloor}{2} = r + \frac{3}{2p} + \frac{1}{4}.$$

Proof. By (18) we know that $f \in \mathscr{SCC}_{p,r}$ can be written as

$$f(x) = \sum_{j \in \mathbb{Z}} \sum_{|k| \le \alpha^{j/2}} \sum_{l \in \mathbb{Z}^2} c(j,k,l) \alpha^{\frac{3}{4}j} \psi(\alpha^j x_1 - \alpha^{j/2} k x_2 - l_1, \alpha^{j/2} x_2 - l_2),$$

where we can choose ψ compactly supported in $[-D,D] \times [-D,D]$ for some $D > 1$ and normalized such that the derivatives of order $0 \le |\gamma| \le K := \max\{K_1, K_2\}$, $K_1 := 1 + \lfloor \sigma_1 \rfloor$, $K_2 := 1 + \lfloor \sigma_2 \rfloor$ are not larger than 1.

We split $f \in \mathscr{SCC}_{p,r}$ as in (19) and (20) into f_1 and f_2. Then, we obtain with the index transform $l_1 = r_1 - kl_2$ that

$$f_1(x) = \sum_{j \ge 0} \sum_{|k| \le \alpha^{j/2}} \sum_{l_2 \in \mathbb{Z}} \sum_{n_1 \in \mathbb{Z}} \sum_{r_1 \in I(j,n_1)} c(j,k,r_1 - kl_2, l_2) \alpha^{\frac{3}{4}j}$$

$$\times \psi(\alpha^j x_1 - \alpha^{j/2} k x_2 - r_1 + kl_2, \alpha^{j/2} x_2 - l_2)$$

where $I(j,n_1) := \{r \in \mathbb{Z} : \alpha^{j/2}(n_1 - 1) < r \le \alpha^{j/2} n_1\}$.

For $j \ge 0$ we set

$$a_{j,n_1,l_2}(x) := \lambda(j,n_1,l_2)^{-1} \alpha^{\frac{3+2K_1}{4}j} \sum_{|k| \le \alpha^{j/2}} \sum_{r_1 \in I(j,n_1)} c(j,k,r_1 - kl_2, l_2)$$

$$\times \alpha^{-K_1 j/2} \psi(\alpha^j x_1 - \alpha^{j/2} k x_2 - r_1 + kl_2, \alpha^{j/2} x_2 - l_2),$$

if $\lambda(j,n_1,l_2) \ne 0$ and $a_{j,n_1,l_2}(x) = 0$ otherwise, where

$$\lambda(j,n_1,l_2) := \alpha^{\frac{3+2K_1}{4}j} \sum_{|k| \le \alpha^{j/2}} \sum_{r_1 \in I(j,n_1)} |c(j,k,r_1 - kl_2, l_2)|.$$

By the support assumption on ψ, the functions appearing in the definition of a_{j,n_1,m_2} are only nonzero if the following conditions are satisfied:

$$-D \le \alpha^{j/2} x_2 - l_2 \le D, \qquad \alpha^{-j/2}(l_2 - D) \le x_2 \le \alpha^{-j/2}(l_2 + D)$$

and

$$-D \le \alpha^j x_1 - \alpha^{j/2} k x_2 - r_1 + kl_2 \le D,$$

$$\alpha^{-j} r_1 + \alpha^{-j} k(\alpha^{j/2} x_2 - l_2) - \alpha^{-j} D \le x_1 \le \alpha^{-j} r_1 + \alpha^{-j} k(\alpha^{j/2} x_2 - l_2) + \alpha^{-j} D,$$

$$\alpha^{-j} r_1 - \alpha^{-j/2}(2D) \le x_1 \le \alpha^{-j} r_1 + \alpha^{-j/2}(2D),$$

$$\alpha^{-j/2} n_1 - \alpha^{-j/2}(3D) \le x_1 \le \alpha^{-j/2} n_1 + \alpha^{-j/2}(2D).$$

Thus, a_{j,n_1,l_2} is supported in $3DQ_{j,n_1,l_2}$, where the cube is considered with respect to $\sqrt{\alpha}$. The appropriate bounds $|D^\gamma a_{j,n_1,l_2}| \le \alpha^{\frac{1}{2}|\gamma|}$, $|\gamma| \le K_1$ can be derived as in the previous proof. Hence, the functions a_{j,n_1,l_2} are K_1-atoms.

Now we obtain for

$$f_1(x) = \sum_{j \ge 0} \sum_{l_2 \in \mathbb{Z}} \sum_{n_1 \in \mathbb{Z}} \lambda(j,n_1,l_2) a_{j,n_1,l_2}(x)$$

that

$$\|f_1\|_{B_{p,p}^{\sigma_1}}^p \lesssim \sum_{j\in\mathbb{Z}} \alpha^{\frac{j}{2}(\sigma_1-\frac{2}{p})p} \sum_{l_2\in\mathbb{Z}}\sum_{n_1\in\mathbb{Z}} |\lambda(j,n_1,l_2)|^p$$

$$= \sum_{j\in\mathbb{Z}} \alpha^{\frac{j}{2}(\sigma_1-\frac{2}{p})p} \alpha^{\frac{j}{2}(\frac{3+2K_1}{2})p} \sum_{l_2\in\mathbb{Z}}\sum_{n_1\in\mathbb{Z}} \left| \sum_{|k|\leq\alpha^{j/2}}\sum_{r_1\in I(j,n_1)} |c(j,k,r_1-kl_2,l_2)| \right|^p$$

$$\leq \sum_{j\in\mathbb{Z}} \alpha^{\frac{j}{2}p(\sigma_1+\frac{7}{2}+K_1-\frac{4}{p})} \sum_{l_2\in\mathbb{Z}}\sum_{n_1\in\mathbb{Z}}\sum_{|k|\leq\alpha^{j/2}}\sum_{r_1\in I(j,n_1)} |c(j,k,r_1-kl_2,l_2)|^p$$

$$= \sum_{j\in\mathbb{Z}} \alpha^{\frac{j}{2}p(\sigma_1+\frac{9}{2}+\lfloor\sigma_1\rfloor-\frac{4}{p})} \sum_{|k|\leq\alpha^{j/2}}\sum_{l_1\in\mathbb{Z}}\sum_{l_2\in\mathbb{Z}} |c(j,k,l_1,l_2)|^p$$

$$\lesssim \|f\|_{\mathscr{SC}_{p,r}}^p.$$

In the case $j<0$ we obtain with $J(j,n_2):=\{r: \alpha^{-j/2}(n_2-1)<r\leq\alpha^{-j/2}n_2\}$ that

$$f_2(x) = \sum_{j<0}\sum_{l_1\in\mathbb{Z}}\sum_{l_2\in\mathbb{Z}} c(j,0,l_1,l_2)\alpha^{\frac{3}{4}j}\psi(\alpha^j x_1-l_1,\alpha^{j/2}x_2-l_2)$$

$$= \sum_{j<0}\sum_{l_1\in\mathbb{Z}}\sum_{n_2\in\mathbb{Z}}\sum_{r_2\in J(j,n_2)} c(j,0,l_1,r_2)\alpha^{\frac{3}{4}j}\psi(\alpha^j x_1-l_1,\alpha^{j/2}x_2-r_2)$$

$$= \sum_{j<0}\sum_{l_1\in\mathbb{Z}}\sum_{n_2\in\mathbb{Z}} \lambda(j,l_1,n_2)a_{j,l_1,n_2}(x),$$

where

$$a_{j,l_1,n_2}(x) := \lambda(j,l_1,n_2)^{-1}\alpha^{\frac{3-2K_2}{4}j} \sum_{r_2\in J(j,n_2)} c(j,0,l_1,r_2)\alpha^{\frac{jK_2}{2}}\psi(\alpha^j x_1-l_1,\alpha^{j/2}x_2-r_2),$$

$$\lambda(j,l_1,n_2) := \alpha^{\frac{3-2K_2}{4}j} \sum_{r_2\in J(j,n_2)} |c(j,0,l_1,r_2)|$$

and $a_{j,l_1,n_2}(x):=0$ if $\lambda_{j,l_1,n_2}=0$. By the support assumption on ψ we get

$$\alpha^{-j}(l_1-D)\leq x_1\leq\alpha^{-j}(l_1+D),$$
$$\alpha^{-j/2}(r_2-D)\leq x_2\leq\alpha^{-j/2}(r_2+D), \text{ i.e., } \alpha^{-j}(n_2-2D)\leq x_2\leq\alpha^{-j}(n_2+D).$$

Consequently, a_{j,l_1,n_2} is supported in $2DQ_{j,l_1,n_2}$. Since $1\geq\alpha^{j|\gamma|/2}\geq\alpha^{j|\gamma|}\geq\alpha^{jK_2}$ for $0\leq|\gamma|\leq K_2$ and $j<0$ we obtain further that $|D^\gamma a_{j,n_1,l_2}|\leq\alpha^{jK_2/2}\alpha^{j|\gamma|/2}\leq\alpha^{j|\gamma|}$ so that a_{j,l_1,n_2} are K_2-atoms. Thus,

$$\|f_2\|_{B_{p,p}^{\sigma_2}}^p \lesssim \sum_{j\in\mathbb{Z}} \alpha^{j(\sigma_2-\frac{2}{p})p} \sum_{l_1\in\mathbb{Z}}\sum_{n_2\in\mathbb{Z}} |\lambda(j,l_1,n_2)|^p$$

$$\leq \sum_{j<0} \alpha^{j(\sigma_2-\frac{2}{p}+\frac{3-2K_2}{4})p} \sum_{l_1\in\mathbb{Z}}\sum_{n_2\in\mathbb{Z}} \left| \sum_{r_2\in J(j,n_2)} c(j,0,l_1,r_2) \right|^p$$

$$\leq \sum_{j<0} \alpha^{j(\sigma_2 - \frac{3}{2p} + \frac{1}{4} - \frac{K_2}{2})p} \sum_{l \in \mathbb{R}^2} |c(j,0,l)|^p$$

$$\leq \sum_{j \in \mathbb{Z}} \alpha^{jpr} \sum_{k \in \mathbb{Z}} \sum_{l \in \mathbb{R}^2} |c(j,k,l)|^p$$

$$\lesssim \|f\|^p_{\mathscr{SC}_{p,r}},$$

where $r = \sigma_2 - \frac{3}{2p} - \frac{1}{4} - \frac{\lfloor \sigma_2 \rfloor}{2}$. $\square$

Remark 6. Embedding results in Besov spaces have also been shown for the curvelet setting by Borup and Nielsen [2]. However, the technique used by these authors is completely different. In contrast to our approach they work in the frequency domain. We prefer to consider the time domain with flexible atomic decompositions for the following reasons. As already outlined above time domain techniques provide a very natural way to derive trace theorems which might be very difficult or even impossible in the Fourier domain. Moreover, since we are working with compactly supported atoms the treatment of shearlet coorbit spaces on bounded domains, including again embedding and trace theorems, seems to be manageable.

It has also turned out that our approach provides some advantages for higher dimensions. Trace theorems for shearlet coorbit spaces on $\mathbb{R}^d$, $d \geq 3$ to higher dimensional hyperplanes are not straightforward since it is not clear that these traces will also be contained in Besov spaces. One natural conjecture would be that the traces of shearlet coorbit spaces on $\mathbb{R}^3$ with respect to two-dimensional hyperplanes are again shearlet coorbit spaces. This conjecture was proved in [8]. As expected flexible atomic and molecular decomposition techniques for shearlet coorbit spaces can be applied.

6 Analysis of Singularities

In this section, we deal with the decay of the shearlet transform at hyperplane singularities and at special simplex singularities in $\mathbb{R}^d$. For the behavior of the shearlet transform at singularities in $\mathbb{R}^2$ we refer to [32, 38].

6.1 Hyperplane Singularities

We consider $(d-m)$-dimensional hyperplanes in $\mathbb{R}^d$, $m = 1, \ldots, d-1$ through the origin given by

$$\underbrace{\begin{pmatrix} x_1 \\ \vdots \\ x_m \end{pmatrix}}_{x_A} + P \underbrace{\begin{pmatrix} x_{m+1} \\ \vdots \\ x_d \end{pmatrix}}_{x_E} = \begin{pmatrix} 0 \\ \vdots \\ 0 \end{pmatrix}, \quad P := \begin{pmatrix} p_1^{\mathsf{T}} \\ \vdots \\ p_m^{\mathsf{T}} \end{pmatrix} \in \mathbb{R}^{m, d-m}. \tag{21}$$

Note that this setting excludes some special hyperplanes, e.g., for $d = 3$ and $m = 1$ planes containing the x_1-axis and for $d = 3$ and $m = 2$ lines contained within the x_1x_2-plane. To detect such hyperplane singularities one has to perform a simple variable exchange in the shearlet stetting or to define "cone adapted shearlets" similar to [32].

Let δ denote the Delta distribution. Then, we obtain for

$$v_m := \delta(x_A + Px_E)$$

that

$$\hat{v}_m(\omega) = \int_{\mathbb{R}^d} \delta(x_A + Px_E) e^{-2\pi i(\langle x_A, \omega_A \rangle + \langle x_E, \omega_E \rangle)} \, dx$$

$$= \int_{\mathbb{R}^{d-m}} e^{-2\pi i(-\langle Px_E, \omega_A \rangle + \langle x_E, \omega_E \rangle)} \, dx_E$$

$$= \delta(\omega_E - P^{\mathrm{T}} \omega_A). \tag{22}$$

The following theorem describes the decay of the shearlet transform at hyperplane singularities. We use the notation $\mathscr{SH}_\psi f(a,s,t) \sim |a|^r$ as $a \to 0$, if there exists constants $0 < c \leq C < \infty$ such that

$$c|a|^r \leq |\mathscr{SH}_\psi f(a,s,t)| \leq C|a|^r \quad \text{as } a \to 0.$$

Theorem 12. *Let $\psi \in L_2(\mathbb{R}^d)$ be a shearlet satisfying $\hat{\psi} \in C^\infty(\mathbb{R}^d)$. Assume further that $\hat{\psi}(\omega) = \hat{\psi}_1(\omega_1)\hat{\psi}_2(\tilde{\omega}/\omega_1)$, where $\operatorname{supp} \hat{\psi}_1 \in [-a_1, -a_0] \cup [a_0, a_1]$ for some $a_1 > a_0 \geq \alpha > 0$ and $\operatorname{supp} \hat{\psi}_2 \in Q_b$. If*

$$(s_m, \ldots, s_{d-1}) = (-1, s_1, \ldots, s_{m-1})P \quad \text{and} \quad (t_1, \ldots, t_m) = -(t_{m+1}, \ldots, t_d)P^{\mathrm{T}}, \tag{23}$$

then

$$\mathscr{SH}_\psi v_m(a,s,t) \sim |a|^{\frac{1-2m}{2d}} \quad \text{as } a \to 0. \tag{24}$$

Otherwise, the shearlet transform $\mathscr{SH}_\psi v_m$ decays rapidly, i.e., faster than any polynomial, as $a \to 0$.

The condition (23) requires that the shearlet is aligned with the hyperplane (21) and that t lies within the hyperplane. The condition on $\hat{\psi}_1$ and $\hat{\psi}_2$ can be relaxed toward a rapid decay of the functions.

Proof. An application of Plancherel's theorem for tempered distribution together with (22) and (7) yields

$$\mathscr{SH}_\psi v_m(a,s,t) := \langle v_m, \psi_{a,s,t} \rangle = \langle \hat{v}_m, \hat{\psi}_{a,s,t} \rangle$$

$$= \int_{\mathbb{R}^d} \delta(\omega_E - P^{\mathrm{T}}\omega_A)|a|^{1-\frac{1}{2d}} e^{2\pi i \langle t, \omega \rangle} \, \bar{\hat{\psi}}\left(a\omega_1, \operatorname{sgn}(a)|a|^{\frac{1}{d}}(\omega_1 s + \tilde{\omega})\right) d\omega$$

$$= |a|^{1-\frac{1}{2d}} \int_{\mathbb{R}^m} e^{2\pi i \langle t_A + Pt_E, \omega_A \rangle} \, \bar{\hat{\psi}}\left(a\omega_1, \operatorname{sgn}(a)|a|^{\frac{1}{d}}(\omega_1 s + \begin{pmatrix} \tilde{\omega}_A \\ P^{\mathrm{T}}\omega_A \end{pmatrix})\right) d\omega_A$$

with $\tilde{\omega}_A = (\omega_2, \ldots, \omega_m)^{\mathrm{T}}$ for $m \geq 2$. In the following, we restrict our attention to the case $m \geq 2$. If $m = 1$, we can simply neglect $\tilde{\omega}_A$ and the assertion follows in a similar way. By definition of $\hat{\psi}$ this can be rewritten as

$$|a|^{1-\frac{1}{2d}} \int_{\mathbb{R}^m} e^{2\pi i \langle t_A + P t_E, \omega_A \rangle} \bar{\hat{\psi}}_1(a\omega_1) \bar{\hat{\psi}}_2\left(|a|^{\frac{1}{d}-1}\left(s + \frac{1}{\omega_1}\begin{pmatrix} \tilde{\omega}_A \\ P^{\mathrm{T}}\omega_A \end{pmatrix}\right)\right) \mathrm{d}\omega_A.$$

Substituting $\tilde{\xi}_A = (\xi_2, \ldots, \xi_m)^{\mathrm{T}} := \tilde{\omega}_A/\omega_1$, i.e., $\mathrm{d}\tilde{\omega}_A = |\omega_1|^{m-1}\,\mathrm{d}\tilde{\xi}_A$, we get

$$\mathscr{SH}_\psi V_m(a,s,t) = |a|^{1-\frac{1}{2d}} \int_{\mathbb{R}} \int_{\mathbb{R}^{m-1}} e^{2\pi i \omega_1 \langle t_A + P t_E, (1, \tilde{\xi}_A^{\mathrm{T}})^{\mathrm{T}} \rangle} \bar{\hat{\psi}}_1(a\omega_1)|\omega_1|^{m-1}$$

$$\times \bar{\hat{\psi}}_2\left(|a|^{\frac{1}{d}-1}\left(s + \begin{pmatrix} \tilde{\xi}_A \\ P^{\mathrm{T}}(1, \tilde{\xi}_A^{\mathrm{T}})^{\mathrm{T}} \end{pmatrix}\right)\right) \mathrm{d}\tilde{\xi}_A\,\mathrm{d}\omega_1$$

and further by substituting $\xi_1 := a\omega_1$

$$\mathscr{SH}_\psi V_m(a,s,t) = |a|^{1-m-\frac{1}{2d}} \int_{\mathbb{R}^{m-1}} \int_{\mathbb{R}} e^{2\pi i \frac{\xi_1}{a} \langle t_A + P t_E, (1, \tilde{\xi}_A^{\mathrm{T}})^{\mathrm{T}} \rangle} |\xi_1|^{m-1} \bar{\hat{\psi}}_1(\xi_1)\,\mathrm{d}\xi_1$$

$$\times \bar{\hat{\psi}}_2\left(|a|^{\frac{1}{d}-1}\left(s + \begin{pmatrix} \tilde{\xi}_A \\ P^{\mathrm{T}}(1, \tilde{\xi}_A^{\mathrm{T}})^{\mathrm{T}} \end{pmatrix}\right)\right) \mathrm{d}\tilde{\xi}_A.$$

Finally, by substituting $\tilde{\omega}_A := |a|^{\frac{1}{d}-1}(\tilde{\xi}_A + s_a)$, where $s_a := (s_1, \ldots, s_{m-1})^{\mathrm{T}}$ and $s_e := (s_m, \ldots, s_{d-1})^{\mathrm{T}}$, we obtain

$$\mathscr{SH}_\psi V_m(a,s,t) = |a|^{\frac{1-2m}{2d}} \int_{\mathbb{R}^{m-1}} \int_{\mathbb{R}} e^{2\pi i \frac{\xi_1}{a} \langle t_A + P t_E, (1, |a|^{1-1/d}\tilde{\omega}_A^{\mathrm{T}} - s_a^{\mathrm{T}}) \rangle} |\xi_1|^{m-1} \bar{\hat{\psi}}_1(\xi_1)\,\mathrm{d}\xi_1$$

$$\times \bar{\hat{\psi}}_2\left(|a|^{\frac{1}{d}-1}\left(s_e - P^{\mathrm{T}}\begin{pmatrix} -1 \\ s_a \end{pmatrix}\right) + P^{\mathrm{T}}\begin{pmatrix} 0 \\ \tilde{\omega}_A \end{pmatrix}\right) \mathrm{d}\tilde{\omega}_A.$$

If the vector

$$s_e - P^{\mathrm{T}}\begin{pmatrix} -1 \\ s_a \end{pmatrix} \neq 0_{d-m} \tag{25}$$

then at least one component of its product with $|a|^{1/d-1}$ becomes arbitrary large as $a \to 0$. On the other hand, by the support property of $\hat{\psi}_2$, we conclude that $\hat{\psi}_2(\tilde{\omega}_A, \cdot)$ becomes zero if $\tilde{\omega}_A$ is not in $Q_{(b_1, \ldots, b_{m-1})} \subset \mathbb{R}^{m-1}$. But for all $\tilde{\omega}_A \in Q_{(b_1, \ldots, b_{m-1})}$ at least one component of

$$|a|^{\frac{1}{d}-1}\left(s_e - P^{\mathrm{T}}\begin{pmatrix} 1 \\ s_a \end{pmatrix}\right) + P^{\mathrm{T}}\begin{pmatrix} 0 \\ \tilde{\omega}_A \end{pmatrix}$$

is not within the support of $\hat{\psi}_2$ for a sufficiently small so that $\hat{\psi}_2$ becomes zero again. Assume now that we have equality in (25). Then

$$\mathscr{SH}_\psi V_m(a,s,t) = |a|^{\frac{1-2m}{2d}} \int_{\mathbb{R}^{m-1}} \int_{\mathbb{R}} e^{2\pi i \frac{\xi_1}{a} \langle t_A + P t_E, (1, |a|^{1-1/d}\tilde{\omega}_A^{\mathrm{T}} - s_a^{\mathrm{T}}) \rangle} |\xi_1|^{m-1} \bar{\hat{\psi}}_1(\xi_1)\,\mathrm{d}\xi_1$$

$$\times \bar{\tilde{\psi}}_2 \left(P^{\mathrm{T}} \begin{pmatrix} \tilde{\omega}_A \\ 0 \\ \tilde{\omega}_A \end{pmatrix} \right) \mathrm{d}\tilde{\omega}_A$$

$$= C|a|^{\frac{1-2m}{2d}} \int_{\mathbb{R}^{m-1}} \tilde{\psi}_1^{(m-1)} \left(\langle t_A + Pt_E, (1,|a|^{1-1/d} \tilde{\omega}_A^{\mathrm{T}} - s_a^{\mathrm{T}}) \rangle /a \right)$$

$$\times \bar{\tilde{\psi}}_2 \left(P^{\mathrm{T}} \begin{pmatrix} \tilde{\omega}_A \\ 0 \\ \tilde{\omega}_A \end{pmatrix} \right) \mathrm{d}\tilde{\omega}_A,$$

$$\mathscr{SH}_\psi v_m(a,s,t) = C|a|^{\frac{1-2m}{2d}} \int_{\mathbb{R}^{m-1}} \tilde{\psi}_1^{(m-1)} \left(\langle t_A + Pt_E, \left(\frac{|a|^{1/d-1}}{\tilde{\omega}_A^{\mathrm{T}} - |a|^{1/d-1} s_a} \right) \rangle |a|^{-\frac{1}{d}} \right)$$

$$\times \bar{\tilde{\psi}}_2 \left(P^{\mathrm{T}} \begin{pmatrix} \tilde{\omega}_A \\ 0 \\ \tilde{\omega}_A \end{pmatrix} \right) \mathrm{d}\tilde{\omega}_A,$$

where $\tilde{\psi}_1$ has the Fourier transform $\hat{\tilde{\psi}}_1(\xi_1) := \hat{\tilde{\psi}}_1(\xi_1)$ for $\xi_1 \geq 0$ and $\hat{\tilde{\psi}}_1(\xi_1) := -\hat{\tilde{\psi}}_1(\xi_1)$ for $\xi_1 < 0$. Since by our assumptions the support of $\hat{\psi}_1$ is bounded away from the origin, we see that $\hat{\tilde{\psi}}_1$ is again in $C^\infty(\mathbb{R})$. If $t_A + Pt_E \neq 0_m$, then, since $\hat{\tilde{\psi}}_1 \in C^\infty$ the function $\tilde{\psi}_1^{(m-1)}$ decays rapidly as $a \to 0$ for all $\tilde{\omega}_A$ in the bounded domain, where $\hat{\psi}_2$ does not become zero. Consequently, the value of the shearlet transform decays rapidly. If $t_A + Pt_E = 0_m$, then

$$\mathscr{SH}_\psi v_m(a,s,t) = C|a|^{\frac{1-2m}{2d}} \tilde{\psi}_1^{(m-1)}(0) \int_{\mathbb{R}^{m-1}} \bar{\tilde{\psi}}_2 \left(P^{\mathrm{T}} \begin{pmatrix} \tilde{\omega}_A \\ 0 \\ \tilde{\omega}_A \end{pmatrix} \right) \mathrm{d}\tilde{\omega}_A \sim |a|^{\frac{1-2m}{2d}}.$$

This finishes the proof. □

Remark 7. Other choices of the dilation matrix are possible, e.g.,

$$A_a := \begin{pmatrix} a & 0_{d-1}^{\mathrm{T}} \\ 0_{d-1} & \mathrm{sgn}(a)\sqrt{|a|} I_{d-1} \end{pmatrix}.$$

Then, we have to replace (24) by $|a|^{\frac{d-2m-1}{4}}$ which increases for $d < 2m+1$ as $a \to 0$. Therefore, we prefer our choice.

6.2 Tetrahedron Singularities

In the following, we deal with the cone $\mathscr{C}$ in the first octant of $\mathbb{R}^3$ given by

$$\mathscr{C} := \{x = Ct : t \geq 0 \text{ componentwise}\},$$

where

$$C := (p\,q\,r) = \begin{pmatrix} 1 & 1 & 1 \\ p_1 & q_1 & r_1 \\ p_2 & q_2 & r_2 \end{pmatrix}, \quad p_j, q_j, r_j > 0, \; j = 1,2$$

and the vectors p, q, r are linearly independent. The vector

$$n_{pq} := \left(1, \frac{p_2 - q_2}{p_1 q_2 - p_2 q_1}, \frac{q_1 - p_1}{p_1 q_2 - p_2 q_1}\right)^{\mathrm{T}} = \left(1, \tilde{n}_{pq}^{\mathrm{T}}\right)^{\mathrm{T}}$$

is a multiple of the normal vector of the plane spanned by p and q. Similarly, we use the notation n_{pr}, n_{qr} for the corresponding vectors perpendicular to the pr- and qr-plane. Let $\chi_{\mathscr{C}}$ denote the characteristic function of the cone $\mathscr{C}$. Since the Fourier transform of the Heavyside function H is

$$\hat{H}(\omega) = \frac{1}{2\pi i} \mathrm{pv}\left(\frac{1}{\omega}\right) + \sqrt{\frac{\pi}{2}} \delta(\omega),$$

see [22, p. 340], we obtain that

$$\hat{\chi}_{\mathscr{C}}(\omega) = \int_{\mathscr{C}} e^{-2\pi i \langle x, \omega \rangle} \, dx = |\det C| \int_{\mathbb{R}^3_+} e^{-2\pi i \langle t, C^{\mathrm{T}}\omega \rangle} \, dt$$

$$= c_1 \left(\frac{1}{p^{\mathrm{T}}\omega} \frac{1}{q^{\mathrm{T}}\omega} \frac{1}{r^{\mathrm{T}}\omega}\right)$$

$$+ c_2 \left(\frac{1}{p^{\mathrm{T}}\omega} \frac{1}{q^{\mathrm{T}}\omega} \delta(r^{\mathrm{T}}\omega) + \frac{1}{p^{\mathrm{T}}\omega} \frac{1}{r^{\mathrm{T}}\omega} \delta(q^{\mathrm{T}}\omega) + \frac{1}{q^{\mathrm{T}}\omega} \frac{1}{r^{\mathrm{T}}\omega} \delta(p^{\mathrm{T}}\omega)\right)$$

$$+ c_3 \left(\frac{1}{p^{\mathrm{T}}\omega} \delta(q^{\mathrm{T}}\omega)\delta(r^{\mathrm{T}}\omega) + \frac{1}{q^{\mathrm{T}}\omega} \delta(p^{\mathrm{T}}\omega)\delta(r^{\mathrm{T}}\omega) + \frac{1}{r^{\mathrm{T}}\omega} \delta(p^{\mathrm{T}}\omega)\delta(q^{\mathrm{T}}\omega)\right)$$

$$+ c_4 \left(\delta(p^{\mathrm{T}}\omega)\delta(q^{\mathrm{T}}\omega)\delta(r^{\mathrm{T}}\omega)\right) \tag{26}$$

with nonzero constants c_j, $j = 1,2,3,4$. We have omitted the pv to simplify the notation. This can be used to prove the following theorem.

Theorem 13. *Let $\psi \in L_2(\mathbb{R}^3)$ be a shearlet satisfying $\hat{\psi} \in C^\infty(\mathbb{R}^3)$. Assume further that $\hat{\psi}(\omega) = \hat{\psi}_1(\omega_1)\hat{\psi}_2(\tilde{\omega}/\omega_1)$, where $\mathrm{supp}\,\hat{\psi}_1 \in [-a_1, -a_0] \cup [a_0, a_1]$ for some $a_1 > a_0 \geq \alpha > 0$ and $\mathrm{supp}\,\hat{\psi}_2 \in Q_b$. Let $a > 0$. If*

$$1 - p_1 s_1 - p_2 s_2 \neq 0, \; 1 - q_1 s_1 - q_2 s_2 \neq 0, \; 1 - r_1 s_1 - r_2 s_2 \neq 0 \quad \text{and} \quad t = (0,0,0)^{\mathrm{T}},$$

then

$$\mathscr{SH}_\psi \chi_{\mathscr{C}}(a, s, t) \sim a^{13/9}.$$

If

$$1 - p_1 s_1 - p_2 s_2 = 0, \; 1 - q_1 s_1 - q_2 s_2 \neq 0, \; 1 - r_1 s_1 - r_2 s_2 \neq 0 \quad \text{or}$$
$$1 - q_1 s_1 - q_2 s_2 = 0, \; 1 - p_1 s_1 - p_2 s_2 \neq 0, \; 1 - r_1 s_1 - r_2 s_2 \neq 0 \quad \text{or}$$
$$1 - r_1 s_1 - r_2 s_2 = 0, \; 1 - p_1 s_1 - p_2 s_2 \neq 0, \; 1 - q_1 s_1 - q_2 s_2 \neq 0$$

and $t_1 - t_2 s_1 - t_3 s_2 = 0$ which is in particular the case if $t = cp$, $t = cq$ or $t = cr$, respectively, then

$$\mathscr{SH}_{\psi}\chi_{\mathscr{C}}(a,s,t) \lesssim a^{3/2}.$$

If

$$s = -\tilde{n}_{pq}, \; n_{pq}^{\mathrm{T}}t = 0 \quad \text{or} \quad s = -\tilde{n}_{pr}, \; n_{pr}^{\mathrm{T}}t = 0 \quad \text{or} \quad s = -\tilde{n}_{qr}, \; n_{qr}^{\mathrm{T}}t = 0$$

then

$$\mathscr{SH}_{\psi}\chi_{\mathscr{C}}(a,s,t) \lesssim a^{5/6}.$$

Otherwise, the shearlet transform $\mathscr{SH}_{\psi}\chi_{\mathscr{C}}(a,s,t)$ decays rapidly as $a \to 0$.

Figure 2 illustrates the decay of the shearlet transform.

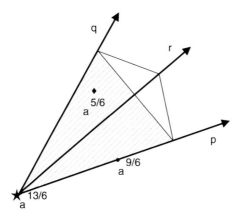

Fig. 2 Decay of the shearlet transform of the characteristic function of the cone $\mathscr{C}$: we have $\star$ if $\hat{s} \not\perp p,q,r$, $\bullet$ if $\hat{s} \perp p$ but $\hat{s} \not\perp q,r$ and $\diamond$ if $\hat{s} \perp p,q$, where $\hat{s} := (1, -s_1, -s_2)^{\mathrm{T}}$ is orthogonal to the plane containing the largest shearlet value

Proof. To determine the decay of $\mathscr{SH}_{\psi}\chi_{\mathscr{C}}(a,s,t) = \langle \hat{\chi}_{\mathscr{C}}, \hat{\psi}_{a,s,t} \rangle$ as $a \to 0$, we consider the four parts of (26) separately.

1. Since p,q,r are linearly independent, we have by the support of $\hat{\psi}$ that

$$\langle \delta(p^{\mathrm{T}}\cdot)\delta(q^{\mathrm{T}}\cdot)\delta(r^{\mathrm{T}}\cdot), \hat{\psi}_{a,s,t} \rangle = \hat{\psi}_{a,s,t}(0) = 0.$$

2. Next we obtain

$$\langle \delta(p^{\mathrm{T}}\cdot)\delta(q^{\mathrm{T}}\cdot)\frac{1}{r^{\mathrm{T}}\cdot}, \hat{\psi}_{a,s,t} \rangle$$

$$= a^{5/6}\frac{1}{r^{\mathrm{T}}n_{pq}}\int_{\mathbb{R}} e^{2\pi i \omega_1 \langle t,n_{pq}\rangle} \frac{\bar{\hat{\psi}}_1(a\omega_1)}{\omega_1} \bar{\hat{\psi}}_2\left(a^{-2/3}(s+\tilde{n}_{pq})\right) d\omega_1$$

$$\sim a^{5/6}\bar{\hat{\psi}}_2\left(a^{-2/3}(s+\tilde{n}_{pq})\right)\int_{\mathbb{R}} e^{2\pi i \xi_1 \langle t,n_{pq}\rangle/a} \frac{\bar{\hat{\psi}}_1(\xi_1)}{\xi_1} d\xi_1. \qquad (27)$$

If $s \neq -\tilde{n}_{pq}$, then (27) becomes zero for sufficiently small a since $\hat{\tilde{\psi}}_2$ is compactly supported. If $s = -\tilde{n}_{pq}$, then

$$\langle \delta(p^{\mathsf{T}} \cdot) \delta(q^{\mathsf{T}} \cdot) \frac{1}{r^{\mathsf{T}} \cdot}, \hat{\psi}_{a,s,t} \rangle \sim a^{5/6} \phi_1(\langle t, n_{pq} \rangle / a),$$

where ϕ_1 defined by $\hat{\phi}_1(\xi) := \hat{\tilde{\psi}}_1(\xi)/\xi \in \mathscr{S}$ is rapidly decaying. Thus, the above expression decays rapidly as $a \to 0$ except for $n_{pq}^{\mathsf{T}} t = 0$, i.e., t is in the pq-plane, where the decay is $a^{5/6}$.

3. For $I_3 := \langle \delta(p^{\mathsf{T}} \cdot) \frac{1}{q^{\mathsf{T}} \cdot} \frac{1}{r^{\mathsf{T}} \cdot}, \hat{\psi}_{a,s,t} \rangle$ we get with $\omega_3 = -(\omega_1 + p_1 \omega_2)/p_2$ that

$$I_3 = a^{5/6} \int_{\mathbb{R}^2} e^{2\pi i \langle t, \omega \rangle} \hat{\tilde{\psi}}_1(a\omega_1) \hat{\tilde{\psi}}_2 \left(a^{-2/3} \left(s + \frac{1}{\omega_1} \binom{\omega_2}{\omega_3} \right) \right) \frac{1}{q^{\mathsf{T}} \omega} \frac{1}{r^{\mathsf{T}} \omega} \, d\omega_1 d\omega_2.$$

Substituting first $\xi_2 := a^{-2/3}(s_1 + \omega_2/\omega_1)$ and then $\xi_1 := a\omega_1$ this becomes

$$I_3 = a^{3/2} \int_{\mathbb{R}^2} e^{2\pi i \xi_1 \left(t_1 - \frac{t_3}{p_2} - s_1(t_2 - \frac{p_1 t_3}{p_2}) \right)/a} e^{2\pi i \xi_1 \xi_2 (t_2 - \frac{p_1 t_3}{p_2})/a^{1/3}} \frac{\hat{\tilde{\psi}}_1(\xi_1)}{\xi_1}$$

$$\times \hat{\tilde{\psi}}_2 \left(a^{-2/3} (-\frac{1}{p_2} + \frac{p_1}{p_2} s_1 + s_2) - \frac{p_1}{p_2} \xi_2 \right) \frac{1}{g_{pq}(\xi_2)} \frac{1}{g_{pr}(\xi_2)} \, d\xi_1 d\xi_2,$$

where $g_{pq}(\xi_2) := 1 - \frac{q_2}{p_2} - s_1(q_1 - \frac{p_1 q_2}{p_2}) + a^{2/3} \xi_2(q_1 - \frac{p_1 q_2}{p_2})$. If $1 - p_1 s_1 - p_2 s_2 \neq 0$, then $\hat{\tilde{\psi}}_2 \left((\xi_2, a^{-2/3}(-\frac{1}{p_2} + \frac{p_1}{p_2} s_1 + s_2) - \frac{p_1}{p_2} \xi_2)^{\mathsf{T}} \right)$ becomes zero for sufficiently small a by the support property of $\hat{\psi}_2$.

Let $1 - p_1 s_1 - p_2 s_2 = 0$.

3.1. If $1 - \frac{q_2}{p_2} - s_1(q_1 - \frac{p_1 q_2}{p_2}) \neq 0$, i.e., $s_1 \neq -\frac{p_2 - q_2}{p_1 q_2 - p_2 q_1}$ and $1 - \frac{r_2}{p_2} - s_1(r_1 - \frac{p_1 r_2}{p_2})$ $\neq 0$, i.e., $s_1 \neq -\frac{p_2 - r_2}{p_1 r_2 - p_2 r_1}$, then the function ϕ_2 defined by $\hat{\phi}_2 := \frac{\hat{\tilde{\psi}}_2 \left(\xi_2(1, -\frac{p_1}{p_2})^{\mathsf{T}} \right)}{g_{pq}(\xi_2) g_{pr}(\xi_2)} \in \mathscr{S}$ is rapidly decaying and we obtain

$$I_3 = a^{3/2} \int_{\mathbb{R}^1} e^{2\pi i \xi_1 \left(t_1 - \frac{t_3}{p_2} - s_1(t_2 - \frac{p_1 t_3}{p_2}) \right)/a} \frac{\hat{\tilde{\psi}}_1(\xi_1)}{\xi_1} \phi_2 \left(\frac{\xi_1(t_2 p_2 - p_1 t_3)}{p_2 a^{1/3}} \right) d\xi_1.$$

If $t_2 p_2 - p_1 t_3 \neq 0$, then

$$\phi_2 \left(\frac{\xi_1(t_2 p_2 - p_1 t_3)}{p_2 a^{1/3}} \right) \leq C \frac{a^{2r/3}}{a^{2r/3} + \|\xi_1(t_2 - p_1 t_3/p_2)\|^{2r}} \quad \forall r \in \mathbb{N}$$

and since $\hat{\tilde{\psi}}_1(\xi_1) = 0$ for $\xi_1 \in [-a_0, a_0]$, we see that I_3 is rapidly decaying as $a \to 0$. If $t_2 p_2 - p_1 t_3 = 0$, then

$$I_3 \sim a^{3/2} \phi_1 \left(\frac{t_1 - \frac{t_3}{p_2}}{a} \right),$$

which decays rapidly as $a \to 0$ except for $t_1 p_2 = t_3$. Now $t_2 p_2 - p_1 t_3 = 0$ and $t_1 p_2 = t_3$ imply that $t = cp$, $c \in \mathbb{R}$. In this case we have that $I_3 \sim a^{3/2}$.

3.2. If $s_1 = -\frac{p_2 - q_2}{p_1 q_2 - p_2 q_1}$ and consequently $s = -\tilde{n}_{pq}$, then

$$I_3 \sim a^{5/6} \int_{\mathbb{R}^2} e^{2\pi i \xi_1 \left(t_1 - \frac{t_3}{p_2} - s_1 (t_2 - \frac{p_1 t_3}{p_2})\right)/a} \, e^{2\pi i \xi_1 \xi_2 (t_2 - \frac{p_1 t_3}{p_2})/a^{1/3}} \, \frac{\bar{\hat{\psi}}_1(\xi_1)}{\xi_1}$$

$$\times \frac{\bar{\hat{\psi}}_2\left(\xi_2 (1, -p_1/p_2)^{\mathsf{T}}\right)}{g_{pr}(\xi_2)} \frac{1}{\xi_2} \, d\xi_2 d\xi_1$$

$$\sim a^{5/6} \int_{\mathbb{R}} e^{2\pi i \xi_1 \left(t_1 - \frac{t_3}{p_2} - s_1 (t_2 - \frac{p_1 t_3}{p_2})\right)/a} \frac{\bar{\hat{\psi}}_1(\xi_1)}{\xi_1} \, (\phi_2 * \mathrm{sgn}) \left(\frac{\xi_1 (p_2 t_2 - p_1 t_3)}{p_2 a^{1/3}}\right) d\xi_1$$

$$\lesssim a^{5/6} \phi_1 \left(\frac{t_1 - \frac{t_3}{p_2} - s_1 (t_2 - \frac{p_1 t_3}{p_2})}{a}\right),$$

where ϕ_2 and ϕ_1 are defined by $\hat{\phi}_2(\xi_2) := \frac{\bar{\hat{\psi}}_2\left(\xi_2 (1, -p_1/p_2)^{\mathsf{T}}\right)}{g_{pr}(\xi_2)} \in \mathscr{S}$ and $\hat{\phi}_1(\xi_1) := \frac{\bar{\hat{\psi}}_1(\xi_1)}{\xi_1} \in \mathscr{S}$. The last expression decays rapidly as $a \to 0$ except for $t_1 - \frac{t_3}{p_2} - s_1 (t_2 - \frac{p_1 t_3}{p_2}) = 0$, where $I_3 \lesssim a^{5/6}$. Together with the conditions on s the later is the case if $n_{pq}^{\mathsf{T}} t = 0$.

4. Finally, we examine $I_4 := \langle \frac{1}{p^{\mathsf{T}}} \frac{1}{q^{\mathsf{T}}} \frac{1}{r^{\mathsf{T}}}, \hat{\psi}_{a,s,t} \rangle$. We obtain

$$I_4 = a^{5/6} \int_{\mathbb{R}^3} e^{2\pi i \langle t, \omega \rangle} \bar{\hat{\psi}}_1(a\omega_1) \bar{\hat{\psi}}_2 \left(a^{-2/3} \left(s + \frac{1}{\omega_1} \begin{pmatrix} \omega_2 \\ \omega_3 \end{pmatrix}\right)\right) \frac{1}{p^{\mathsf{T}}\omega} \frac{1}{q^{\mathsf{T}}\omega} \frac{1}{r^{\mathsf{T}}\omega} \, d\omega$$

and further by substituting $\xi_j := a^{-2/3}(s_{j-1} + \omega_j/\omega_1)$, $j = 2, 3$ and $\xi_1 := a\omega_1$

$$I_4 = a^{13/6} \int_{\mathbb{R}^3} e^{2\pi i \xi_1 \left(t_1 + t_2 (a^{2/3}\xi_2 - s_1) + t_3 (a^{2/3}\xi_3 - s_2)\right)/a} \frac{\bar{\hat{\psi}}_1(\xi_1)}{\xi_1}$$

$$\times \frac{\bar{\hat{\psi}}_2 \left((\xi_2, \xi_3)^{\mathsf{T}}\right)}{g_p(\xi_2, \xi_3) g_q(\xi_2, \xi_3) g_r(\xi_2, \xi_3)} \, d\xi,$$

where $g_p(\xi_2, \xi_3) := 1 - p_1 s_1 - p_2 s_2 + a^{2/3}(\xi_2 p_1 + \xi_3 p_2)$.

4.1. If $1 - p_1 s_1 - p_2 s_2 \neq 0$, $1 - q_1 s_1 - q_2 s_2 \neq 0$ and $1 - r_1 s_1 - r_2 s_2 \neq 0$, then ϕ_2 defined by $\hat{\phi}_2(\xi_2, \xi_3) := \frac{\bar{\hat{\psi}}_1\left((\xi_2, \xi_3)^{\mathsf{T}}\right)}{g_p(\xi_2, \xi_3) g_q(\xi_2, \xi_3) g_r(\xi_2, \xi_3)} \in \mathscr{S}$ is rapidly decaying and

$$I_4 = a^{13/6} \int_{\mathbb{R}} e^{2\pi i \xi_1 (t_1 - t_2 s_1 - t_3 s_2)/a} \frac{\bar{\hat{\psi}}_1(\xi_1)}{\xi_1} \phi_2(\xi_1 (t_2, t_3)/a^{1/3}) d\xi_1.$$

Similarly as before, we see that I_4 decays rapidly as $a \to 0$ if $(t_2, t_3) \neq (0, 0)$. For $t_2 = t_3 = 0$ we conclude that $I_4 \sim a^{13/6} \phi_1 \left((t_1 - t_2 s_1 - t_3 s_2)/a\right)$. The right-hand side is rapidly decaying as $a \to 0$ except for $t_1 - t_2 s_1 - t_3 s_2 = 0$, i.e., for $t = (0,0,0)^{\mathrm{T}}$, where $I_4 \sim a^{13/6}$.

4.2. If $1 - p_1 s_1 - p_2 s_2 = 0$ and $1 - q_1 s_1 - q_2 s_2 \neq 0$, $1 - r_1 s_1 - r_2 s_2 \neq 0$, we obtain with $\hat{\phi}_2(\xi_2, \xi_3) := \dfrac{\bar{\hat{\psi}}_1\big(((\xi_2,\xi_3)^{\mathrm{T}}\big)}{g_q(\xi_2,\xi_3)\, g_r(\xi_2,\xi_3)} \in \mathscr{S}$ that

$$
I_4 = a^{3/2} \int_{\mathbb{R}^3} e^{2\pi \mathrm{i}\xi_1(t_1 - t_2 s_1 - t_3 s_2)/a}\, e^{2\pi \mathrm{i}\xi_1(t_2\xi_2 + t_3\xi_3)/a^{1/3}}\, \frac{\bar{\hat{\psi}}_1(\xi_1)}{\xi_1}
$$

$$
\times\, \hat{\phi}_2(\xi_2,\xi_3)\, \frac{1}{p_1\xi_2 + p_2\xi_3}\, \mathrm{d}\xi
$$

$$
\sim a^{3/2} \int_{\mathbb{R}} e^{2\pi \mathrm{i}\xi_1(t_1 - t_2 s_1 - t_3 s_2)/a}\, \frac{\bar{\hat{\psi}}_1(\xi_1)}{\xi_1}\, (\phi_2 * h)(\xi_1(t_2,t_3)/a^{1/3})\, \mathrm{d}\xi_1
$$

$$
\lesssim a^{3/2} \phi_1(t_1 - t_2 s_1 - t_3 s_2)/a),
$$

where $h(u,v) := \operatorname{sgn}(-v/p_2)\, \delta(t_2 - p_1 t_3/p_2)$. Thus, I_4 decays rapidly as $a \to 0$ except for $t_1 - t_2 s_1 - t_3 s_2 = 0$.

4.3. Let $1 - p_1 s_1 - p_2 s_2 = 0$ and $1 - q_1 s_1 - q_2 s_2 = 0$, i.e., $s = -\tilde{n}_{pq}$. Then, we obtain with $\hat{\phi}_2(\xi_2,\xi_3) := \dfrac{\bar{\hat{\psi}}_1\big(((\xi_2,\xi_3)^{\mathrm{T}}\big)}{g_r(\xi_2,\xi_3)} \in \mathscr{S}$ that

$$
I_4 = a^{5/6} \int_{\mathbb{R}^3} e^{2\pi \mathrm{i}\xi_1(t_1 - t_2 s_1 - t_3 s_2)/a}\, e^{2\pi \mathrm{i}\xi_1(t_2\xi_2 + t_3\xi_3)/a^{1/3}}\, \frac{\bar{\hat{\psi}}_1(\xi_1)}{\xi_1}
$$

$$
\times\, \hat{\phi}_2(\xi_2,\xi_3)\, \frac{1}{p_1\xi_2 + p_2\xi_3}\, \frac{1}{q_1\xi_2 + q_2\xi_3}\, \mathrm{d}\xi
$$

$$
= a^{5/6} \int_{\mathbb{R}} e^{2\pi \mathrm{i}\xi_1(t_1 - t_2 s_1 - t_3 s_2)/a}\, \frac{\bar{\hat{\psi}}_1(\xi_1)}{\xi_1}\, (\phi_2 * h)(\xi_1(t_2,t_3)/a^{1/3})\, \mathrm{d}\xi_1
$$

$$
\lesssim a^{5/6} \phi_1(t_1 - t_2 s_1 - t_3 s_2)/a),
$$

where $h(u,v) := \operatorname{sgn}\dfrac{p_2 u - p_1 v}{p_1 q_2 - q_1 p_2}\operatorname{sgn}\dfrac{q_2 u - q_1 v}{p_1 q_2 - q_1 p_2}$. If $t_1 - t_2 s_1 - t_3 s_2 = 0$, i.e., $n_{pq}^{\mathrm{T}} t = 0$, then $I_4 \lesssim a^{5/6}$, otherwise we have a rapid decay as $a \to 0$. This finishes the proof.   □

References

1. R.A. Adams, *Sobolev Spaces*, Academic Press, Now York, 1975.
2. L. Borup and M. Nielsen, *Frame decomposition of decomposition spaces*, J. Fourier Anal. Appl. **13** (2007), 39 - 70.
3. E. J. Candès and D. L. Donoho, *Ridgelets: a key to higher-dimensional intermittency?*, Phil. Trans. R. Soc. Lond. A. **357** (1999), 2495 - 2509.
4. E. J. Candès and D. L. Donoho, *Curvelets - A surprisingly effective nonadaptive representation for objects with edges*, in Curves and Surfaces, L. L. Schumaker et al., eds., Vanderbilt University Press, Nashville, TN (1999).

5. E. J. Candès and D. L. Donoho, *Continuous curvelet transform: I. Resolution of the wavefront set*, Appl. Comput. Harmon. Anal. **19** (2005), 162 - 197.

6. E. Cordero, F. De Mari, K. Nowak and A. Tabacco, *Analytic features of reproducing groups for the metaplectic representation*. J. Fourier Anal. Appl., **12**(2) (2006), 157 - 180.

7. S. Dahlke, M. Fornasier, H. Rauhut, G. Steidl, and G. Teschke, *Generalized coorbit theory, Banach frames, and the relations to alpha–modulation spaces*, Proc. Lond. Math. Soc. **96** (2008), 464 - 506.

8. S. Dahlke, S. Häuser, G. Steidl, and G. Teschke, *Coorbit spaces: traces and embeddings in higher dimensions*, Preprint 11-2, Philipps Universität Marburg (2011).

9. S. Dahlke, S. Häuser, and G. Teschke, *Coorbit space theory for the Toeplitz shearlet transform*, to appear in Int. J. Wavelets Multiresolut. Inf. Process.

10. S. Dahlke, G. Kutyniok, P. Maass, C. Sagiv, H.-G. Stark, and G. Teschke, *The uncertainty principle associated with the continuous shearlet transform*, Int. J. Wavelets Multiresolut. Inf. Process. **6** (2008), 157 - 181.

11. S. Dahlke, G. Kutyniok, G. Steidl, and G. Teschke, *Shearlet coorbit spaces and associated Banach frames*, Appl. Comput. Harmon. Anal. **27/2** (2009), 195 - 214.

12. S. Dahlke, G. Steidl, and G. Teschke, *The continuous shearlet transform in arbitrary space dimensions*, J. Fourier Anal. Appl. **16** (2010), 340 - 354.

13. S. Dahlke, G. Steidl and G. Teschke, *Shearlet Coorbit Spaces: Compactly Supported Analyzing Shearlets, Traces and Embeddings*, J. Fourier Anal. Appl., DOI10.1007/s00041-011-9181-6.

14. S. Dahlke and G. Teschke, *The continuous shearlet transform in higher dimensions: Variations of a theme*, in Group Theory: Classes, Representations and Connections, and Applications (C. W. Danelles, Ed.), Nova Publishers, p. 167 - 175, 2009.

15. R. DeVore, *Nonlinear Approximation*, Acta Numerica **7** (1998), 51 - 150.

16. R. DeVore and V.N. Temlyakov, *Some remarks on greedy algorithms*, Adv. in Comput.Math. **5** (1996), 173 - 187.

17. M. N. Do and M. Vetterli, *The contourlet transform: an efficient directional multiresolution image representation*, IEEE Transactions on Image Processing **14**(12) (2005), 2091 - 2106.

18. H. G. Feichtinger and K. Gröchenig, *A unified approach to atomic decompositions via integrable group representations*, Proc. Conf. "Function Spaces and Applications", Lund 1986, Lecture Notes in Math. **1302** (1988), 52 - 73.

19. H. G. Feichtinger and K. Gröchenig, *Banach spaces related to integrable group representations and their atomic decomposition I*, J. Funct. Anal. **86** (1989), 307 - 340.

20. H. G. Feichtinger and K. Gröchenig, *Banach spaces related to integrable group representations and their atomic decomposition II*, Monatsh. Math. **108** (1989), 129 - 148.

21. H. G. Feichtinger and K. Gröchenig, *Non–orthogonal wavelet and Gabor expansions and group representations*, in: Wavelets and Their Applications, M.B. Ruskai et.al. (eds.), Jones and Bartlett, Boston, 1992, 353 - 376.

22. G. B. Folland, *Fourier Analysis and its Applications*, Brooks/Cole Publ. Company, Boston, 1992.

23. M. Frazier and B. Jawerth, *Decomposition of Besov sapces*, Indiana University Mathematics Journal **34/4** (1985), 777 - 799.

24. K. Gröchenig, *Describing functions: Atomic decompositions versus frames*, Monatsh. Math. **112** (1991), 1 - 42.

25. K. Gröchenig, *Foundations of Time–Frequency Analysis*, Birkhäuser, Boston, Basel, Berlin, 2001.

26. K. Gröchenig, E. Kaniuth and K.F. Taylor, *Compact open sets in duals and projections in L_1-algebras of certain semi-direct product groups*, Math. Proc. Camb. Phil. Soc. **111** (1992), 545 - 556.

27. K. Gröchenig and S. Samarah, *Nonlinear approximation with local Fourier bases*, Constr. Approx. **16** (2000), 317 - 331.

28. K. Guo, W. Lim, D. Labate, G. Weiss, and E. Wilson, *Wavelets with composite dilations and their MRA properties*, Appl. Comput. Harmon. Anal. **20** (2006), 220 - 236.

29. K. Guo, G. Kutyniok, and D. Labate, *Sparse multidimensional representations using anisotropic dilation und shear operators,* in Wavelets und Splines (Athens, GA, 2005), G. Chen und M. J. Lai, eds., Nashboro Press, Nashville, TN (2006), 189 - 201.

30. L.I. Hedberg and Y. Netrusov, *An axiomatic approach to function spaces, spectral synthesis, and Luzin approximation,* Memoirs of the American Math. Soc. **188**, 1- 97 (2007).

31. P. Kittipoom, G. Kutyniok, and W.-Q Lim, *Construction of compactly supported shearlet frames,* Preprint, 2009.

32. G. Kutyniok and D. Labate, *Resolution of the wavefront set using continuous shearlets,* Trans. Amer. Math. Soc. **361** (2009), 2719 - 2754.

33. G. Kutyniok, J. Lemvig, and W.-Q. Lim, *Compactly supported shearlets,* Approximation Theory XIII (San Antonio, TX, 2010), Springer, to appear.

34. R. S. Laugesen, N. Weaver, G. L. Weiss and E. N. Wilson, *A characterization of the higher dimensional groups associated with continuous wavelets,* The Journal of Geom. Anal. **12/1** (2002), 89 - 102.

35. Y. Lu and M.N. Do, *Multidimensional directional filterbanks and surfacelets,* IEEE Trans. Image Process. **16** (2007), 918 - 931.

36. C. Schneider, *Besov spaces of positive smoothness,* PhD thesis, University of Leipzig, 2009.

37. H. Triebel, *Function Spaces I,* Birkhäuser, Basel - Boston - Berlin, 2006

38. S. Yi, D. Labate, G. R. Easley, and H. Krim, *A shearlet approach to edge analysis and detection,* IEEE Trans. Image Process. **16** (2007), 918 - 931.

Shearlets and Optimally Sparse Approximations

Gitta Kutyniok, Jakob Lemvig, and Wang-Q Lim

Abstract Multivariate functions are typically governed by anisotropic features such as edges in images or shock fronts in solutions of transport-dominated equations. One major goal both for the purpose of compression and for an efficient analysis is the provision of optimally sparse approximations of such functions. Recently, cartoon-like images were introduced in 2D and 3D as a suitable model class, and approximation properties were measured by considering the decay rate of the L^2 error of the best N-term approximation. Shearlet systems are to date the only representation system, which provide optimally sparse approximations of this model class in 2D as well as 3D. Even more, in contrast to all other directional representation systems, a theory for compactly supported shearlet frames was derived which moreover also satisfy this optimality benchmark. This chapter shall serve as an introduction to and a survey about sparse approximations of cartoon-like images by band-limited and also compactly supported shearlet frames as well as a reference for the state of the art of this research field.

Key words: Anisotropic features, Band-limited shearlets, Cartoon-like images, Compactly supported shearlets, Linear and nonlinear approximations, Multidimensional data, Sparse approximations

G. Kutyniok
Department of Mathematics, Technische Universität Berlin, 10623 Berlin, Germany
e-mail: kutyniok@math.tu-berlin.de

J. Lemvig
Department of Mathematics, Technical University of Denmark, 2800 Kgs. Lyngby, Denmark
e-mail: j.lemvig@mat.dtu.dk

W.-Q Lim
Institut für Mathematik, Technische Universität Berlin, 10623 Berlin, Germany
e-mail: lim@math.tu-berlin.de

G. Kutyniok and D. Labate (eds.), *Shearlets: Multiscale Analysis for Multivariate Data,*   145
Applied and Numerical Harmonic Analysis, DOI 10.1007/978-0-8176-8316-0_5,
© Springer Science+Business Media, LLC 2012

1 Introduction

Scientists face a rapidly growing deluge of data, which requires highly sophisticated methodologies for analysis and compression. Simultaneously, the data itself are becoming increasingly complex and higher dimensional. One of the most prominent features of data is singularities. This statement is justified, for instance, by the observation from neuroscientists that the human eye is most sensitive to smooth geometric areas divided by sharp edges. Intriguingly, already the step from univariate to multivariate data causes a significant change in the behavior of singularities. While one-dimensional (1D) functions can only exhibit point singularities, singularities of two-dimensional (2D) functions can already be of both point and curvilinear type. In fact, multivariate functions are typically governed by *anisotropic phenomena*. Think, for instance, of edges in digital images or evolving shock fronts in solutions of transport-dominated equations. These two exemplary situations also show that such phenomena occur even for both explicitly and implicitly given data.

One major goal both for the purpose of compression and for an efficient analysis is the introduction of representation systems for "good" approximation of anisotropic phenomena, more precisely, of multivariate functions governed by anisotropic features. This raises the following fundamental questions:

(P1) What is a suitable model for functions governed by anisotropic features?
(P2) How do we measure "good" approximation and what is a benchmark for optimality?
(P3) Is the step from 1D to 2D already the crucial step or how does this framework scale with increasing dimension?
(P4) Which representation system behaves optimally?

Let us now first debate these questions on a higher and more intuitive level, and later on delve into the precise mathematical formalism.

1.1 Choice of Model for Anisotropic Features

Each model design has to face the trade-off between closeness to the true situation versus sufficient simplicity to enable analysis of the model. The suggestion of a suitable model for functions governed by anisotropic features in [6] solved this problem in the following way. As a model for an image, it first of all requires the $L^2(\mathbb{R}^2)$ functions serving as a model to be supported on the unit square $[0,1]^2$. These functions shall then consist of the minimal number of smooth parts, namely two. To avoid artificial problems with a discontinuity ending at the boundary of $[0,1]^2$, the boundary curve of one of the smooth parts is entirely contained in $(0,1)^2$. It now remains to decide upon the regularity of the smooth parts of the model functions and of the boundary curve, which were chosen to both be C^2. Thus, concluding, a possible suitable model for functions governed by anisotropic features are 2D functions which are supported on $[0,1]^2$ and C^2 apart from a closed C^2 discontinuity curve; these are typically referred to as *cartoon-like images* (cf. "Introduction of this book!"). This provides an answer to (P1). Extensions of this 2D model to piecewise smooth curves were then suggested in [1], and extensions to three-dimensional (3D) as well as to different types of regularity were introduced in [8, 12].

1.2 Measure for Sparse Approximation and Optimality

The quality of the performance of a representation system with respect to cartoon-like images is typically measured by taking a nonlinear approximation viewpoint. More precisely, given a cartoon-like image and a representation system which forms an orthonormal basis, the chosen measure is the asymptotic behavior of the L^2 error of the best N-term (nonlinear) approximation in the number of terms N. This intuitively measures how fast the ℓ^2 norm of the tail of the expansion decays as more and more terms are used for the approximation. A slight subtlety has to be observed if the representation system does not form an orthonormal basis, but a frame. In this case, the N-term approximation using the N largest coefficients is considered, which, in case of an orthonormal basis, is the same as the best N-term approximation, but not in general. The term "optimally sparse approximation" is then awarded to those representation systems which deliver the fastest possible decay rate in N for all cartoon-like images, where we consider log-factors as negligible, thereby providing an answer to (P2).

1.3 Why is 3D the Crucial Dimension?

We already identified the step from 1D to 2D as crucial for the appearance of anisotropic features at all. Hence, one might ask: Is it sufficient to consider only the 2D situation, and higher dimensions can be treated similarly? Or, does each dimension causes its own problems? To answer these questions, let us consider the step from 2D to 3D which shows an interesting phenomenon. A 3D function can exhibit point ($=$0D), curvilinear ($=$1D), and surface ($=$2D) singularities. Thus, suddenly anisotropic features appear in two different dimensions: as 1D and as 2D features. Hence, the 3D situation has to be analyzed with particular care. It is not at all clear whether two different representation systems are required for optimally approximating both types of anisotropic features simultaneously, or whether one system will suffice. This shows that the step from 2D to 3D can justifiably be also coined "crucial." Once it is known how to handle anisotropic features of different dimensions, the step from 3D to 4D can be dealt with in a similar way as also the extension to even higher dimensions. Thus, answering (P3), we conclude that the two crucial dimensions are 2D and 3D with higher dimensional situations deriving from the analysis of those.

1.4 Performance of Shearlets and Other Directional Systems

Within the framework we just briefly outlined, it can be shown that wavelets do not provide optimally sparse approximations of cartoon-like images. This initiated a flurry of activity within the applied harmonic analysis community with the aim to develop so-called directional representation systems which satisfy this benchmark, certainly besides other desirable properties depending on the application at hand. In 2004, Candés and Donoho were the first to introduce the tight curvelet frames, a directional representation system, which provides provably optimally sparse approximations of cartoon-like images in the sense we discussed. One year later, contourlets were introduced by Do and Vetterli [4], which similarly derived an optimal approximation rate. The first analysis of the performance of (band-limited) tight shearlet frames was undertaken by Guo and Labate in [7], who proved that these shearlets also do satisfy this benchmark. In the situation of (band-limited) shearlets the analysis was then driven even further, and very recently Guo and Labate proved a similar result for 3D cartoon-like images which in this case are defined as a function which is C^2 apart from a C^2 discontinuity surface, i.e., focusing on only one of the types of anisotropic features we are facing in 3D.

1.5 Band-Limited Versus Compactly Supported Systems

The results mentioned in the previous subsection only concerned band-limited systems. Even in the contourlet case, although compactly supported contourlets seem to be included, the proof for optimal sparsity only works for band-limited generators due to the requirement of infinite directional vanishing moments. However, for various applications compactly supported generators are essential, wherefore already in the wavelet case the introduction of compactly supported wavelets was a major advance. Prominent examples of such applications can be found in imaging sciences, when denoising images while avoiding a smoothing of the edges, and in the theory of partial differential equations, when the shearlets act as a generating system for a trial space and the compact support ensures fast computational realizations.

So far, shearlets are the only system for which a theory for compactly supported generators has been developed and compactly supported shearlet frames have been constructed [10], see also the survey paper [13]. It should though be mentioned that these frames are somehow close to being tight, but at this point it is not clear whether also compactly supported *tight* shearlet frames can be constructed. Interestingly, it was proved in [14] that this class of shearlet frames also delivers optimally sparse approximations of the 2D cartoon-like image model class with a very different proof than [7] now adapted to the particular nature of compactly supported generators. And with [12] the 3D situation is now also fully understood, even taking the two different types of anisotropic features—curvilinear and surface singularities—into account.

1.6 Outline

In Sect. 2, we introduce the 2D and 3D cartoon-like image model class. Optimality of sparse approximations of this class is then discussed in Sect. 3. Section 4 is concerned with the introduction of 3D shearlet systems with both band-limited and compactly supported generators, which are shown to provide optimally sparse approximations within this class in the final Sect. 5.

2 Cartoon-Like Image Class

We start by making the definition of cartoon-like images, which was intuitively derived in the introduction of this chapter, mathematically precise. We start with the most basic definition of this class which was also historically first stated in [6]. We allow ourselves to state this together with its 3D version from [8] by remarking that d could be either $d = 2$ or $d = 3$.

For fixed $\mu > 0$, the *class* $\mathscr{E}^2(\mathbb{R}^d)$ *of cartoon-like image* shall be the set of functions $f : \mathbb{R}^d \to \mathbb{C}$ of the form

$$f = f_0 + f_1 \chi_B,$$

where $B \subset [0,1]^d$ and $f_i \in C^2(\mathbb{R}^d)$ with $\operatorname{supp} f_0 \subset [0,1]^d$ and $\|f_i\|_{C^2} \leq \mu$ for each $i = 0, 1$. For dimension $d = 2$, we assume that ∂B is a closed C^2-curve with curvature bounded by ν, and, for $d = 3$, the discontinuity ∂B shall be a closed C^2-surface with principal curvatures bounded by ν. An indiscriminately chosen cartoon-like function $f = \chi_B$, where the discontinuity surface ∂B is a deformed sphere in $\mathbb{R}^3$, is depicted in Fig. 1.

Since objects in images often have sharp corners, in [1] for 2D and in [12] for 3D also less regular images were allowed, where ∂B is only assumed to be *piecewise* C^2-smooth. To be precise, ∂B is assumed to be a union of finitely many pieces $\partial B_1, \dots, \partial B_L$ which do not overlap except at their boundaries, and each patch ∂B_i can be represented in parametric form by a C^2-smooth function and has curvature (for $d = 2$) or principal curvatures (for $d = 3$) bounded by μ. Letting $L \in \mathbb{N}$ denote the number of C^2 pieces, we speak of the extended class of *cartoon-like images* $\mathscr{E}_L^2(\mathbb{R}^d)$ as consisting of cartoon-like images having C^2-smoothness apart from a piecewise C^2 discontinuity curve in the 2D setting and a piecewise C^2 discontinuity surface in the 3D setting. We note that this concept is also essential for being able to analyze the behavior of a system with respect to the two different types of anisotropic features appearing in 3D; see the discussion in Sect. 1.3. Indeed, in the 3D setting, besides the C^2 discontinuity surfaces, this model exhibits curvilinear C^2 singularities as well as point singularities, e.g., the cartoon-like image $f = \chi_B$ in Fig. 2 exhibits a discontinuity surface $\partial B \subset \mathbb{R}^3$ consisting of *three* C^2-smooth surfaces with point and curvilinear singularities where these surfaces meet.

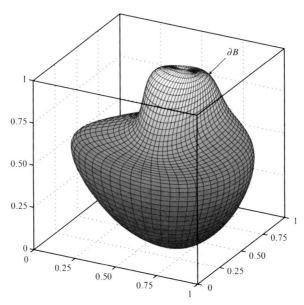

Fig. 1 A simple cartoon-like image $f = \chi_B \in \mathscr{E}_L^2(\mathbb{R}^3)$ with $L = 1$ for dimension $d = 3$, where the discontinuity surface ∂B is a deformed sphere

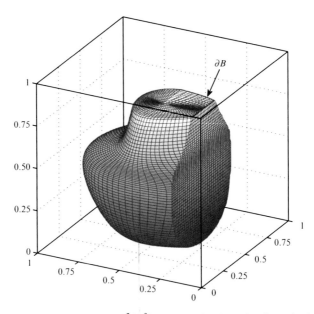

Fig. 2 A cartoon-like image $f = \chi_B \in \mathscr{E}_L^2(\mathbb{R}^3)$ with $L = 3$, where the discontinuity surface ∂B is piecewise C^2-smooth

The model in [12] goes even one step further and considers a different regularity for the smooth parts, say being in C^β, and for the smooth pieces of the discontinuity, say being in C^α with $1 \le \alpha \le \beta \le 2$. This very general class of cartoon-like images is then denoted by $\mathcal{E}^\beta_{\alpha,L}(\mathbb{R}^d)$, with the agreement that $\mathcal{E}^2_L(\mathbb{R}^d) = \mathcal{E}^\beta_{\alpha,L}(\mathbb{R}^d)$ for $\alpha = \beta = 2$.

For the purpose of clarity, in the sequel we will focus on the first most basic cartoon-like model where $\alpha = \beta = 2$, and add hints on generalizations when appropriate (in particular, in Sect. 5.2.4).

3 Sparse Approximations

After having clarified the model situation, we will now discuss which measure for the accuracy of approximation by representation systems we choose, and what optimality means in this case.

3.1 (Nonlinear) N-term Approximations

Let $\mathcal{C}$ denote a given class of elements in a separable Hilbert space $\mathcal{H}$ with norm $\|\cdot\| = \langle \cdot, \cdot \rangle^{1/2}$ and $\Phi = (\phi_i)_{i \in I}$ a dictionary for $\mathcal{H}$, i.e., $\overline{\text{span}}\Phi = \mathcal{H}$, with indexing set I. The Φ plays the role of our representation system. Later $\mathcal{C}$ will be chosen to be the class of cartoon-like images and Φ a shearlet frame, but for now we will assume this more general setting. We now seek to approximate each single element of $\mathcal{C}$ with elements from Φ by "few" terms of this system. Approximation theory provides us with the concept of best N-term approximation which we now introduce; for a general introduction to approximation theory, we refer to [3].

For this, let $f \in \mathcal{C}$ be arbitrarily chosen. Since Φ is a complete system, for any $\varepsilon > 0$ there exists a finite linear combination of elements from Φ of the form

$$g = \sum_{i \in F} c_i \phi_i \quad \text{with } F \subset I \text{ finite, i.e., } \#|F| < \infty$$

such that $\|f - g\| \le \varepsilon$. Moreover, if Φ is a frame with countable indexing set I, there exists a sequence $(c_i)_{i \in I} \in \ell_2(I)$ such that the representation

$$f = \sum_{i \in I} c_i \phi_i$$

holds with convergence in the Hilbert space norm $\|\cdot\|$. The reader should notice that if Φ does not form a basis, this representation of f is certainly not the only possible one. Letting now $N \in \mathbb{N}$, we aim to approximate f by only N terms of Φ, i.e., by

$$\sum_{i \in I_N} c_i \phi_i \quad \text{with } I_N \subset I, \#|I_N| = N,$$

which is termed *N-term approximation* to f. This approximation is typically *nonlinear* in the sense that if f_N is an N-term approximation to f with indices I_N and g_N is an N-term approximation to some $g \in \mathscr{C}$ with indices J_N, then $f_N + g_N$ is only an N-term approximation to $f + g$ in case $I_N = J_N$.

But certainly we would like to pick the best approximation with the accuracy of approximation measured in the Hilbert space norm. We define the *best N-term approximation* to f by the N-term approximation

$$f_N = \sum_{i \in I_N} c_i \phi_i,$$

which satisfies that for all $I_N \subset I$, $\# |I_N| = N$, and for all scalars $(c_i)_{i \in I}$,

$$\|f - f_N\| \le \left\| f - \sum_{i \in I_N} c_i \phi_i \right\|.$$

Let us next discuss the notion of best N-term approximation for the special cases of Φ forming an orthonormal basis, a tight frame, and a general frame alongside an error estimate for the accuracy of this approximation.

3.1.1 Orthonormal bases

Let Φ be an orthonormal basis for $\mathscr{H}$. In this case, we can actually write down the best N-term approximation $f_N = \sum_{i \in I_N} c_i \phi_i$ for f. Since in this case

$$f = \sum_{i \in I} \langle f, \phi_i \rangle \phi_i,$$

and this representation is unique, we obtain

$$
\begin{aligned}
\|f - f_N\|_{\mathscr{H}} &= \left\| \sum_{i \in I} \langle f, \phi_i \rangle \phi_i - \sum_{i \in I_N} c_i \phi_i \right\| \\
&= \left\| \sum_{i \in I_N} [\langle f, \phi_i \rangle - c_i] \phi_i + \sum_{i \in I \setminus I_N} \langle f, \phi_i \rangle \phi_i \right\| \\
&= \| (\langle f, \phi_i \rangle - c_i)_{i \in I_N} \|_{\ell^2} + \| (\langle f, \phi_i \rangle)_{i \in I \setminus I_N} \|_{\ell^2}.
\end{aligned}
$$

The first term $\| (\langle f, \phi_i \rangle - c_i)_{i \in I_N} \|_{\ell^2}$ can be minimized by choosing $c_i = \langle f, \phi_i \rangle$ for all $i \in I_N$. And the second term $\| (\langle f, \phi_i \rangle)_{i \in I \setminus I_N} \|_{\ell^2}$ can be minimized by choosing I_N to be the indices of the N largest coefficients $\langle f, \phi_i \rangle$ in magnitude. Notice that this does not uniquely determine f_N since some coefficients $\langle f, \phi_i \rangle$ might have the same magnitude. But it characterizes the set of best N-term approximations to some $f \in \mathscr{C}$ precisely. Even more, we have complete control of the error of best N-term approximation by

$$\|f - f_N\| = \| (\langle f, \phi_i \rangle)_{i \in I \setminus I_N} \|_{\ell^2}. \tag{1}$$

3.1.2 Tight frames

Assume now that Φ constitutes a tight frame with bound $A = 1$ for $\mathcal{H}$. In this situation, we still have

$$f = \sum_{i \in I} \langle f, \phi_i \rangle \phi_i,$$

but this expansion is now not unique anymore. Moreover, the frame elements are not orthogonal. Both conditions prohibit an analysis of the error of best N-term approximation as in the previously considered situation of an orthonormal basis. And in fact, examples can be provided to show that selecting the N largest coefficients $\langle f, \phi_i \rangle$ in magnitude does not always lead to the *best* N-term approximation, but merely to *an* N-term approximation. To be able to still analyze the approximation error, one typically—as will be also our choice in the sequel—chooses the N-term approximation provided by the indices I_N associated with the N largest coefficients $\langle f, \phi_i \rangle$ in magnitude with these coefficients, i.e.,

$$f_N = \sum_{i \in I_N} \langle f, \phi_i \rangle \phi_i.$$

This selection also allows for some control of the approximation in the Hilbert space norm, which we will defer to the next subsection in which we consider the more general case of arbitrary frames.

3.1.3 General Frames

Let now Φ form a frame for $\mathcal{H}$ with frame bounds A and B, and let $(\tilde{\phi}_i)_{i \in I}$ denote the canonical dual frame. We then consider the expansion of f in terms of this dual frame, i.e.,

$$f = \sum_{i \in I} \langle f, \phi_i \rangle \tilde{\phi}_i. \qquad (2)$$

Notice that we could also consider

$$f = \sum_{i \in I} \langle f, \tilde{\phi}_i \rangle \phi_i.$$

Let us explain why the first form is of more interest to us in this chapter. By definition, we have $(\langle f, \tilde{\phi}_i \rangle)_{i \in I} \in \ell^2(I)$ as well as $(\langle f, \phi_i \rangle)_{i \in I} \in \ell^2(I)$. Since we only consider expansions of functions f belonging to a subset $\mathcal{C}$ of $\mathcal{H}$, this can, at least, potentially improve the decay rate of the coefficients so that they belong to $\ell^p(I)$ for some $p < 2$. This is exactly what is understood by *sparse approximation* (also called *compressible approximations* in the context of inverse problems). We hence aim to analyze shearlets with respect to this behavior, i.e., the decay rate of shearlet coefficients. This then naturally leads to form (2). We remark that in case of a tight frame, there is no distinction necessary, since then $\tilde{\phi}_i = \phi_i$ for all $i \in I$.

As in the tight frame case, it is not possible to derive a usable, explicit form for the best N-term approximation. We therefore again crudely approximate the best N-term approximation by choosing the N-term approximation provided by the indices I_N associated with the N largest coefficients $\langle f, \phi_i \rangle$ in magnitude with these coefficients, i.e.,

$$f_N = \sum_{i \in I_N} \langle f, \phi_i \rangle \tilde{\phi}_i.$$

Surprisingly, even with this rather crude greedy selection procedure, we obtain very strong results for the approximation rate of shearlets as we will see in Sect. 5.

The following result shows how the N-term approximation error can be bounded by the tail of the square of the coefficients c_i. The reader might want to compare this result with the error in case of an orthonormal basis stated in (1).

Lemma 1. *Let $(\phi_i)_{i \in I}$ be a frame for $\mathcal{H}$ with frame bounds A and B, and let $(\tilde{\phi}_i)_{i \in I}$ be the canonical dual frame. Let $I_N \subset I$ with $\#|I_N| = N$, and let f_N be the N-term approximation $f_N = \sum_{i \in I_N} \langle f, \phi_i \rangle \tilde{\phi}_i$. Then*

$$\|f - f_N\|^2 \leq \frac{1}{A} \sum_{i \notin I_N} |\langle f, \phi_i \rangle|^2. \tag{3}$$

Proof. Recall that the canonical dual frame satisfies the frame inequality with bounds B^{-1} and A^{-1}. At first hand, it therefore might look as if the estimate (3) should follow directly from the frame inequality for the canonical dual. However, since the sum in (3) does not run over the entire index set $i \in I$, but only $I \setminus I_N$, this is not the case. So, to prove the lemma, we first consider

$$\|f - f_N\| = \sup \{|\langle f - f_N, g \rangle| : g \in \mathcal{H}, \|g\| = 1\}$$
$$= \sup \left\{ \left| \sum_{i \notin I_N} \langle f, \phi_i \rangle \langle \tilde{\phi}_i, g \rangle \right| : g \in \mathcal{H}, \|g\| = 1 \right\}. \tag{4}$$

Using Cauchy–Schwarz' inequality, we then have that

$$\left| \sum_{i \notin I_N} \langle f, \phi_i \rangle \langle \tilde{\phi}_i, g \rangle \right|^2 \leq \sum_{i \notin I_N} |\langle f, \phi_i \rangle|^2 \sum_{i \notin I_N} |\langle \tilde{\phi}_i, g \rangle|^2 \leq A^{-1} \|g\|^2 \sum_{i \notin I_N} |\langle f, \phi_i \rangle|^2,$$

where we have used the upper frame inequality for the dual frame $(\tilde{\phi}_i)_i$ in the second step. We can now continue (4) and arrive at

$$\|f - f_N\|^2 \leq \sup \left\{ \frac{1}{A} \|g\|^2 \sum_{i \notin I_N} |\langle f, \phi_i \rangle|^2 : g \in \mathcal{H}, \|g\| = 1 \right\} = \frac{1}{A} \sum_{i \notin I_N} |\langle f, \phi_i \rangle|^2.$$

$\square$

Relating to the previous discussion about the decay of coefficients $\langle f, \phi_i \rangle$, let c^* denote the non-increasing (in modulus) rearrangement of $c = (c_i)_{i \in I} = (\langle f, \phi_i \rangle)_{i \in I}$, e.g., c_n^* denotes the nth largest coefficient of c in modulus. This rearrangement corresponds to a bijection $\pi : \mathbb{N} \to I$ that satisfies

$$\pi : \mathbb{N} \to I, \quad c_{\pi(n)} = c_n^* \text{ for all } n \in \mathbb{N}.$$

Strictly speaking, the rearrangement (and hence the mapping π) might not be unique; we will simply take c^* to be one of these rearrangements. Since $c \in \ell^2(I)$, also $c^* \in \ell^2(\mathbb{N})$. Suppose further that $|c_n^*|$ even decays as

$$|c_n^*| \lesssim n^{-(\alpha+1)/2} \qquad \text{for} \quad n \to \infty$$

for some $\alpha > 0$, where the notation $h(n) \lesssim g(n)$ means that there exists a $C > 0$ such that $h(n) \leq Cg(n)$, i.e., $h(n) = O(g(n))$. Clearly, we then have $c^* \in \ell^p(\mathbb{N})$ for $p \geq \frac{2}{\alpha+1}$. By Lemma 1, the N-term approximation error will therefore decay as

$$\|f - f_N\|^2 \leq \frac{1}{A} \sum_{n>N} |c_n^*|^2 \lesssim \sum_{n>N} n^{-\alpha+1} \asymp N^{-\alpha},$$

where f_N is the N-term approximation of f by keeping the N largest coefficients, that is,

$$f_N = \sum_{n=1}^{N} c_n^* \tilde{\phi}_{\pi(n)}. \tag{5}$$

The notation $h(n) \asymp g(n)$, also written $h(n) = \Theta(g(n))$, used above means that h is bounded both above and below by g asymptotically as $n \to \infty$, that is, $h(n) = O(g(n))$ and $g(n) = O(h(n))$.

3.2 A Notion of Optimality

We now return to the setting of functions spaces $\mathcal{H} = L^2(\mathbb{R}^d)$, where the subset $\mathcal{C}$ will be the class of cartoon-like images, that is, $\mathcal{C} = \mathcal{E}_L^2(\mathbb{R}^d)$. We then aim for a benchmark, i.e., an optimality statement for sparse approximation of functions in $\mathcal{E}_L^2(\mathbb{R}^d)$. For this, we will again only require that our representation system Φ is a dictionary, that is, we assume only that $\Phi = (\phi_i)_{i \in I}$ is a complete family of functions in $L^2(\mathbb{R}^d)$ with I not necessarily being countable. Without loss of generality, we can assume that the elements ϕ_i are normalized, i.e., $\|\phi_i\|_{L^2} = 1$ for all $i \in I$. For $f \in \mathcal{E}_L^2(\mathbb{R}^d)$, we then consider expansions of the form

$$f = \sum_{i \in I_f} c_i \phi_i,$$

where $I_f \subset I$ is a countable selection from I that may depend on f. Relating to the previous subsection, the first N elements of $\Phi_f := \{\phi_i\}_{i \in I_f}$ could for instance be the N terms from Φ selected for the best N-term approximation of f.

Since artificial cases shall be avoided, this selection procedure has the following natural restriction which is usually termed *polynomial depth search*: The nth term in Φ_f is obtained by only searching through the first $q(n)$ elements of the list Φ_f, where q is a polynomial. Moreover, the selection rule may *adaptively* depend on f, and the nth element may also be modified adaptively and depend on the first $(n-1)$th chosen elements. We shall denote any sequence of coefficients c_i chosen according to these restrictions by $c(f) = (c_i)_i$. The role of the polynomial q is to limit how deep or how far down in the listed dictionary Φ_f we are allowed to search for the next element ϕ_i in the approximation. Without such a depth search limit, one could choose Φ to be a countable, dense subset of $L^2(\mathbb{R}^d)$ which would yield arbitrarily good sparse approximations, but also infeasible approximations in practice.

Using information theoretic arguments, it was then shown in [5, 12] that almost no matter what selection procedure we use to find the coefficients $c(f)$, we cannot have $\|c(f)\|_{\ell^p}$ bounded for $p < \frac{2(d-1)}{d+1}$ for $d = 2, 3$.

Theorem 1 ([5, 12]). *Retaining the definitions and notations in this subsection and allowing only polynomial depth search, we obtain*

$$\max_{f \in \mathscr{E}_L^2(\mathbb{R}^d)} \|c(f)\|_{\ell^p} = +\infty, \qquad \text{for} \qquad p < \frac{2(d-1)}{d+1}.$$

In case Φ is an orthonormal basis for $L^2(\mathbb{R}^d)$, the norm $\|c(f)\|_{\ell^p}$ is trivially bounded for $p \geq 2$ since we can take $c(f) = (c_i)_{i \in I} = (\langle f, \phi_i \rangle)_{i \in I}$. Although not explicitly stated, the proof can be straightforwardly extended from 3D to higher dimensions as also the definition of cartoon-like images can be similarly extended. It is then intriguing to analyze the behavior of $\frac{2(d-1)}{d+1}$ from Theorem 1. In fact, as $d \to \infty$, we observe that $\frac{2(d-1)}{d+1} \to 2$. Thus, the decay of any $c(f)$ for cartoon-like images becomes slower as d grows and approaches ℓ^2, which—as we just mentioned—is actually the rate guaranteed for *all* $f \in L^2(\mathbb{R}^d)$.

Theorem 1 is truly a statement about the optimal achievable sparsity level: No representation system—up to the restrictions described above—can deliver approximations for $\mathscr{E}_L^2(\mathbb{R}^d)$ with coefficients satisfying $c(f) \in \ell_p$ for $p < \frac{2(d-1)}{d+1}$. This implies that one *cannot* find a $\beta > \frac{d+1}{2(d-1)}$ such that $|c(f)_n^*| \lesssim n^{-\beta}$ holds, where $c(f)^* = (c(f)_n^*)_{n \in \mathbb{N}}$ is a decreasing (in modulus) arrangement of the coefficients $c(f)$. Hence, the best decay of $c(f)^*$ we can hope for is

$$|c(f)_n^*| \lesssim n^{-\frac{d+1}{2(d-1)}} = \begin{cases} n^{-3/2} & : \quad d = 2, \\ n^{-1} & : \quad d = 3. \end{cases}$$

Assume for a moment that we have such an "optimal" dictionary Φ at hand that delivers this decay rate, and assume further that it is also a frame. Let $\alpha := \frac{2}{d-1}$. Then $\frac{2(d-1)}{d+1} = \frac{\alpha+1}{2}$, hence $|c(f)_n^*| \lesssim n^{-\frac{\alpha+1}{2}}$. As we saw in Sect. 3.1.3, this implies that

$$\|f - f_N\|_{L^2}^2 \lesssim N^{-\alpha} = N^{-\frac{2}{d-1}} \qquad \text{as } N \to \infty,$$

where f_N is the N-term approximation of f by keeping the N largest coefficients.

On the other hand, assume toward a contradiction that

$$\|f - f_N\|_{L^2}^2 \lesssim N^{-\alpha'} \qquad \text{as } N \to \infty$$

holds for some $\alpha' > \alpha$. In the following, we need to make a stronger assumption on Φ, namely, that it is a Riesz basis. Then the rearranged frame coefficients $c(f)^*$ satisfy

$$N |c(f)_{2N}^*|^2 \leq \sum_{n > N} |c(f)_n^*|^2 \lesssim \|f - f_N\|_{L^2}^2 \lesssim N^{-\alpha'} \qquad \text{for } N \in \mathbb{N},$$

which implies

$$|c(f)_n^*|^2 \lesssim n^{-(\alpha'+1)} \qquad \text{for } n \in \mathbb{N}.$$

Hence, $|c(f)_n^*| \lesssim n^{-\beta}$ for $\beta = \frac{\alpha'+1}{2} > \frac{\alpha+1}{2}$, but this contradicts what we concluded above. This shows that the best N-term approximation error $\|f - f_N\|_{L^2}^2$ behaves asymptotically as $N^{-\frac{2}{d-1}}$ or worse. Thus, the optimally achievable rate is, at best, $N^{-\frac{2}{d-1}}$. This optimal rate can be used as a benchmark for measuring the sparse approximation ability of cartoon-like images of different representation systems. Let us phrase this formally.

Definition 1. Let $\Phi = (\phi_i)_{i \in I}$ be a frame for $L^2(\mathbb{R}^d)$ with $d = 2$ or $d = 3$. We say that Φ *provides optimally sparse approximations* of cartoon-like images if, for each $f \in \mathscr{E}_L^2(\mathbb{R}^d)$, the associated N-term approximation f_N (cf. (5)) by keeping the N largest coefficients of $c = c(f) = (\langle f, \phi_i \rangle)_{i \in I}$ satisfies

$$\|f - f_N\|_{L^2}^2 \lesssim N^{-\frac{2}{d-1}} \qquad \text{as } N \to \infty, \tag{6}$$

and

$$|c_n^*| \lesssim n^{-\frac{d+1}{2(d-1)}} \qquad \text{as } n \to \infty, \tag{7}$$

where we ignore log-factors.

Note that, for frames Φ, the bound $|c_n^*| \lesssim n^{-\frac{d+1}{2(d-1)}}$ automatically implies that $\|f - f_N\|^2 \lesssim N^{-\frac{2}{d-1}}$ whenever f_N is chosen as in (5). This follows from Lemma 1 and the estimate

$$\sum_{n > N} |c_n^*|^2 \lesssim \sum_{n > N} n^{-\frac{d+1}{d-1}} \lesssim \int_N^\infty x^{-\frac{d+1}{d-1}} dx \leq C \cdot N^{-\frac{2}{d-1}}, \tag{8}$$

where we have used that $-\frac{d+1}{d-1} + 1 = -\frac{2}{d-1}$. Hence, we are searching for a representation system Φ which forms a frame and delivers decay of $c = (\langle f, \phi_i \rangle)_{i \in I}$ as (up to log-factors)

$$|c_n^*| \lesssim n^{-\frac{d+1}{2(d-1)}} = \begin{cases} n^{-3/2} & : \quad d = 2, \\ n^{-1} & : \quad d = 3. \end{cases} \tag{9}$$

as $n \to \infty$ for any cartoon-like image.

3.3 Approximation by Fourier Series and Wavelets

We will next study two examples of more traditional representation systems—the Fourier basis and wavelets—with respect to their ability to meet this benchmark. For this, we choose the function $f = \chi_B$, where "B" is a ball contained in $[0,1]^d$ with $d = 2$ or $d = 3$, as a simple cartoon-like image in $\mathcal{E}_L^2(\mathbb{R}^d)$ with $L = 1$. We then analyze the error $\|f - f_N\|^2$ for f_N being the N-term approximation by the N largest coefficients and compare with the optimal decay rate stated in Definition 1. It will, however, turn out that these systems are far from providing optimally sparse approximations of cartoon-like images, thus underlining the pressing need to introduce representation systems delivering this optimal rate; and we already now refer to Sect. 5 in which shearlets will be proven to satisfy this property.

Since Fourier series and wavelet systems are orthonormal bases (or more generally, Riesz bases), the best N-term approximation is found by keeping the N largest coefficients as discussed in Sect. 3.1.1.

3.3.1 Fourier series

The error of the best N-term Fourier series approximation of a typical cartoon-like image decays asymptotically as $N^{-1/d}$. The following proposition shows this behavior in the case of a very simple cartoon-like image: the characteristic function on a ball.

Proposition 1. Let $d \in \mathbb{N}$, and let $\Phi = (e^{2\pi i k x})_{k \in \mathbb{Z}^d}$. Suppose $f = \chi_B$, where B is a ball contained in $[0,1]^d$. Then

$$\|f - f_N\|_{L^2}^2 \asymp N^{-1/d} \qquad for \ N \to \infty,$$

where f_N is the best N-term approximation from Φ.

Proof. We fix a new origin as the center of the ball B. Then f is a radial function $f(x) = h(\|x\|_2)$ for $x \in \mathbb{R}^d$. The Fourier transform of f is also a radial function and can be expressed explicitly by Bessel functions of first kind [11, 15]:

$$\hat{f}(\xi) = r^{d/2} \frac{J_{d/2}(2\pi r \|\xi\|_2)}{\|\xi\|_2^{d/2}},$$

where r is the radius of the ball B. Since the Bessel function $J_{d/2}(x)$ decays like $x^{-1/2}$ as $x \to \infty$, the Fourier transform of f decays like $|\hat{f}(\xi)| \asymp \|\xi\|_2^{-(d+1)/2}$ as $\|\xi\|_2 \to \infty$. Letting $I_N = \{k \in \mathbb{Z}^d : \|k\|_2 \leq N\}$ and f_{I_N} be the partial Fourier sum with terms from I_N, we obtain

$$\|f - f_{I_N}\|_{L^2}^2 = \sum_{k \notin I_N} |\hat{f}(k)|^2 \asymp \int_{\|\xi\|_2 > N} \|\xi\|_2^{-(d+1)} d\xi$$

$$= \int_N^\infty r^{-(d+1)} r^{(d-1)} dr = \int_N^\infty r^{-2} dr = N^{-1}.$$

The conclusion now follows from the cardinality of $\# |I_N| \asymp N^d$ as $N \to \infty$. $\square$

3.3.2 Wavelets

Since wavelets are designed to deliver sparse representations of point singularities—
see "Introduction of this book!"—we expect this system to outperform the Fourier
approach. This will indeed be the case. However, the optimal rate will still by far
be missed. The best N-term approximation of a typical cartoon-like image using
a wavelet basis performs only slightly better than Fourier series with asymptotic
behavior as $N^{-1/(d-1)}$. This is illustrated by the following result.

Proposition 2. *Let $d = 2,3$, and let Φ be a wavelet basis for $L^2(\mathbb{R}^d)$ or $L^2([0,1]^d)$.
Suppose $f = \chi_B$, where B is a ball contained in $[0,1]^d$. Then*

$$\|f - f_N\|_{L^2}^2 \asymp N^{-\frac{1}{d-1}} \qquad \text{for } N \to \infty,$$

where f_N is the best N-term approximation from Φ.

Proof. Let us first consider wavelet approximation by the Haar tensor wavelet basis
for $L^2([0,1]^d)$ of the form

$$\left\{ \phi_{0,k} : |k| \leq 2^J - 1 \right\} \cup \left\{ \psi_{j,k}^1, \ldots, \psi_{j,k}^{2d-1} : j \geq J, |k| \leq 2^{j-J} - 1 \right\},$$

where $J \in \mathbb{N}$, $k \in \mathbb{N}_0^d$, and $g_{j,k} = 2^{jd/2}g(2^j \cdot -k)$ for $g \in L^2(\mathbb{R}^d)$. There are only a
finite number of coefficients of the form $\langle f, \phi_{0,k}\rangle$; hence, we do not need to consider
these for our asymptotic estimate. For simplicity, we take $J = 0$. At scale $j \geq 0$ there
exist $\Theta(2^{j(d-1)})$ nonzero wavelet coefficients, since the surface area of ∂B is finite
and the wavelet elements are of size $2^{-j} \times \cdots \times 2^{-j}$.

To illustrate the calculations leading to the sought approximation error rate,
we will first consider the case where B is a cube in $[0,1]^d$. For this, we first
consider the nonzero coefficients associated with the face of the cube contain-
ing the point $(b,c,\ldots,c)$. For scale j, let k be such that $\operatorname{supp} \psi_{j,k}^1 \cap \operatorname{supp} f \neq \emptyset$,
where $\psi^1(x) = h(x_1)p(x_2)\cdots p(x_d)$ and h and p are the Haar wavelet and scal-
ing function, respectively. Assume that b is located in the first half of the interval
$[2^{-j}k_1, 2^{-j}(k_1+1)]$; the other case can be handled similarly. Then

$$|\langle f, \psi_{j,k}^1\rangle| = \int_{2^{-j}k_1}^b 2^{jd/2}dx_1 \prod_{i=2}^d \int_{2^{-j}k_i}^{2^{-j}(k_i+1)} dx_i = (b-2^{-j}k_1)2^{-j(d-1)}2^{jd/2} \asymp 2^{-jd/2},$$

where we have used that $(b-2^{-j}k_1)$ will typically be of size $\frac14 2^{-j}$. Note that for
the chosen j and k above, we also have that $\langle f, \psi_{j,k}^l\rangle = 0$ for all $l = 2,\ldots,2^d-1$.

There will be $2 \cdot \lceil 2c2^{j(d-1)}\rceil$ nonzero coefficients of size $2^{-jd/2}$ associated with
the wavelet ψ^1 at scale j. The same conclusion holds for the other wavelets ψ^l, $l =
2,\ldots,2^d-1$. To summarize, at scale j there will be $C2^{j(d-1)}$ nonzero coefficients
of size $C2^{-jd/2}$. On the first j_0 scales, that is $j = 0,1,\ldots,j_0$, we therefore have
$\sum_{j=0}^{j_0} 2^{j(d-1)} \asymp 2^{j_0(d-1)}$ nonzero coefficients. The nth largest coefficient c_n^* is of size
$n^{-\frac{d}{2(d-1)}}$ since, for $n = 2^{j(d-1)}$, we have

$$2^{-j\frac{d}{2}} = n^{-\frac{d}{2(d-1)}}.$$

Therefore,

$$\|f - f_N\|_{L^2}^2 = \sum_{n>N} |c_n^*|^2 \asymp \sum_{n>N} n^{-\frac{d}{d-1}} \asymp \int_N^\infty x^{-\frac{d}{d-1}} \, dx = \frac{d}{d-1} N^{-\frac{1}{d-1}}.$$

Hence, for the best N-term approximation f_N of f using a wavelet basis, we obtain the asymptotic estimates

$$\|f - f_N\|_{L^2}^2 = \Theta(N^{-\frac{1}{d-1}}) = \begin{cases} \Theta(N^{-1}), & \text{if } d = 2, \\ \Theta(N^{-1/2}), & \text{if } d = 3, \end{cases}$$

Let us now consider the situation that B is a ball. In fact, in this case we can do similar (but less transparent) calculations leading to the same asymptotic estimates as above. We will not repeat these calculations here, but simply remark that the upper asymptotic bound in $|\langle f, \psi_{j,k}^l \rangle| \asymp 2^{-jd/2}$ can be seen by the following general argument:

$$|\langle f, \psi_{j,k}^l \rangle| \le \|f\|_{L^\infty} \|\psi_{j,k}^l\|_{L^1} \le \|f\|_{L^\infty} \|\psi^l\|_{L^1} 2^{-jd/2} \le C 2^{-jd/2},$$

which holds for each $l = 1, \ldots, 2^d - 1$.

Finally, we can conclude from our calculations that choosing another wavelet basis will not improve the approximation rate.   □

Remark 1. We end this subsection with a remark on *linear* approximations. For a linear wavelet approximation of f, one would use

$$f \approx \langle f, \phi_{0,0} \rangle \phi_{0,0} + \sum_{l=1}^{2^d-1} \sum_{j=0}^{j_0} \sum_{|k| \le 2^j - 1} \langle f, \psi_{j,k}^l \rangle \psi_{j,k}^l$$

for some $j_0 > 0$. If restricting to linear approximations, the summation order is not allowed to be changed, and we therefore need to include all coefficients from the first j_0 scales. At scale $j \ge 0$, there exist a total of 2^{jd} coefficients, which by our previous considerations can be bounded by $C \cdot 2^{-jd/2}$. Hence, we include 2^j times as many coefficients as in the nonlinear approximation on each scale. This implies that the error rate of the linear N-term wavelet approximation is $N^{-1/d}$, which is the *same* rate as obtained by Fourier approximations.

3.3.3 Key problem

The key problem of the suboptimal behavior of Fourier series and wavelet bases is the fact that these systems are not generated by anisotropic elements. Let us illustrate this for 2D in the case of wavelets. Wavelet elements are isotropic due to the scaling matrix $\mathrm{diag}\, 2^j, 2^j$. However, already intuitively, approximating a curve with isotropic elements requires many more elements than if the analyzing elements would be anisotropic themselves, see Fig. 3a, b.

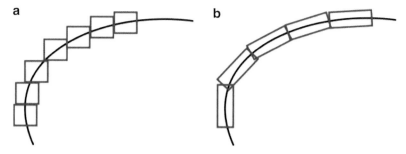

Fig. 3 (**a**) Isotropic elements capturing a discontinuity curve. (**b**) Rotated, anisotropic elements capturing a discontinuity curve

Considering wavelets with anisotropic scaling will not remedy the situation since within one fixed scale one cannot control the direction of the (now anisotropically shaped) elements. Thus, to capture a discontinuity curve as in Fig. 3b, one needs not only anisotropic elements but also a location parameter to locate the elements on the curve and a rotation parameter to align the elongated elements in the direction of the curve.

Let us finally remark why a parabolic scaling matrix diag $2^j, 2^{j/2}$ will be natural to use as anisotropic scaling. Since the discontinuity curves of cartoon-like images are C^2-smooth with bounded curvature, we may write the curve locally by a Taylor expansion. Let us assume it has the form $(s, E(s))$ for some $E : \mathbb{R} \to \mathbb{R}$. Then, by our assumptions on discontinuity curves of cartoon-like images,

$$E(s) = E(s') + E'(s')s + \tfrac{1}{2}E''(t)s^2$$

near $s = s'$ for some t between s' and s. Clearly, the translation parameter will be used to position the anisotropic element near $(s', E(s'))$, and the orientation parameter to align with $(1, E'(s')s)$. If the length of the element is l, then, due to the term $E''(t)s^2$, the most beneficial height would be l^2. And, in fact, parabolic scaling yields precisely this relation, i.e.,

$$\text{height} \approx \text{length}^2.$$

Hence, the main idea in the following will be to design a system which consists of anisotropically shaped elements together with a directional parameter to achieve the optimal approximation rate for cartoon-like images.

4 Pyramid-Adapted Shearlet Systems

After we have set our benchmark for directional representation systems in terms of an optimality criteria for sparse approximations of the cartoon-like image class $\mathscr{E}_L^2(\mathbb{R}^d)$, we next introduce classes of shearlet systems we claim behave optimally. As already mentioned in the introduction of this chapter, optimally sparse approximations were proven for a class of band-limited and of compactly supported shearlet frames. For the definition of cone-adapted discrete shearlets and, in particular, classes of band-limited as well as of compactly supported shearlet frames leading to optimally sparse approximations, we refer to "Introduction of this book!". In this section, we present the definition of discrete shearlets in 3D, from which the mentioned definitions in the 2D situation can also be directly derived. As special cases, we then introduce particular classes of band-limited as well as of compactly supported shearlet frames, which will be shown to provide optimally approximations of $\mathscr{E}_L^2(\mathbb{R}^3)$ and, with a slight modification which we will elaborate on in Sect. 5.2.4, also for $\mathscr{E}_{\alpha,L}^{\beta}(\mathbb{R}^3)$ with $1 < \alpha \leq \beta \leq 2$.

4.1 General Definition

The first step in the definition of cone-adapted discrete 2D shearlets was a partitioning of 2D frequency domain into two pairs of high-frequency cones and one low-frequency rectangle. We mimic this step by partitioning 3D frequency domain into the three pairs of *pyramids* given by

$$\mathscr{P} = \{(\xi_1,\xi_2,\xi_3) \in \mathbb{R}^3 : |\xi_1| \geq 1, |\xi_2/\xi_1| \leq 1, |\xi_3/\xi_1| \leq 1\},$$
$$\tilde{\mathscr{P}} = \{(\xi_1,\xi_2,\xi_3) \in \mathbb{R}^3 : |\xi_2| \geq 1, |\xi_1/\xi_2| \leq 1, |\xi_3/\xi_2| \leq 1\},$$
$$\check{\mathscr{P}} = \{(\xi_1,\xi_2,\xi_3) \in \mathbb{R}^3 : |\xi_3| \geq 1, |\xi_1/\xi_3| \leq 1, |\xi_2/\xi_3| \leq 1\},$$

and the centered cube

$$\mathscr{C} = \{(\xi_1,\xi_2,\xi_3) \in \mathbb{R}^3 : \|(\xi_1,\xi_2,\xi_3)\|_{\infty} < 1\}.$$

This partition is illustrated in Fig. 4 which depicts the three pairs of pyramids and Fig. 5 depicting the centered cube surrounded by the three pairs of pyramids $\mathscr{P}$, $\tilde{\mathscr{P}}$, and $\check{\mathscr{P}}$.

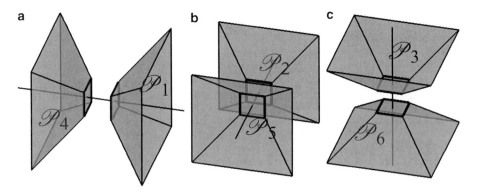

Fig. 4 The partition of the frequency domain: the "top" of the six pyramids

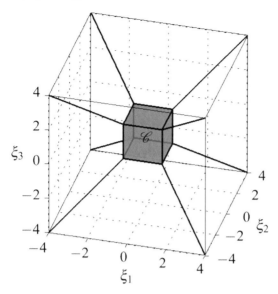

Fig. 5 The partition of the frequency domain: the centered cube $\mathscr{C}$. The arrangement of the six pyramids is indicated by the *"diagonal" lines*. See Fig. 4 for a sketch of the pyramids

The partitioning of frequency space into pyramids allows us to restrict the range of the shear parameters. Without such a partitioning as, e.g., in shearlet systems arising from the shearlet group, one must allow arbitrarily large shear parameters, which leads to a treatment biased toward one axis. The defined partition, however, enables restriction of the shear parameters to $[-\lceil 2^{j/2} \rceil, \lceil 2^{j/2} \rceil]$ like the definition of cone-adapted discrete shearlet systems. We would like to emphasize that this approach is key to provide an almost uniform treatment of different directions in a sense of a good approximation to rotation.

Pyramid-adapted discrete shearlets are scaled according to the *paraboloidal scaling matrices* $A_{2^j}, \tilde{A}_{2^j}$ or $\check{A}_{2^j}, j \in \mathbb{Z}$ defined by

$$A_{2^j} = \begin{pmatrix} 2^j & 0 & 0 \\ 0 & 2^{j/2} & 0 \\ 0 & 0 & 2^{j/2} \end{pmatrix}, \quad \tilde{A}_{2^j} = \begin{pmatrix} 2^{j/2} & 0 & 0 \\ 0 & 2^j & 0 \\ 0 & 0 & 2^{j/2} \end{pmatrix}, \quad \text{and} \quad \check{A}_{2^j} = \begin{pmatrix} 2^{j/2} & 0 & 0 \\ 0 & 2^{j/2} & 0 \\ 0 & 0 & 2^j \end{pmatrix},$$

and directionality is encoded by the *shear matrices* $S_k, \tilde{S}_k$, or $\check{S}_k, k = (k_1, k_2) \in \mathbb{Z}^2$, given by

$$S_k = \begin{pmatrix} 1 & k_1 & k_2 \\ 0 & 1 & 0 \\ 0 & 0 & 1 \end{pmatrix}, \quad \tilde{S}_k = \begin{pmatrix} 1 & 0 & 0 \\ k_1 & 1 & k_2 \\ 0 & 0 & 1 \end{pmatrix}, \quad \text{and} \quad \check{S}_k = \begin{pmatrix} 1 & 0 & 0 \\ 0 & 1 & 0 \\ k_1 & k_2 & 1 \end{pmatrix},$$

respectively. The reader should note that these definitions are (discrete) special cases of the general setup in the chapter on "Multivariate Shearlet Transform, Shearlet Coorbit Spaces and Their Structural Properties". The translation lattices will be defined through the following matrices: $M_c = \text{diag}(c_1, c_2, c_2)$, $\tilde{M}_c = \text{diag}(c_2, c_1, c_2)$, and $\check{M}_c = \text{diag}(c_2, c_2, c_1)$, where $c_1 > 0$ and $c_2 > 0$.

We are now ready to introduce 3D shearlet systems, for which we will make use of the vector notation $|k| \leq K$ for $k = (k_1, k_2)$ and $K > 0$ to denote $|k_1| \leq K$ and $|k_2| \leq K$.

Definition 2. For $c = (c_1, c_2) \in (\mathbb{R}_+)^2$, the *pyramid-adapted discrete shearlet system* $SH(\phi, \psi, \tilde{\psi}, \check{\psi}; c)$ generated by $\phi, \psi, \tilde{\psi}, \check{\psi} \in L^2(\mathbb{R}^3)$ is defined by

$$SH(\phi, \psi, \tilde{\psi}, \check{\psi}; c) = \Phi(\phi; c_1) \cup \Psi(\psi; c) \cup \tilde{\Psi}(\tilde{\psi}; c) \cup \check{\Psi}(\check{\psi}; c),$$

where

$$\Phi(\phi; c_1) = \left\{ \phi_m = \phi(\cdot - m) : m \in c_1 \mathbb{Z}^3 \right\},$$

$$\Psi(\psi; c) = \left\{ \psi_{j,k,m} = 2^j \psi(S_k A_{2^j} \cdot - m) : j \geq 0, |k| \leq \lceil 2^{j/2} \rceil, m \in M_c \mathbb{Z}^3 \right\},$$

$$\tilde{\Psi}(\tilde{\psi}; c) = \left\{ \tilde{\psi}_{j,k,m} = 2^j \tilde{\psi}(\tilde{S}_k \tilde{A}_{2^j} \cdot - m) : j \geq 0, |k| \leq \lceil 2^{j/2} \rceil, m \in \tilde{M}_c \mathbb{Z}^3 \right\},$$

and

$$\check{\Psi}(\check{\psi}; c) = \left\{ \check{\psi}_{j,k,m} = 2^j \check{\psi}(\check{S}_k \check{A}_{2^j} \cdot - m) : j \geq 0, |k| \leq \lceil 2^{j/2} \rceil, m \in \check{M}_c \mathbb{Z}^3 \right\},$$

where $j \in \mathbb{N}_0$ and $k \in \mathbb{Z}^2$. For the sake of brevity, we will sometimes also use the notation ψ_λ with $\lambda = (j, k, m)$.

We now focus on two different special classes of pyramid-adapted discrete shearlets leading to the class of band-limited shearlets, and the class of compactly supported shearlets for which optimality of their approximation properties with respect to cartoon-like images will be proven in Sect. 5.

4.2 Band-Limited 3D Shearlets

Let the shearlet generator $\psi \in L^2(\mathbb{R}^3)$ be defined by

$$\hat{\psi}(\xi) = \hat{\psi}_1(\xi_1) \hat{\psi}_2\left(\frac{\xi_2}{\xi_1}\right) \hat{\psi}_2\left(\frac{\xi_3}{\xi_1}\right), \tag{10}$$

where ψ_1 and ψ_2 satisfy the following assumptions:

(a) $\hat{\psi}_1 \in C^\infty(\mathbb{R})$, supp $\hat{\psi}_1 \subset [-4, -\frac{1}{2}] \cup [\frac{1}{2}, 4]$, and

$$\sum_{j \geq 0} \left| \hat{\psi}_1(2^{-j}\xi) \right|^2 = 1 \quad \text{for } |\xi| \geq 1, \xi \in \mathbb{R}. \tag{11}$$

(b) $\hat{\psi}_2 \in C^\infty(\mathbb{R})$, supp $\hat{\psi}_2 \subset [-1, 1]$, and

$$\sum_{l=-1}^{1} \left| \hat{\psi}_2(\xi + l) \right|^2 = 1 \quad \text{for } |\xi| \leq 1, \xi \in \mathbb{R}. \tag{12}$$

Thus, in frequency domain, the band-limited function $\psi \in L^2(\mathbb{R}^3)$ is almost a tensor product of one wavelet with two "bump" functions, thereby a canonical generalization of the classical band-limited 2D shearlets, see also "Introduction of this book!". This implies the support in frequency domain to have a needle-like shape with the wavelet acting in radial direction ensuring high directional selectivity, see also Fig. 6. The derivation from being a tensor product, i.e., the substitution of ξ_2

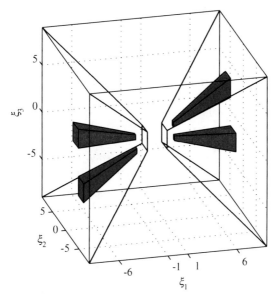

Fig. 6 Support of two shearlet elements $\psi_{j,k,m}$ in the frequency domain. The two shearlet elements have the same scale parameter $j = 2$, but different shearing parameters $k = (k_1, k_2)$

and ξ_3 by the quotients ξ_2/ξ_1 and ξ_3/ξ_1, respectively, in fact ensures a favorable behavior with respect to the shearing operator, and thus a tiling of frequency domain which leads to a tight frame for $L^2(\mathbb{R}^3)$.

A first step toward this result is the following observation.

Theorem 2 ([8]). *Let ψ be a band-limited shearlet defined as in this subsection. Then the family of functions $P_{\mathscr{P}} \Psi(\psi)$ forms a tight frame for $\check{L}^2(\mathscr{P}) := \{f \in L^2(\mathbb{R}^3) : \operatorname{supp} \hat{f} \subset \mathscr{P}\}$, where $P_{\mathscr{P}}$ denotes the orthogonal projection onto $\check{L}^2(\mathscr{P})$ and*

$$\Psi(\psi) = \{\psi_{j,k,m} : j \geq 0, |k| \leq \lceil 2^{j/2} \rceil, m \in \tfrac{1}{8}\mathbb{Z}^3\}.$$

Proof. For each $j \geq 0$, (12) implies that

$$\sum_{k=-\lceil 2^{j/2} \rceil}^{\lceil 2^{j/2} \rceil} |\hat{\psi}_2(2^{j/2}\xi + k)|^2 = 1, \quad \text{for } |\xi| \leq 1.$$

Hence, using (11), we obtain

$$
\sum_{j\geq 0}\sum_{k_1,k_2=-\lceil 2^{j/2}\rceil}^{\lceil 2^{j/2}\rceil}|\hat{\psi}(S_k^T A_{2^j}^{-1}\xi)|^2
$$

$$
=\sum_{j\geq 0}|\hat{\psi}_1(2^{-j}\xi_1)|^2|\sum_{k_1=-\lceil 2^{j/2}\rceil}^{\lceil 2^{j/2}\rceil}|\hat{\psi}_2(2^{j/2}\tfrac{\xi_2}{\xi_1}+k_1)|^2\sum_{k_2=-\lceil 2^{j/2}\rceil}^{\lceil 2^{j/2}\rceil}|\hat{\psi}_2(2^{j/2}\tfrac{\xi_2}{\xi_1}+k_2)|^2
$$

$$
=1,
$$

for $\xi=(\xi_1,\xi_2,\xi_3)\in\mathscr{P}$. Using this equation together with the fact that $\hat{\psi}$ is supported inside $[-4,4]^3$ proves the theorem.   □

By Theorem 2 and a change of variables, we can construct shearlet tight frames for $\check{L}^2(\mathscr{P})$, $\check{L}^2(\tilde{\mathscr{P}})$, and $\check{L}^2(\breve{\mathscr{P}})$, respectively. Furthermore, wavelet theory provides us with many choices of $\phi\in L^2(\mathbb{R}^3)$ such that $\Phi(\phi;\tfrac{1}{8})$ forms a tight frame for $\check{L}^2(\mathscr{R})$. Since $\mathbb{R}^3=\mathscr{R}\cup\mathscr{P}\cup\tilde{\mathscr{P}}\cup\breve{\mathscr{P}}$ as a disjoint union, we can express any function $f\in L^2(\mathbb{R}^3)$ as $f=P_{\mathscr{R}}f+P_{\mathscr{P}}f+P_{\tilde{\mathscr{P}}}f+P_{\breve{\mathscr{P}}}f$, where P_C denotes the orthogonal projection onto the closed subspace $\check{L}^2(C)$ for some measurable set $C\subset\mathbb{R}^3$. We then expand the projection $P_{\mathscr{P}}f$ in terms of the corresponding tight frame $P_{\mathscr{P}}\Psi(\psi)$ and similar for the other three projections. Finally, our representation of f will then be the sum of these four expansions. We remark that the projection of f and the shearlet frame elements onto the four subspaces can lead to artificially slow decaying shearlet coefficients; this will, e.g., be the case if f is in the Schwartz class. This problem does, in fact, not occur in the construction of compactly supported shearlets presented in the next subsection.

4.3 Compactly Supported 3D Shearlets

It is easy to see that the general form (10) does never lead to a function which is compactly supported in spatial domain. Thus, we need to modify the function by now taking indeed exact tensor products as our shearlet generators, which has the additional benefit of leading to fast algorithmic realizations. This, however, causes the problem that the shearlets do not behave as favorable with respect to the shearing operator as in the previous subsection, and the question arises whether they actually do lead to at least a frame for $L^2(\mathbb{R}^3)$. The next results shows this to be true for an even much more general form of shearlet generators including compactly supported separable generators. The attentive reader will notice that this theorem even covers the class of band-limited shearlets introduced in Sect. 4.2.

Theorem 3 ([12]). *Let $\phi,\psi\in L^2(\mathbb{R}^3)$ be functions such that*

$$
|\hat{\phi}(\xi)|\leq C_1\min\{1,|\xi_1|^{-\gamma}\}\cdot\min\{1,|\xi_2|^{-\gamma}\}\cdot\min\{1,|\xi_3|^{-\gamma}\},
$$

and

$$|\hat{\psi}(\xi)| \le C_2 \cdot \min\{1, |\xi_1|^\delta\} \cdot \min\{1, |\xi_1|^{-\gamma}\} \cdot \min\{1, |\xi_2|^{-\gamma}\} \cdot \min\{1, |\xi_3|^{-\gamma}\},$$

for some constants $C_1, C_2 > 0$ and $\delta > 2\gamma > 6$. Define $\tilde{\psi}(x) = \psi(x_2, x_1, x_3)$ and $\breve{\psi}(x) = \psi(x_3, x_2, x_1)$ for $x = (x_1, x_2, x_3) \in \mathbb{R}^3$. Then there exists a constant $c_0 > 0$ such that the shearlet system $\mathrm{SH}(\phi, \psi, \tilde{\psi}, \breve{\psi}; c)$ forms a frame for $L^2(\mathbb{R}^3)$ for all $c = (c_1, c_2)$ with $c_2 \le c_1 \le c_0$ provided that there exists a positive constant $M > 0$ such that

$$|\hat{\phi}(\xi)|^2 + \sum_{j \ge 0} \sum_{k_1, k_2 \in K_j} |\hat{\psi}(S_k^T A_{2^j} \xi)|^2 + |\hat{\tilde{\psi}}(\tilde{S}_k^T \tilde{A}_{2^j} \xi)|^2 + |\hat{\breve{\psi}}(\breve{S}_k^T \breve{A}_{2^j} \xi)|^2 > M \quad (13)$$

for a.e $\xi \in \mathbb{R}^3$, where $K_j := \left[-\lceil 2^{j/2} \rceil, \lceil 2^{j/2} \rceil \right]$.

We next provide an example of a family of compactly supported shearlets satisfying the assumptions of Theorem 3. However, for applications, one is typically not only interested in whether a system forms a frame, but in the ratio of the associated frame bounds. In this regard, these shearlets also admit a theoretically derived estimate for this ratio. Numerical estimates of the frame bound ratio show that the ratio can often be improved by a factor 20 over the theoretically derived bounds. Nevertheless, the theoretically derived bounds show that the frame bound ratio behaves in a controlled manner.

Example 1. Let $K, L \in \mathbb{N}$ be such that $L \ge 10$ and $\frac{3L}{2} \le K \le 3L - 2$, and define a shearlet $\psi \in L^2(\mathbb{R}^3)$ by

$$\hat{\psi}(\xi) = m_1(4\xi_1)\hat{\phi}(\xi_1)\hat{\phi}(2\xi_2)\hat{\phi}(2\xi_3), \quad \xi = (\xi_1, \xi_2, \xi_3) \in \mathbb{R}^3, \quad (14)$$

where the function m_0 is the low pass filter satisfying

$$|m_0(\xi_1)|^2 = \cos^{2K}(\pi\xi_1)) \sum_{n=0}^{L-1} \binom{K-1+n}{n} \sin^{2n}(\pi\xi_1),$$

for $\xi_1 \in \mathbb{R}$, the function m_1 is the associated bandpass filter defined by

$$|m_1(\xi_1)|^2 = |m_0(\xi_1 + 1/2)|^2, \quad \xi_1 \in \mathbb{R},$$

and ϕ the scaling function is given by

$$\hat{\phi}(\xi_1) = \prod_{j=0}^{\infty} m_0(2^{-j}\xi_1), \quad \xi_1 \in \mathbb{R}.$$

In [10, 12], it is shown that ϕ and ψ indeed are compactly supported. Moreover, we have the following result.

Table 1 Frame bound ratio for the shearlet frame from Example 1 with parameters $K = 39, L = 19$

Theoretical (B/A)	Numerical (B/A)	Translation constants (c_1, c_2)
345.7	13.42	(0.9, 0.25)
226.6	13.17	(0.9, 0.20)
226.4	13.16	(0.9, 0.15)
226.4	13.16	(0.9, 0.10)

Theorem 4 ([12]). *Suppose $\psi \in L^2(\mathbb{R}^3)$ is defined as in (14). Then there exists a sampling constant $c_0 > 0$ such that the shearlet system $\Psi(\psi; c)$ forms a frame for $\check{L}^2(\mathscr{P})$ for any translation matrix M_c with $c = (c_1, c_2) \in (\mathbb{R}_+)^2$ and $c_2 \leq c_1 \leq c_0$.*

Proof (sketch). Using upper and lower estimates of the absolute value of the trigonometric polynomial m_0 (cf. [2, 10]), one can show that ψ satisfies the hypothesis of Theorem 3 as well as

$$\sum_{j \geq 0} \sum_{k_1, k_2 \in K_j} |\hat{\psi}(S_k^T A_{2^j} \xi)|^2 > M \qquad \text{for all } \xi \in \mathscr{P},$$

where $M > 0$ is a constant, for some sufficiently small $c_0 > 0$. We note that this inequality is an analog to (13) for the pyramid $\mathscr{P}$. Hence, by a result similar to Theorem 3, but for the case, where we restrict to the pyramid $\check{L}^2(\mathscr{P})$, it then follows that $\Psi(\psi; c)$ is a frame.

To obtain a frame for all of $L^2(\mathbb{R}^3)$, we simply set $\tilde{\psi}(x) = \psi(x_2, x_1, x_3)$ and $\check{\psi}(x) = \psi(x_3, x_2, x_1)$ as in Theorem 3, and choose $\phi(x) = \phi(x_1)\phi(x_2)\phi(x_3)$ as scaling function for $x = (x_1, x_2, x_3) \in \mathbb{R}^3$. Then the corresponding shearlet system $\text{SH}(\phi, \psi, \tilde{\psi}, \check{\psi}; c, \alpha)$ forms a frame for $L^2(\mathbb{R}^3)$. The proof basically follows from Daubechies' classical estimates for wavelet frames in [2, Sect. 3.3.2] and the fact that anisotropic and sheared windows obtained by applying the scaling matrix A_{2^j} and the shear matrix S_k^T to the effective support[1] of $\hat{\psi}$ cover the pyramid $\mathscr{P}$ in the frequency domain. The same arguments can be applied to each of shearlet generators ψ, $\tilde{\psi}$, and $\check{\psi}$ as well as the scaling function ϕ to show a covering of the entire frequency domain and thereby the frame property of the pyramid-adapted shearlet system for $L^2(\mathbb{R}^3)$. We refer to [12] for the detailed proof.

Theoretical and numerical estimates of frame bounds for a particular parameter choice are shown in Table 1. We see that the theoretical estimates are overly pessimistic since they are a factor 20 larger than the numerical estimated frame bound ratios. We mention that in 2D the estimated frame bound ratios are approximately $1/10$ of the ratios found in Table 1.

---

[1] Loosely speaking, we say that $f \in L^2(\mathbb{R}^d)$ has *effective* support on B if the ratio $\|f\chi_B\|_{L^2} / \|f\|_{L^2}$ is "close" to 1.

4.4 *Some Remarks on Construction Issues*

The supports of the compactly supported shearlets $\psi_{j,k,m}$ from Example 1 are, in spatial domain, of size $2^{-j/2}$ times $2^{-j/2}$ times 2^{-j} due to the scaling matrix A_{2^j}. This reveals that the shearlet elements will become "plate-like" as $j \to \infty$. For an illustration, we refer to Fig. 7. Band-limited shearlets, on the other hand, do not have compactly support, but their effective support (the region where the energy of the function is concentrated) in spatial domain will likewise be of size $2^{-j/2}$ times $2^{-j/2}$ times 2^{-j} owing to their smoothness in frequency domain. Considering the fact that

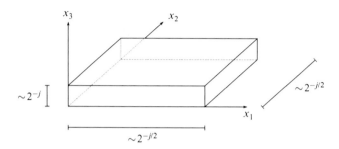

Fig. 7 Support of a shearlet $\check{\psi}_{j,0,m}$ from Example 1

intuitively such shearlet elements should provide sparse approximations of surface singularities, one could also think of using the scaling matrix $A_{2^j} = \operatorname{diag} 2^j, 2^j, 2^{j/2}$ with similar changes for $\tilde{A}_{2^j}$ and $\check{A}_{2^j}$ to derive "needle-like" shearlet elements in space domain. These would intuitively behave favorable with respect to the other type of anisotropic features occurring in 3D, that is curvilinear singularities. Surprisingly, we will show in Sect. 5.2 that for optimally sparse approximation platelike shearlets, i.e., shearlets associated with scaling matrix $A_{2^j} = \operatorname{diag} 2^j, 2^{j/2}, 2^{j/2}$, and similarly $\tilde{A}_{2^j}$ and $\check{A}_{2^j}$ are sufficient.

Let us also mention that, more generally, non-paraboloidal scaling matrices of the form $A_j = \operatorname{diag}\left(2^j, 2^{a_1 j}, 2^{a_2 j}\right)$ for $0 < a_1, a_2 \leq 1$ can be considered. The parameters a1 and a2 allow precise control of the aspect ratio of the shearlet elements, ranging from very plate-like to very needle-like, and they should be chosen according to the application at hand, i.e., choosing the shearlet shape which best matches the geometric characteristics of the considered data. The case $a_i < 1$ is covered by the setup of the multidimensional shearlet transform explained in the chapter on "Multivariate Shearlet Transform, Shearlet Coorbit Spaces and Their Structural Properties".

Let us finish this section with a general thought on the construction of bandlimited (not separable) tight shearlet frames versus compactly supported (nontight, but separable) shearlet frames. It seems that there is a trade-off between *compact support* of the shearlet generators, *tightness* of the associated frame, and *separability* of the shearlet generators. In fact, even in 2D, all known constructions of tight shearlet frames do not use separable generators, and these constructions can be shown to *not* be applicable to compactly supported generators. Presumably, tightness is difficult to obtain while allowing for compactly supported generators, but we can gain separability which leads to fast algorithmic realizations, see the chapter on "Digital Shearlet Transforms". If we though allow non-compactly supported generators, tightness is possible as shown in Sect. 4.2, but separability seems to be out of reach, which causes problems for fast algorithmic realizations.

5 Optimal Sparse Approximations

In this section, we will show that shearlets—both band-limited and compactly supported as defined in Sect. 4—indeed provide the optimal sparse approximation rate for cartoon-like images from Sect. 3.2. Thus, letting $(\psi_\lambda)_\lambda = (\psi_{j,k,m})_{j,k,m}$ denote the band-limited shearlet frame from Sect. 4.2 and the compactly supported shearlet frame from Sect. 4.3 in both 2D and 3D (see the chapter on "Introduction to Shearlets") and $d \in \{2,3\}$, we aim to prove that

$$\|f - f_N\|_{L^2}^2 \lesssim N^{-\frac{2}{d-1}} \quad \text{for all } f \in \mathcal{E}_L^2(\mathbb{R}^d),$$

where—as debated in Sect. 3.1—f_N denotes the N-term approximation using the N largest coefficients as in (5). Hence, in 2D we aim for the rate N^{-2} and in 3D we aim for the rate N^{-1} with ignoring log-factors. As mentioned in Sect. 3.2, see (9), in order to prove these rate, it suffices to show that the nth largest shearlet coefficient c_n^* decays as

$$|c_n^*| \lesssim n^{-\frac{d+1}{2(d-1)}} = \begin{cases} n^{-3/2} & : \quad d = 2, \\ n^{-1} & : \quad d = 3. \end{cases}$$

According to Definition 1, this will show that among all adaptive and nonadaptive representation systems shearlet frames behave optimal with respect to sparse approximation of cartoon-like images. That one is able to obtain such an optimal approximation error rate might seem surprising since the shearlet system as well as the approximation procedure will be nonadaptive.

To present the necessary hypotheses, illustrate the key ideas of the proofs, and debate the differences between the arguments for band-limited and compactly supported shearlets, we first focus on the situation of 2D shearlets. We then discuss the 3D situation, with a *sketchy* proof, mainly discussing the essential differences to the proof for 2D shearlets and highlighting the crucial nature of this case (cf. Sect. 1.3).

5.1 Optimal Sparse Approximations in 2D

As discussed in the previous section, in the case $d = 2$, we aim for the estimates $|c_n^*| \lesssim n^{-3/2}$ and $\|f - f_N\|_{L^2}^2 \lesssim N^{-2}$ (up to log-factors). In Sect. 5.1.1, we will first provide a heuristic analysis to argue that shearlet frames indeed can deliver these rates. In Sects. 5.1.2 and 5.1.3, we then discuss the required hypotheses and state the main optimality result. The subsequent subsections are then devoted to proving the main result.

5.1.1 A heuristic analysis

We start by giving a heuristic argument (inspired by a similar argument for curvelets in [1]) on why the error $\|f - f_N\|_{L^2}^2$ satisfies the asymptotic rate N^{-2}. We emphasize that this heuristic argument applies to both the band-limited and the compactly supported case.

For simplicity we assume $L = 1$, and let $f \in \mathscr{E}_L^2(\mathbb{R}^2)$ be a 2D cartoon-like image. The main concern is to derive the estimate (18) for the shearlet coefficients $\langle f, \mathring{\psi}_{j,k,m} \rangle$, where $\mathring{\psi}$ denotes either ψ or $\tilde{\psi}$. We consider only the case $\mathring{\psi} = \psi$ since the other case can be handled similarly. For compactly supported shearlet, we can think of our generators having the form $\psi(x) = \eta(x_1)\phi(x_2)$, $x = (x_1, x_2)$, where η is a wavelet and ϕ a bump (or a scaling) function. It will become important that the wavelet "points" in the x_1-axis direction, which corresponds to the "short" direction of the shearlet. For band-limited generators, we can think of our generators having the form $\hat{\psi}(\xi) = \hat{\eta}(\xi_2/\xi_1)\hat{\phi}(\xi_2)$ for $\xi = (\xi_1, \xi_2)$. We, moreover, restrict our analysis to shearlets $\psi_{j,k,m}$ since the frame elements $\tilde{\psi}_{j,k,m}$ can be handled in a similar way.

We now consider three cases of coefficients $\langle f, \psi_{j,k,m} \rangle$:

(a) Shearlets $\psi_{j,k,m}$ whose support does not overlap with the boundary ∂B.
(b) Shearlets $\psi_{j,k,m}$ whose support overlaps with ∂B and is nearly tangent.
(c) Shearlets $\psi_{j,k,m}$ whose support overlaps with ∂B, but not tangentially.

It turns out that only coefficients from case (b) will be significant. Case (b) is, loosely speaking, the situation, where the wavelet η crosses the discontinuity curve over the entire "height" of the shearlet, see Fig. 8.

Case (a). Since f is C^2-smooth away from ∂B, the coefficients $|\langle f, \psi_{j,k,m} \rangle|$ will be sufficiently small owing to the approximation property of the wavelet η. The situation is sketched in Fig. 8.

Case (b). At scale $j > 0$, there are about $O(2^{j/2})$ coefficients since the shearlet elements are of length $2^{-j/2}$ (and "thickness" 2^{-j}) and the length of ∂B is finite. By Hölder's inequality, we immediately obtain

$$\left| \langle f, \psi_{j,k,m} \rangle \right| \leq \|f\|_{L^\infty} \|\psi_{j,k,m}\|_{L^1} \leq C_1 \, 2^{-3j/4} \|\psi\|_{L^1} \leq C_2 \cdot 2^{-3j/4}$$

for some constants $C_1, C_2 > 0$. In other words, we have $O(2^{j/2})$ coefficients bounded by $C_2 \cdot 2^{-3j/4}$. Assuming the case (a) and (c) coefficients are negligible, the nth largest coefficient c_n^* is then bounded by

$$|c_n^*| \leq C \cdot n^{-3/2},$$

which was what we aimed to show; compare to (7) in Definition 1. This in turn implies (cf. estimate (8)) that

$$\sum_{n>N} |c_n^*|^2 \leq \sum_{n>N} C \cdot n^{-3} \leq C \cdot \int_N^\infty x^{-3} dx \leq C \cdot N^{-2}.$$

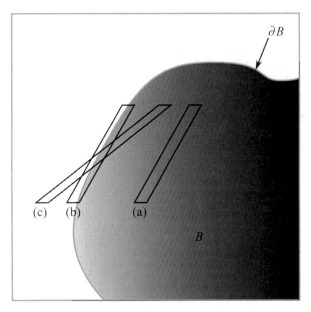

Fig. 8 Sketch of the three cases: (**a**) the support of $\psi_{j,k,m}$ does not overlap with ∂B, (**b**) the support of $\psi_{j,k,m}$ does overlap with ∂B and is nearly tangent, (**c**) the support of $\psi_{j,k,m}$ does overlap with ∂B, but not tangentially. Note that only a section of the discontinuity curve ∂B is shown, and that for the case of band-limited shearlets only the effective support is shown

By Lemma 1, as desired it follows that

$$\|f - f_N\|_{L^2}^2 \leq \frac{1}{A} \sum_{n>N} |c_n^*|^2 \leq C \cdot N^{-2},$$

where A denotes the lower frame bound of the shearlet frame.

Case (c). Finally, when the shearlets are sheared away from the tangent position in case (b), they will again be small. This is due to the frequency support of f and ψ_λ as well as to the directional vanishing moment conditions assumed in Setup 1 or 2, which will be formally introduced in the next subsection.

Summarizing our findings, we have argued, at least heuristically, that shearlet frames provide optimal sparse approximation of cartoon-like images as defined in Definition 1.

5.1.2 Required hypotheses

After having build up some intuition on why the optimal sparse approximation rate is achievable using shearlets, we will now go into more detail and discuss the hypotheses required for the main result. This will along the way already highlight some differences between the band-limited and compactly supported case.

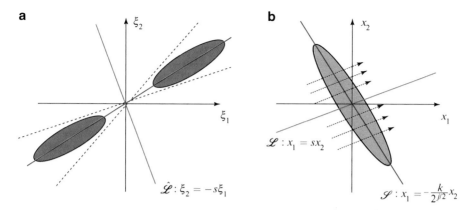

Fig. 9 (a) *Shaded region*: the effective part of supp $\hat{\psi}_{j,k,m}$ in the frequency domain. (b) *Shaded region*: the effective part of supp $\psi_{j,k,m}$ in the spatial domain. *Dashed lines*: the direction of line integration $I(t)$

For this discussion, we restrict our attention to the horizontal cone $\mathscr{C}$ since the vertical cone can be treated in the same way by a change of variables. Furthermore, we assume that $f \in L^2(\mathbb{R}^2)$ is piecewise C^{L+1}-smooth with a discontinuity on the line $\mathscr{L} : x_1 = sx_2$, $s \in \mathbb{R}$, so that the function f is well approximated by two 2D polynomials of degree $L > 0$, one polynomial on either side of $\mathscr{L}$, and denote this piecewise polynomial $q(x_1, x_2)$. We denote the restriction of q to lines $x_1 = sx_2 + t$, $t \in \mathbb{R}$, by $p_t(x_2) = q(sx_2 + t, x_2)$. Hence, p_t is a 1D polynomial along lines parallel to $\mathscr{L}$ going through $(x_1, x_2) = (t, 0)$; these lines are marked by dashed lines in Fig. 9b.

We now aim at estimating the absolute value of a shearlet coefficient $\langle f, \psi_{j,k,m} \rangle$ by

$$|\langle f, \psi_{j,k,m} \rangle| \le |\langle q, \psi_{j,k,m} \rangle| + |\langle (q - f), \psi_{j,k,m} \rangle|. \tag{15}$$

We first observe that $|\langle f, \psi_{j,k,m} \rangle|$ will be small depending on the approximation quality of the (piecewise) polynomial q and the decay of ψ in the spatial domain. Hence, it suffices to focus on estimating $|\langle q, \psi_{j,k,m} \rangle|$.

For this, let us consider the line integration along the direction $(x_1, x_2) = (s, 1)$ as follows: For $t \in \mathbb{R}$ fixed, define integration of $q\psi_{j,k,m}$ along the lines $x_1 = sx_2 + t$, $x_2 \in \mathbb{R}$, as

$$I(t) = \int_{\mathbb{R}} p_t(x_2)\psi_{j,k,m}(sx_2 + t, x_2)dx_2,$$

Observe that $|\langle q, \psi_{j,k,m} \rangle| = 0$ is equivalent to $I \equiv 0$. For simplicity, let us now assume $m = (0,0)$. Then

$$I(t) = 2^{\frac{3}{4}j} \int_{\mathbb{R}} p_t(x_2)\psi(S_k A_{2^j}(sx_2 + t, x_2))dx_2$$

$$= 2^{\frac{3}{4}j} \sum_{\ell=0}^{L} c_\ell \int_{\mathbb{R}} x_2^\ell \psi(S_k A_{2^j}(sx_2 + t, x_2))dx_2$$

$$= 2^{\frac{3}{4}j} \sum_{\ell=0}^{L} c_\ell \int_{\mathbb{R}} x_2^\ell \psi(A_{2^j}S_{k/2^{j/2}+s}(t, x_2))dx_2,$$

and, by the Fourier slice theorem [9] (see also (27)), it follows that

$$|I(t)| = 2^{\frac{3}{4}j} \left| \sum_{\ell=0}^{L} \frac{2^{-\frac{\ell}{2}j}}{(2\pi)^{\ell}} c_\ell \int_{\mathbb{R}} \left(\frac{\partial}{\partial \xi_2}\right)^{\ell} \hat{\psi}(A_{2^j}^{-1} S_{k/2^{j/2}+s}^{-T}(\xi_1,0)) \mathscr{e}^{2\pi i \xi_1 t} d\xi_1 \right|.$$

Note that

$$\int_{\mathbb{R}} \left(\frac{\partial}{\partial \xi_2}\right)^{\ell} \hat{\psi}(A_{2^j}^{-1} S_{k/2^{j/2}+s}^{-T}(\xi_1,0)) e^{2\pi i \xi_1 t} d\xi_1 = 0 \quad \text{for almost all } t \in \mathbb{R}$$

if and only if

$$\left(\frac{\partial}{\partial \xi_2}\right)^{\ell} \hat{\psi}(A_{2^j}^{-1} S_{k/2^{j/2}+s}^{-T}(\xi_1,0)) = 0 \quad \text{for almost all } \xi_1 \in \mathbb{R}.$$

Therefore, to ensure $I(t) = 0$ for any 1D polynomial p_t of degree $L > 0$, we require the following condition:

$$\left(\frac{\partial}{\partial \xi_2}\right)^{\ell} \hat{\psi}_{j,k,0}(\xi_1, -s\xi_1) = 0 \quad \text{for almost all } \xi_1 \in \mathbb{R} \text{ and } \ell = 0,\dots,L.$$

These are the so-called *directional vanishing moments* (cf. [4]) in the direction $(s,1)$. We now consider the two cases, band-limited shearlets and compactly supported shearlets, separately.

If ψ is a band-limited shearlet generator, we automatically have

$$\left(\frac{\partial}{\partial \xi_2}\right)^{\ell} \hat{\psi}_{j,k,m}(\xi_1, -s\xi_1) = 0 \quad \text{for } \ell = 0,\dots,L \quad \text{if } \left| s + \frac{k}{2^{j/2}} \right| \geq 2^{-j/2}, \quad (16)$$

since supp $\hat{\psi} \subset \mathscr{D}$, where $\mathscr{D} = \{\xi \in \mathbb{R}^2 : |\xi_2/\xi_1| \leq 1\}$ as discussed in the chapter on "Introduction to Shearlets". Observe that the "direction" of supp $\psi_{j,k,m}$ is determined by the line $\mathscr{S} : x_1 = -\frac{k}{2^{j/2}}x_2$. Hence, equation (16) implies that if the direction of supp $\psi_{j,k,m}$, i.e., of $\mathscr{S}$ is *not* close to the direction of $\mathscr{L}$ in the sense that $|s + \frac{k}{2^{j/2}}| \geq 2^{-j/2}$, then

$$|\langle q, \psi_{j,k,m}\rangle| = 0.$$

However, if ψ is a compactly supported shearlet generator, equation (16) can never hold since it requires that supp $\hat{\psi} \subset \mathscr{D}$. Therefore, for compactly supported generators, we will assume that $(\frac{\partial}{\partial \xi_2})^l \hat{\psi}$, $l = 0,1$, has sufficient decay in $\mathscr{D}^c$ to force $I(t)$ and hence $|\langle q, \psi_{j,k,m}\rangle|$ to be sufficiently small. It should be emphasized that the drawback that $I(t)$ will only be "small" for compactly supported shearlets (due to the lack of exact directional vanishing moments) will be compensated by the perfect localization property which still enables optimal sparsity.

Thus, the developed conditions ensure that both terms on the right-hand side of (15) can be effectively bounded.

This discussion gives naturally rise to the following hypotheses for optimal sparse approximation. Let us start with the hypotheses for the band-limited case.

Setup 1. The generators $\phi, \psi, \tilde{\psi} \in L^2(\mathbb{R}^2)$ are band-limited and C^∞ in the frequency domain. Furthermore, the shearlet system $SH(\phi, \psi, \tilde{\psi}; c)$ forms a frame for $L^2(\mathbb{R}^2)$ (cf. the construction in the chapter on "Introduction to Shearlets" or Sect. 4.2).

In contrast to this, the conditions for the compactly supported shearlets are as follows:

Setup 2. The generators $\phi, \psi, \tilde{\psi} \in L^2(\mathbb{R}^2)$ are compactly supported, and the shearlet system $SH(\phi, \psi, \tilde{\psi}; c)$ forms a frame for $L^2(\mathbb{R}^2)$. Furthermore, for all $\xi = (\xi_1, \xi_2) \in \mathbb{R}^2$, the function ψ satisfies

(i) $|\hat{\psi}(\xi)| \leq C \cdot \min\{1, |\xi_1|^\delta\} \cdot \min\{1, |\xi_1|^{-\gamma}\} \cdot \min\{1, |\xi_2|^{-\gamma}\}$, and

(ii) $\left| \frac{\partial}{\partial \xi_2} \hat{\psi}(\xi) \right| \leq |h(\xi_1)| \left(1 + \frac{|\xi_2|}{|\xi_1|} \right)^{-\gamma}$,

where $\delta > 6$, $\gamma \geq 3$, $h \in L^1(\mathbb{R})$, and C a constant, and $\tilde{\psi}$ satisfies analogous conditions with the obvious change of coordinates (cf. the construction in Sect. 4.3).

Conditions (i) and (ii) in Setup 2 are exactly the decay assumptions on $\left(\frac{\partial}{\partial \xi_2}\right)^l \hat{\psi}$, $l = 0, 1$, discussed above that guarantees control of the size of $I(t)$.

5.1.3 Main result

We are now ready to present the main result, which states that under Setup 1 or Setup 2 shearlets provide optimally sparse approximations for cartoon-like images.

Theorem 5 ([7, 14]). *Assume Setup 1 or 2. Let $L \in \mathbb{N}$. For any $v > 0$ and $\mu > 0$, the shearlet frame* $SH(\phi, \psi, \tilde{\psi}; c)$ *provides optimally sparse approximations of functions* $f \in \mathcal{E}_L^2(\mathbb{R}^2)$ *in the sense of Definition 1, i.e.,*

$$\|f - f_N\|_{L^2}^2 = O(N^{-2}(\log N)^3), \qquad as \ N \to \infty, \tag{17}$$

and

$$|c_n^*| \lesssim n^{-3/2}(\log n)^{3/2}, \qquad as \ n \to \infty, \tag{18}$$

where $c = \{\langle f, \mathring{\psi}_\lambda \rangle : \lambda \in \Lambda, \mathring{\psi} = \psi \ or \ \mathring{\psi} = \tilde{\psi}\}$ *and* $c^* = (c_n^*)_{n \in \mathbb{N}}$ *is a decreasing (in modulus) rearrangement of* c.

5.1.4 Band-limitedness vs. compactly supportedness

Before we delve into the proof of Theorem 5, we first carefully discuss the main differences between band-limited shearlets and compactly supported shearlets which requires adaptions of the proof.

In the case of compactly supported shearlets, we can consider the two cases $|\operatorname{supp}\mathring{\psi}_\lambda \cap \partial B| \neq 0$ and $|\operatorname{supp}\mathring{\psi}_\lambda \cap \partial B| = 0$. In case the support of the shearlet intersects the discontinuity curve ∂B of the cartoon-like image f, we will estimate each shearlet coefficient $\langle f, \mathring{\psi}_\lambda\rangle$ individually using the decay assumptions on $\hat{\mathring{\psi}}$ in Setup 2, and then apply a simple counting estimate to obtain the sought estimates (17) and (18). In the other case, in which the shearlet does not interact with the discontinuity, we are simply estimating the decay of shearlet coefficients of a C^2 function. The argument here is similar to the approximation of smooth functions using wavelet frames and rely on estimating coefficients at all scales using the frame property.

In the case of band-limited shearlets, it is not allowed to consider two cases $|\operatorname{supp}\mathring{\psi}_\lambda \cap \partial B| = 0$ and $|\operatorname{supp}\mathring{\psi}_\lambda \cap \partial B| \neq 0$ separately since all shearlet elements $\mathring{\psi}_\lambda$ intersect the boundary of the set B. In fact, one needs to first localize the cartoon-like image f by compactly supported smooth window functions associated with dyadic squares using a partition of unity. Letting f_Q denote such a localized version, we then estimate $\langle f_Q, \mathring{\psi}_\lambda\rangle$ instead of directly estimating the shearlet coefficients $\langle f, \mathring{\psi}_\lambda\rangle$. Moreover, in the case of band-limited shearlets, one needs to estimate the sparsity of the sequence of the shearlet coefficients rather than analyzing the decay of individual coefficients.

In the next subsections, we present the proof—first for band-limited, then for compactly supported shearlets—in the case $L = 1$, i.e., when the discontinuity curve in the model of cartoon-like images is smooth. In these proofs, we will repeatedly use that it suffices to consider the case $\mathring{\psi} = \psi$, that is, the horizontal cone since the case $\mathring{\psi} = \tilde{\psi}$, that is, the vertical cone, can be handled similarly. Finally, the extension to $L \neq 1$ will be discussed for both cases simultaneously.

We will first, however, introduce some notation used in the proofs and prove a helpful lemma which will be used in both cases: band-limited and compactly supported shearlets. For a fixed j, we let $\mathcal{Q}_j$ be a collection of dyadic squares defined by

$$\mathcal{Q}_j = \{Q = [\tfrac{l_1}{2^{j/2}}, \tfrac{l_1+1}{2^{j/2}}] \times [\tfrac{l_2}{2^{j/2}}, \tfrac{l_2+1}{2^{j/2}}] : l_1, l_2 \in \mathbb{Z}\}.$$

We let Λ denote the set of all indices (j,k,m) in the shearlet system and define

$$\Lambda_j = \{(j,k,m) \in \Lambda : -\lceil 2^{j/2}\rceil \leq k \leq \lceil 2^{j/2}\rceil, m \in \mathbb{Z}^2\}.$$

For $\varepsilon > 0$, we define the set of "relevant" indices on scale j as

$$\Lambda_j(\varepsilon) = \{\lambda \in \Lambda_j : |\langle f, \psi_\lambda\rangle| > \varepsilon\}$$

and, on all scales, as

$$\Lambda(\varepsilon) = \{\lambda \in \Lambda : |\langle f, \psi_\lambda \rangle| > \varepsilon\}.$$

Lemma 2. *Assume Setup 1 or 2. Let $f \in \mathscr{E}_L^2(\mathbb{R}^2)$. Then the following assertions hold:*

(i) For some constant C, we have

$$\#|\Lambda_j(\varepsilon)| = 0 \quad for \quad j \geq \frac{4}{3}\log_2(\varepsilon^{-1}) + C \tag{19}$$

(ii) If

$$\#|\Lambda_j(\varepsilon)| \lesssim \varepsilon^{-2/3}, \tag{20}$$

for $j \geq 0$, then

$$\#|\Lambda(\varepsilon)| \lesssim \varepsilon^{-2/3}\log_2(\varepsilon^{-1}), \tag{21}$$

which, in turn, implies (17) and (18).

Proof. (i). Since $\psi \in L^1(\mathbb{R}^2)$ for both the band-limited and compactly supported setup, we have that

$$\begin{aligned} |\langle f, \psi_\lambda \rangle| &= \left| \int_{\mathbb{R}^2} f(x) 2^{\frac{3j}{4}} \psi(S_k A_{2^j} x - m) \mathrm{d}x \right| \\ &\leq 2^{\frac{3j}{4}} \|f\|_\infty \int_{\mathbb{R}^2} |\psi(S_k A_{2^j} x - m)| \mathrm{d}x \\ &= 2^{-\frac{3j}{4}} \|f\|_\infty \|\psi\|_1. \end{aligned} \tag{22}$$

As a consequence, there is a scale j_ε such that $|\langle f, \psi_\lambda \rangle| < \varepsilon$ for each $j \geq j_\varepsilon$. It therefore follows from (22) that

$$\#|\Lambda(\varepsilon)| = 0 \quad for \quad j > \frac{4}{3}\log_2(\varepsilon^{-1}) + C.$$

(ii). By assertion (i) and estimate (20), we have that

$$\#|\Lambda(\varepsilon)| \leq C\,\varepsilon^{-2/3}\log_2(\varepsilon^{-1}).$$

From this, the value ε can be written as a function of the total number of coefficients $n = \#|\Lambda(\varepsilon)|$. We obtain

$$\varepsilon(n) \leq C\,n^{-3/2}(\log_2(n))^{3/2} \quad \text{for sufficiently large } n.$$

This implies that

$$|c_n^*| \leq C\,n^{-3/2}(\log_2(n))^{3/2}$$

and

$$\sum_{n>N} |c_n^*|^2 \leq C N^{-2} (\log_2(N))^3 \qquad \text{for sufficiently large } N > 0,$$

where c_n^* as usual denotes the nth largest shearlet coefficient in modulus. $\qquad\qquad$ □

5.1.5 Proof for band-limited shearlets for $L = 1$

Since we assume $L = 1$, we have that $f \in \mathscr{E}_L^2(\mathbb{R}^2) = \mathscr{E}^2(\mathbb{R}^2)$. As mentioned in the previous section, we will now measure the sparsity of the shearlet coefficients $\{\langle f, \mathring{\psi}_\lambda \rangle : \lambda \in \Lambda\}$. For this, we will use the weak ℓ^p quasi norm $\|\cdot\|_{w\ell^p}$ defined as follows. For a sequence $s = (s_i)_{i \in I}$, we let, as usual, s_n^* be the nth largest coefficient in s in modulus. We then define:

$$\|s\|_{w\ell^p} = \sup_{n>0} n^{\frac{1}{p}} |s_n^*| .$$

One can show [16] that this definition is equivalent to

$$\|s\|_{w\ell^p} = \left(\sup\{\# | \{i : |s_i| > \varepsilon\} | \varepsilon^p : \varepsilon > 0\} \right)^{\frac{1}{p}} .$$

As mentioned above, we will only consider the case $\mathring{\psi} = \psi$ since the case $\mathring{\psi} = \tilde{\psi}$ can be handled similarly. To analyze the decay properties of the shearlet coefficients $(\langle f, \psi_\lambda \rangle)_\lambda$ at a given scale parameter $j \geq 0$, we smoothly localize the function f near dyadic squares. Fix the scale parameter $j \geq 0$. For a nonnegative C^∞ function w with support in $[0, 1]^2$, we then define a smooth partition of unity

$$\sum_{Q \in \mathscr{Q}_j} w_Q(x) = 1, \qquad x \in \mathbb{R}^2,$$

where, for each dyadic square $Q \in \mathscr{Q}_j$, $w_Q(x) = w(2^{j/2}x_1 - l_1, 2^{j/2}x_2 - l_2)$. We will then examine the shearlet coefficients of the localized function $f_Q := f w_Q$. With this smooth localization of the function f, we can now consider the two separate cases, $|\operatorname{supp} w_Q \cap \partial B| = 0$ and $|\operatorname{supp} w_Q \cap \partial B| \neq 0$. Let

$$\mathscr{Q}_j = \mathscr{Q}_j^0 \cup \mathscr{Q}_j^1,$$

where the union is disjoint and $\mathscr{Q}_j^0$ is the collection of those dyadic squares $Q \in \mathscr{Q}_j$ such that the edge curve ∂B intersects the support of w_Q. Since each Q has side length $2^{-j/2}$ and the edge curve ∂B has finite length, it follows that

$$\# |\mathscr{Q}_j^0| \lesssim 2^{j/2}. \tag{23}$$

Similarly, since f is compactly supported in $[0,1]^2$, we see that

$$\#|\mathcal{Q}_j^1| \lesssim 2^j. \tag{24}$$

The following theorems analyzes the sparsity of the shearlets coefficients for each dyadic square $Q \in \mathcal{Q}_j$.

Theorem 6 ([7]). *Let $f \in \mathcal{E}^2(\mathbb{R}^2)$. For $Q \in \mathcal{Q}_j^0$, with $j \geq 0$ fixed, the sequence of shearlet coefficients $\{d_\lambda := \langle f_Q, \mathring{\psi}_\lambda \rangle : \lambda \in \Lambda_j\}$ obeys*

$$\left\| (d_\lambda)_{\lambda \in \Lambda_j} \right\|_{w\ell^{2/3}} \lesssim 2^{-\frac{3j}{4}}.$$

Theorem 7 ([7]). *Let $f \in \mathcal{E}^2(\mathbb{R}^2)$. For $Q \in \mathcal{Q}_j^1$, with $j \geq 0$ fixed, the sequence of shearlet coefficients $\{d_\lambda := \langle f_Q, \mathring{\psi}_\lambda \rangle : \lambda \in \Lambda_j\}$ obeys*

$$\left\| (d_\lambda)_{\lambda \in \Lambda_j} \right\|_{w\ell^{2/3}} \lesssim 2^{-\frac{3j}{2}}.$$

As a consequence of these two theorems, we have the following result.

Theorem 8 ([7]). *Suppose $f \in \mathcal{E}^2(\mathbb{R}^2)$. Then, for $j \geq 0$, the sequence of the shearlet coefficients $\{c_\lambda := \langle f, \mathring{\psi}_\lambda \rangle : \lambda \in \Lambda_j\}$ obeys*

$$\left\| (c_\lambda)_{\lambda \in \Lambda_j} \right\|_{w\ell^{2/3}} \lesssim 1.$$

Proof. Using Theorems 6 and 7, by the p-triangle inequality for weak ℓ^p spaces, $p \leq 1$, we have

$$\| \langle f, \mathring{\psi}_\lambda \rangle \|_{w\ell^{2/3}}^{2/3} \leq \sum_{Q \in \mathcal{Q}_j} \| \langle f_Q, \mathring{\psi}_\lambda \rangle \|_{w\ell^{2/3}}^{2/3}$$

$$= \sum_{Q \in \mathcal{Q}_j^0} \| \langle f_Q, \mathring{\psi}_\lambda \rangle \|_{w\ell^{2/3}}^{2/3} + \sum_{Q \in \mathcal{Q}_j^1} \| \langle f_Q, \mathring{\psi}_\lambda \rangle \|_{w\ell^{2/3}}^{2/3}$$

$$\leq C \#|\mathcal{Q}_j^0| \, 2^{-j/2} + C \#|\mathcal{Q}_j^1| \, 2^{-j}.$$

Equations (23) and (24) complete the proof. □

We can now prove Theorem 5 for the band-limited setup.

Proof (Theorem 5 for Setup 1). From Theorem 8, we have that

$$\#|\Lambda_j(\varepsilon)| \leq C\varepsilon^{-2/3},$$

for some constant $C > 0$, which, by Lemma 2, completes the proof. □

5.1.6 Proof for compactly supported shearlets for L = 1

To derive the sought estimates (17) and (18) for dimension $d = 2$, we will study two separate cases: Those shearlet elements $\mathring{\psi}_\lambda$ which do not interact with the discontinuity curve, and those elements which do. We will again only consider the case $\mathring{\psi} = \psi$ since the case $\mathring{\psi} = \tilde{\psi}$ can be handled similarly.

Case 1. The compact support of the shearlet ψ_λ does not intersect the boundary of the set B, i.e., $|\mathrm{supp}\,\psi_\lambda \cap \partial B| = 0$.
Case 2. The compact support of the shearlet ψ_λ does intersect the boundary of the set B, i.e., $|\mathrm{supp}\,\psi_\lambda \cap \partial B| \neq 0$.

For *Case 1*, we will not be concerned with decay estimates of single coefficients $\langle f, \psi_\lambda \rangle$, but with the decay of sums of coefficients over several scales and all shears and translations. The frame property of the shearlet system, the C^2-smoothness of f, and a crude counting argument of the cardinal of the essential indices λ will be enough to provide the needed approximation rate. The proof of this is similar to estimates of the decay of wavelet coefficients for C^2-smooth functions. In fact, shearlet and wavelet frames give the same approximation decay rates in this case. Due to space limitation of this exposition, we will not go into the details of this estimate, but rather focus on the main part of the proof, *Case 2*.

For *Case 2*, we need to estimate each coefficient $\langle f, \psi_\lambda \rangle$ individually and, in particular, how $|\langle f, \psi_\lambda \rangle|$ decays with scale j and shearing k. Without loss of generality, we can assume that $f = f_0 + \chi_B f_1$ with $f_0 = 0$. We let then M denote the area of integration in $\langle f, \psi_\lambda \rangle$, that is,

$$M = \mathrm{supp}\,\psi_\lambda \cap B.$$

Further, let $\mathscr{L}$ be an affine hyperplane (in other and simpler words, a line in $\mathbb{R}^2$) that intersects M and thereby divides M into two sets M_t and M_l, see the sketch in Fig. 10. We thereby have that

$$\langle f, \psi_\lambda \rangle = \langle \chi_M f, \psi_\lambda \rangle = \langle \chi_{M_t} f, \psi_\lambda \rangle + \langle \chi_{M_l} f, \psi_\lambda \rangle. \tag{25}$$

The hyperplane will be chosen in such a way that the area of M_t is sufficiently small. In particular, area (M_t) should be small enough so that the following estimate

$$|\langle \chi_{M_t} f, \psi_\lambda \rangle| \leq \|f\|_{L^\infty} \|\psi_\lambda\|_{L^\infty} \,\mathrm{area}\,(M_t) \leq \mu\, 2^{3j/4}\,\mathrm{area}\,(M_t) \tag{26}$$

do not violate (18). If the hyperplane $\mathscr{L}$ is positioned as indicated in Fig. 10, it can indeed be shown by crudely estimating area (M_t) that (26) does not violate estimate (18). We call estimates of this form, where we have restricted the integration to a small part M_t of M, *truncated* estimates. Hence, in the following we assume that (25) reduces to $\langle f, \psi_\lambda \rangle = \langle \chi_{M_l} f, \psi_\lambda \rangle$.

For the term $\langle \chi_{M_l} f, \psi_\lambda \rangle$, we will have to integrate over a possibly much large part M_l of M. To handle this, we will use that ψ_λ only interacts with the discontinuity of $\chi_{M_l} f$ along a line inside M. This part of the estimate is called the *linearized*

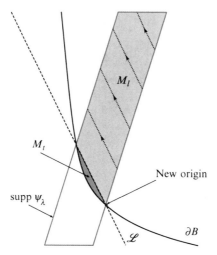

Fig. 10 Sketch of supp ψ_λ, M_l, M_t, and $\mathscr{L}$. The lines of integrations are shown

estimate since the discontinuity curve in $\langle \chi_{M_l} f, \psi_\lambda \rangle$ has been reduced to a line. In $\langle \chi_{M_l} f, \psi_\lambda \rangle$, we are, of course, integrating over two variables, and we will as the inner integration always choose to integrate along lines parallel to the singularity line $\mathscr{L}$, see Fig. 10. The important point here is that along these lines, the function f is C^2-smooth without discontinuities on the entire interval of integration. This is exactly the reason for removing the M_t-part from M. Using the Fourier slice theorem, we will then turn the line integrations along $\mathscr{L}$ in the spatial domain into line integrations in the frequency domain. The argumentation is as follows: consider $g : \mathbb{R}^2 \to \mathbb{C}$ compactly supported and continuous, and let $p : \mathbb{R} \to \mathbb{C}$ be a projection of g onto, say, the x_2 axis, i.e., $p(x_1) = \int_\mathbb{R} g(x_1, x_2) \mathrm{d}x_2$. This immediately implies that $\hat{p}(\xi_1) = \hat{g}(\xi_1, 0)$, which is a simplified version of the Fourier slice theorem. By an inverse Fourier transform, we then have

$$\int_\mathbb{R} g(x_1, x_2) \mathrm{d}x_2 = p(x_1) = \int_\mathbb{R} \hat{g}(\xi_1, 0) e^{2\pi i x_1 \xi_1} \mathrm{d}\xi_1, \tag{27}$$

and hence

$$\int_\mathbb{R} |g(x_1, x_2)| \, \mathrm{d}x_2 = \int_\mathbb{R} |\hat{g}(\xi_1, 0)| \, \mathrm{d}\xi_1. \tag{28}$$

The left-hand side of (28) corresponds to line integrations of g along vertical lines $x_1 = \text{constant}$. By applying shearing to the coordinates $x \in \mathbb{R}^2$, we can transform $\mathscr{L}$ into a line of the form $\{x \in \mathbb{R}^2 : x_1 = \text{constant}\}$, whereby we can apply (28) directly.

We will make this idea more concrete in the proof of the following key estimate for linearized terms of the form $\langle \chi_{M_l} f, \psi_\lambda \rangle$. Since we assume the truncated estimate as negligible, this will in fact allow us to estimate $\langle f, \psi_\lambda \rangle$.

Theorem 9. *Let $\psi \in L^2(\mathbb{R}^2)$ be compactly supported and assume that ψ satisfies the conditions in Setup 2. Further, let λ be such that $\operatorname{supp} \psi_\lambda \cap \partial B \neq \emptyset$. Suppose that $f \in \mathscr{E}(\mathbb{R}^2)$ and that ∂B is linear on the support of ψ_λ in the sense*

$$\operatorname{supp} \psi_\lambda \cap \partial B \subset \mathscr{L}$$

for some affine hyperplane $\mathscr{L}$ of $\mathbb{R}^2$. Then,

(i) if $\mathscr{L}$ has normal vector $(-1, s)$ with $|s| \leq 3$,

$$|\langle f, \psi_\lambda \rangle| \lesssim \frac{2^{-3j/4}}{|k + 2^{j/2}s|^3},$$

(ii) if $\mathscr{L}$ has normal vector $(-1, s)$ with $|s| \geq 3/2$,

$$|\langle f, \psi_\lambda \rangle| \lesssim 2^{-9j/4},$$

(iii) if $\mathscr{L}$ has normal vector $(0, s)$ with $s \in \mathbb{R}$, then

$$|\langle f, \psi_\lambda \rangle| \lesssim 2^{-11j/4}.$$

Proof. Fix λ, and let $f \in \mathscr{E}(\mathbb{R}^2)$. We can without loss of generality assume that f is only nonzero on B.

Cases (i) and (ii). We first consider the cases (i) and (ii). In these cases, the hyperplane can be written as

$$\mathscr{L} = \left\{ x \in \mathbb{R}^2 : \langle x - x_0, (-1, s) \rangle = 0 \right\}$$

for some $x_0 \in \mathbb{R}^2$. We shear the hyperplane by S_{-s} for $s \in \mathbb{R}$ and obtain

$$
\begin{aligned}
S_{-s}\mathscr{L} &= \left\{ x \in \mathbb{R}^2 : \langle S_s x - x_0, (-1, s) \rangle = 0 \right\} \\
&= \left\{ x \in \mathbb{R}^2 : \langle x - S_{-s}x_0, (S_s)^{\mathsf{T}}(-1, s) \rangle = 0 \right\} \\
&= \left\{ x \in \mathbb{R}^2 : \langle x - S_{-s}x_0, (-1, 0) \rangle = 0 \right\} \\
&= \left\{ x = (x_1, x_2) \in \mathbb{R}^2 : x_1 = \hat{x}_1 \right\}, \quad \text{where } \hat{x} = S_{-s}x_0,
\end{aligned}
$$

which is a line parallel to the x_2-axis. Here the power of shearlets comes into play, since it will allow us to only consider line singularities parallel to the x_2-axis. Of course, this requires that we also modify the shear parameter of the shearlet, that is, we will consider the right-hand side of

$$\langle f, \psi_{j,k,m} \rangle = \langle f(S_s \cdot), \psi_{j,\hat{k},m} \rangle$$

with the new shear parameter $\hat{k} = k + 2^{j/2}s$. The integrand in $\langle f(S_s \cdot), \psi_{j,\hat{k},m} \rangle$ has the singularity plane exactly located on the line $x_1 = \hat{x}_1$, i.e., on $S_{-s}\mathscr{L}$.

To simplify the expression for the integration bounds, we will fix a new origin on $S_{-s}\mathcal{L}$, that is, on $x_1 = \hat{x}_1$; the x_2 coordinate of the new origin will be fixed in the next paragraph. Since f is only nonzero of B, the function f will be equal to zero on one side of $S_{-s}\mathcal{L}$, say, $x_1 < \hat{x}_1$. It therefore suffices to estimate

$$\langle f_0(S_s\cdot)\chi_\Omega, \psi_{j,\hat{k},m}\rangle$$

for $f_0 \in C^\beta(\mathbb{R}^2)$ and $\Omega = \mathbb{R}_+ \times \mathbb{R}$. Let us assume that $\hat{k} < 0$. The other case can be handled similarly.

Since ψ is compactly supported, there exists some $c > 0$ such that $\operatorname{supp} \psi \subset [-c,c]^2$. By a rescaling argument, we can assume $c = 1$. Let

$$\mathscr{P}_{j,k} := \left\{ x \in \mathbb{R}^2 : |x_1 + 2^{-j/2}kx_2| \leq 2^{-j}, |x_2| \leq 2^{-j/2} \right\}, \qquad (29)$$

With this notation we have $\operatorname{supp} \psi_{j,k,0} \subset \mathscr{P}_{j,k}$. We say that the shearlet normal direction of the shearlet box $\mathscr{P}_{j,0}$ is $(1,0)$, and thus the shearlet normal of a sheared element $\psi_{j,k,m}$ associated with $\mathscr{P}_{j,k}$ is $(1,k/2^{j/2})$. Now, we fix our origin so that, relative to this new origin, it holds that

$$\operatorname{supp} \psi_{j,\hat{k},m} \subset \mathscr{P}_{j,\hat{k}} + (2^{-j},0) =: \tilde{\mathscr{P}}_{j,k}.$$

Then one face of $\tilde{\mathscr{P}}_{j,\hat{k}}$ intersects the origin.

Next, observe that the parallelogram $\tilde{\mathscr{P}}_{j,k}$ has sides $x_2 = \pm 2^{-j/2}$,

$$2^j x_1 + 2^{j/2}\hat{k}x_2 = 0, \quad \text{and} \quad 2^j x_1 + 2^{j/2}\hat{k}x_2 = 2.$$

As it is only a matter of scaling, we replace the right-hand side of the last equation with 1 for simplicity. Solving the two last equalities for x_2 gives the following lines:

$$L_1: \quad x_2 = -\frac{2^{j/2}}{\hat{k}}x_1, \quad \text{and} \quad L_2: \quad x_2 = -\frac{2^{j/2}}{\hat{k}}x_1 + \frac{2^{-j/2}}{\hat{k}},$$

which show that

$$\left| \left\langle f_0(S_s\cdot)\chi_\Omega, \psi_{j,\hat{k},m} \right\rangle \right| \lesssim \left| \int_0^{K_1} \int_{L_2}^{L_1} f_0(S_s x)\psi_{j,\hat{k},m}(x)\, dx_2 dx_1 \right|, \qquad (30)$$

where the upper integration bound for x_1 is $K_1 = 2^{-j} - 2^{-j}\hat{k}$; this follows from solving L_2 for x_1 and using that $|x_2| \leq 2^{-j/2}$. We remark that the inner integration over x_2 is along lines parallel to the singularity line $\partial\Omega = \{0\} \times \mathbb{R}$; as mentioned, this allows us to better handle the singularity and will be used several times throughout this section.

We consider the 1D Taylor expansion for $f_0(S_s\cdot)$ at each point $x = (x_1,x_2) \in L_2$ in the x_2-direction:

$$f_0(S_s x) = a(x_1) + b(x_1)\left(x_2 + \frac{2^{j/2}}{\hat{k}}x_1\right) + c(x_1,x_2)\left(x_2 + \frac{2^{j/2}}{\hat{k}}x_1\right)^2,$$

where $a(x_1), b(x_1)$ and $c(x_1,x_2)$ are all bounded in absolute value by $C(1+|s|)^2$. Using this Taylor expansion in (30) yields

$$\left|\left\langle f_0(S_s\cdot)\chi_\Omega, \psi_{j,\hat{k},m}\right\rangle\right| \lesssim (1+|s|)^2 \left|\int_0^{K_1} \sum_{l=1}^3 I_l(x_1)\,dx_1\right|, \tag{31}$$

where

$$I_1(x_1) = \left|\int_{L_1}^{L_2} \psi_{j,\hat{k},m}(x)dx_2\right|, \tag{32}$$

$$I_2(x_1) = \left|\int_{L_1}^{L_2} (x_2+K_2)\,\psi_{j,\hat{k},m}(x)dx_2\right|, \tag{33}$$

$$I_3(x_1) = \left|\int_0^{-2^{-j/2}/\hat{k}} (x_2)^2\,\psi_{j,\hat{k},m}(x_1,x_2-K_2)dx_2\right|, \tag{34}$$

and

$$K_2 = \frac{2^{j/2}}{\hat{k}}\,x_1.$$

We next estimate each integral I_1—I_3 separately.

Integral I_1. We first estimate $I_1(x_1)$. The Fourier slice theorem, see also (27), yields directly that

$$I_1(x_1) = \left|\int_{\mathbb{R}} \psi_{j,\hat{k},m}(x)dx_2\right| = \left|\int_{\mathbb{R}^2} \hat{\psi}_{j,\hat{k},m}(\xi_1,0)\,e^{2\pi i x_1 \xi_1}\,d\xi_1\right|.$$

By the assumptions from Setup 2, we have, for all $\xi = (\xi_1,\xi_2,\xi_3) \in \mathbb{R}^2$,

$$|\hat{\psi}_{j,\hat{k},m}(\xi)| \lesssim 2^{-3j/4}|h(2^{-j}\xi_1)|\left(1+\left|\frac{2^{-j/2}\xi_2}{2^{-j}\xi_1}+\hat{k}\right|\right)^{-\gamma}$$

for some $h \in L^1(\mathbb{R})$. Hence, we can continue our estimate of I_1 by

$$I_1(x_1) \lesssim \int_{\mathbb{R}} 2^{-3j/4}|h(2^{-j}\xi_1)|(1+|\hat{k}|)^{-\gamma}d\xi_1,$$

and further, by a change of variables,

$$I_1(x_1) \lesssim \int_{\mathbb{R}} 2^{j/4}|h(\xi_1)|(1+|\hat{k}|)^{-\gamma}d\xi_1 \lesssim 2^{j/4}(1+|\hat{k}|)^{-\gamma}, \tag{35}$$

since $h \in L^1(\mathbb{R})$.

Integral I_2. We start estimating $I_2(x_1)$ by

$$I_2(x_1) \le \left|\int_{\mathbb{R}} x_2\,\psi_{j,\hat{k},m}(x)dx_2\right| + |K_2|\left|\int_{\mathbb{R}} \psi_{j,\hat{k},m}(x)dx_2\right| =: S_1 + S_2.$$

Applying the Fourier slice theorem again and then utilizing the decay assumptions on $\hat{\psi}$ yields

$$S_1 = \left| \int_{\mathbb{R}} x_2 \psi_{j,\hat{k},m}(x) dx_2 \right| \leq \left| \int_{\mathbb{R}} \left(\frac{\partial}{\partial \xi_2} \hat{\psi}_{j,\hat{k},m} \right) (\xi_1, 0) e^{2\pi i x_1 \xi_1} d\xi_1 \right|$$

$$\lesssim \int_{\mathbb{R}} 2^{-j/2} 2^{-3j/4} \left| h(2^{-j}\xi_1) \right| (1 + |\hat{k}|)^{-\gamma} d\xi_1 \lesssim 2^{-j/4}(1 + |\hat{k}|)^{-\gamma}.$$

Since $|x_1| \leq -\hat{k}_1/2^j$, we have $K_2 \leq 2^{-j/2}$. The following estimate of S_2 then follows directly from the estimate of I_1:

$$S_2 \lesssim |K_2| 2^{j/4} (1 + |\hat{k}|)^{-\gamma} \lesssim 2^{-j/4} (1 + |\hat{k}|)^{-\gamma}.$$

From the two last estimates, we conclude that $I_2(x_1) \lesssim 2^{-j/4}(1 + |\hat{k}|)^{-\gamma}$.

Integral I_3. Finally, we estimate $I_3(x_1)$ by

$$I_3(x_1) \leq \left| \int_0^{2^{-j/2}/\hat{k}} (x_2)^2 \| \psi_{j,\hat{k},m} \|_{L^\infty} dx_2 \right|$$

$$\lesssim 2^{3j/4} \left| \int_0^{-2^{-j/2}/\hat{k}} (x_2)^2 dx_2 \right| \lesssim 2^{-3j/4} |\hat{k}|^{-3}. \tag{36}$$

We see that I_2 decays faster than I_1, and hence we can leave I_2 out of our analysis. Applying (35) and (36) to (31), we obtain

$$\left| \left\langle f_0(S_s \cdot) \chi_\Omega, \psi_{j,\hat{k},m} \right\rangle \right| \lesssim (1 + |s|)^2 \left(\frac{2^{-3j/4}}{(1 + |\hat{k}|)^{\gamma-1}} + \frac{2^{-7j/4}}{|\hat{k}|^2} \right). \tag{37}$$

Suppose that $s \leq 3$. Then (37) reduces to

$$|\langle f, \psi_{j,k,m} \rangle| \lesssim \frac{2^{-3j/4}}{(1 + |\hat{k}|)^{\gamma-1}} + \frac{2^{-7j/4}}{|\hat{k}|^2}$$

$$\lesssim \frac{2^{-3j/4}}{(1 + |\hat{k}|)^3},$$

since $\gamma \geq 4$. This proves (i).

On the other hand, if $s \geq 3/2$, then

$$|\langle f, \psi_{j,k,m} \rangle| \lesssim 2^{-9j/4}.$$

To see this, note that

$$\frac{2^{-\frac{3}{4}j}}{(1 + |k + s2^{j/2}|)^3} = \frac{2^{-\frac{9}{4}j}}{(2^{-j/2} + |k/2^{-j/2} + s|)^3} \leq \frac{2^{-\frac{9}{4}j}}{|k/2^{j/2} + s|^3}$$

and

$$|k/2^{j/2} + s| \geq |s| - |k/2^{j/2}| \geq 1/2 - 2^{-j/2} \geq 1/4$$

for sufficiently large $j \geq 0$, since $|k| \leq \lceil 2^{j/2} \rceil \leq 2^{j/2} + 1$, and (ii) is proven.

Case (iii). Finally, we need to consider the case (iii), in which the normal vector of the hyperplane $\mathscr{L}$ is of the form $(0,s)$ for $s \in \mathbb{R}$. For this, let $\tilde{\Omega} = \{x \in \mathbb{R}^2 : x_2 \geq 0\}$. As in the first part of the proof, it suffices to consider coefficients of the form $\langle \chi_{\tilde{\Omega}} f_0, \psi_{j,k,m} \rangle$, where supp $\psi_{j,k,m} \subset \mathscr{P}_{j,k} - (2^{-j}, 0) = \tilde{\mathscr{P}}_{j,k}$ with respect to some new origin. As before, the boundary of $\tilde{\mathscr{P}}_{j,k}$ intersects the origin. By the assumptions in Setup 2, we have that

$$\left(\frac{\partial}{\partial \xi_1} \right)^{\ell} \hat{\psi}(0, \xi_2) = 0 \quad \text{for } \ell = 0, 1,$$

which implies that

$$\int_{\mathbb{R}} x_1^{\ell} \psi(x) dx_1 = 0 \quad \text{for all } x_2 \in \mathbb{R} \text{ and } \ell = 0, 1.$$

Therefore, we have

$$\int_{\mathbb{R}} x_1^{\ell} \psi(S_k x) dx_1 = 0 \quad \text{for all } x_2 \in \mathbb{R}, k \in \mathbb{R}, \text{ and } \ell = 0, 1, \tag{38}$$

since a shearing operation S_k preserves vanishing moments along the x_1 axis. Now, we use Taylor expansion of f_0 in the x_1-direction (i.e., again along the singularity line $\partial \tilde{\Omega}$). By (38) everything but the last term in the Taylor expansion disappears, and we obtain

$$|\langle \chi_{\tilde{\Omega}} f_0, \psi_{j,k,m} \rangle| \lesssim 2^{3j/4} \int_0^{2^{-j/2}} \int_{-2^{-j}}^{2^{-j}} (x_1)^2 \, dx_1 dx_2$$
$$\lesssim 2^{3j/4} 2^{-j/2} 2^{-3j} = 2^{-11j/4},$$

which proves claim (iii). $\square$

We are now ready to show the estimates (20) and (21), which by Lemma 2(ii) completes the proof of Theorem 5.

For $j \geq 0$, fix $Q \in \mathscr{Q}_j^0$, where $\mathscr{Q}_j^0 \subset \mathscr{Q}_j$ is the collection of dyadic squares that intersects $\mathscr{L}$. We then have the following counting estimate:

$$\# |M_{j,k,Q}| \lesssim |k + 2^{j/2} s| + 1 \tag{39}$$

for each $|k| \leq \lceil 2^{j/2} \rceil$, where

$$M_{j,k,Q} := \{ m \in \mathbb{Z}^2 : |\text{supp } \psi_{j,k,m} \cap \mathscr{L} \cap Q| \neq 0 \}$$

To see this claim, note that for a fixed j and k we need to count the number of translates $m \in \mathbb{Z}^2$ for which the support of $\psi_{j,k,m}$ intersects the discontinuity line $\mathscr{L} : x_1 = sx_2 + b$, $b \in \mathbb{R}$, inside Q. Without loss of generality, we can assume that $Q = [0, 2^{-j/2}]^2$, $b = 0$, and supp $\psi_{j,k,0} \subset C \cdot \mathscr{P}_{j,k}$, where $\mathscr{P}_{j,k}$ is defined as in (29). The shearlet $\psi_{j,k,m}$ will therefore be concentrated around the line

$\mathscr{S}_m : x_1 = -\frac{k}{2^{j/2}}x_2 + 2^{-j}m_1 + 2^{-j/2}m_2$, see also Fig. 9b. We will count the number of $m = (m_1, m_2) \in \mathbb{Z}^2$ for which these two lines intersect inside Q since this number, up to multiplication with a constant independent of the scale j, will be equal to $\#|M_{j,k,Q}|$.

First note that since the size of Q is $2^{-j/2} \times 2^{-j/2}$, only a finite number of m_2 translates can make $S_m \cap \mathscr{L} \cap Q \neq \emptyset$ whenever $m_1 \in \mathbb{Z}$ is fixed. For a fixed $m_2 \in \mathbb{Z}$, we then estimate the number of relevant m_1 translates. Equating the x_1 coordinates in $\mathscr{L}$ and $\mathscr{S}_m$ yields

$$\left(\frac{k}{2^{j/2}} + s\right) x_2 = 2^{-j}m_1 + 2^{-j/2}m_2.$$

Without loss of generality, we take $m_2 = 0$ which then leads to

$$2^{-j}|m_1| \leq 2^{-j/2}\left|k + 2^{j/2}s\right| |x_2| \leq 2^{-j}\left|k + 2^{j/2}s\right|,$$

hence $|m_1| \leq |k + 2^{j/2}s|$. This completes the proof of the claim.

For $\varepsilon > 0$, we will consider the shearlet coefficients larger than ε in absolute value. Thus, we define:

$$M_{j,k,Q}(\varepsilon) = \left\{m \in M_{j,k,Q} : |\langle f, \psi_{j,k,m}\rangle| > \varepsilon\right\},$$

where $Q \in \mathscr{Q}_j^0$. Since the discontinuity line $\mathscr{L}$ has finite length in $[0,1]^2$, we have the estimate $\#|\mathscr{Q}_j^0| \lesssim 2^{j/2}$. Assume $\mathscr{L}$ has normal vector $(-1, s)$ with $|s| \leq 3$. Then, by Theorem 9(i), $|\langle f, \psi_{j,k,m}\rangle| > \varepsilon$ implies that

$$|k + 2^{j/2}s| \leq \varepsilon^{-1/3}2^{-j/4}. \tag{40}$$

By Lemma 2(i) and the estimates (39) and (40), we have that

$$\#|\Lambda(\varepsilon)| \lesssim \sum_{j=0}^{\frac{4}{3}\log_2(\varepsilon^{-1})+C} \sum_{Q \in \mathscr{Q}_j^0} \sum_{\{\hat{k}:|\hat{k}|\leq\varepsilon^{-1/3}2^{-j/4}\}} \#|M_{j,k,Q}(\varepsilon)|$$

$$\lesssim \sum_{j=0}^{\frac{4}{3}\log_2(\varepsilon^{-1})+C} \sum_{Q \in \mathscr{Q}_j^0} \sum_{\{\hat{k}:|\hat{k}|\leq\varepsilon^{-1/3}2^{-j/4}\}} (|\hat{k}| + 1)$$

$$\lesssim \sum_{j=0}^{\frac{4}{3}\log_2(\varepsilon^{-1})+C} \#|\mathscr{Q}_j^0| \left(\varepsilon^{-2/3}2^{-j/2}\right)$$

$$\lesssim \varepsilon^{-2/3} \sum_{j=0}^{\frac{4}{3}\log_2(\varepsilon^{-1})+C} 1 \lesssim \varepsilon^{-2/3}\log_2(\varepsilon^{-1}),$$

where, as usual, $\hat{k} = k + s2^{j/2}$. By Lemma 2(ii), this leads to the sought estimates.

On the other hand, if $\mathscr{L}$ has normal vector $(0,1)$ or $(-1,s)$ with $|s| \geq 3$, then $|\langle f, \psi_\lambda \rangle| > \varepsilon$ implies that

$$j \leq \frac{4}{9} \log_2(\varepsilon^{-1}),$$

which follows by assertions (ii) and (iii) in Theorem 9. Hence, we have

$$\# |\Lambda(\varepsilon)| \lesssim \sum_{j=0}^{\frac{4}{9}\log_2(\varepsilon^{-1})} \sum_k \sum_{Q \in \mathscr{Q}_j^0} \# |M_{j,k,Q}(\varepsilon)|.$$

Note that $\# |M_{j,k,Q}| \lesssim 2^{j/2}$, since $\# |\{m \in \mathbb{Z}^2 : |\text{supp } \psi_\lambda \cap Q| \neq 0\}| \lesssim 2^{j/2}$ for each $Q \in \mathscr{Q}_j$, and that the number of shear parameters k for each scale parameter $j \geq 0$ is bounded by $C2^{j/2}$. Therefore,

$$\# |\Lambda(\varepsilon)| \lesssim \sum_{j=0}^{\frac{4}{9}\log_2(\varepsilon^{-1})} 2^{j/2} 2^{j/2} 2^{j/2} = \sum_{j=0}^{\frac{4}{9}\log_2(\varepsilon^{-1})} 2^{3j/2} \lesssim 2^{\frac{4}{9} \cdot \frac{3}{2} \cdot \log_2(\varepsilon^{-1})} \lesssim \varepsilon^{-2/3}.$$

This implies our sought estimate (20) which, together with the estimate for $|s| \leq 3$, completes the proof of Theorem 5 for $L = 1$ under Setup 2.   □

5.1.7 The case L ≠ 1

We now turn to the extended class of cartoon-lime images $\mathscr{E}_L^2(\mathbb{R}^2)$ with $L \neq 1$, i.e., in which the singularity curve is only required to be piecewise C^2. We say that $p \in \mathbb{R}^2$ is a corner point if ∂B is not C^2-smooth in p. The main focus here will be to investigate shearlets that interact with one of the L corner points. We will argue that Theorem 5 also holds in this extended setting. The rest of the proof, that is, for shearlets *not* interacting with corner points, is of course identical to that presented in Sects. 5.1.5 and 5.1.6.

In the compactly supported case, one can simply count the number of shearlets interacting with a corner point at a given scale. Using Lemma 2(i), one then arrives at the sought estimate. On the other hand, for the band-limited case one needs to measure the sparsity of the shearlet coefficients for f localized to each dyadic square. We present the details in the remainder of this section.

Band-limited shearlets. In this case, it is sufficient to consider a dyadic square $Q \in \mathscr{Q}_j^0$ with $j \geq 0$ such that Q contains a singular point of edge curve. Especially, we may assume that j is sufficiently large so that the dyadic square $Q \in \mathscr{Q}_j^0$ contains a single corner point of ∂B. The following theorem analyzes the sparsity of the shearlet coefficients for such a dyadic square $Q \in \mathscr{Q}_j^0$.

Theorem 10. *Let* $f \in \mathscr{E}_L^2(\mathbb{R}^2)$ *and* $Q \in \mathscr{Q}_j^0$ *with* $j \geq 0$ *be a dyadic square containing a singular point of the edge curve. The sequence of shearlet coefficients* $\{d_\lambda := \langle f_Q, \psi_\lambda \rangle : \lambda \in \Lambda_j\}$ *obeys*

$$\left\| (d_\lambda)_{\lambda \in \Lambda_j} \right\|_{w\ell^{2/3}} \leq C.$$

The proof of Theorem 10 is based on a proof of an analog result for curvelets [1]. Although the proof in [1] considers only curvelet coefficients, essentially the same arguments, with modifications to the shearlet setting, can be applied to show Theorem 10.

Finally, we note that the number of dyadic squares $Q \in \mathscr{Q}_j^0$ containing a singular point of ∂B is bounded by a constant not depending on j; one could, e.g., take L as this constant. Therefore, applying Theorem 10 and repeating the arguments in Sect. 5.1.5 complete the proof of Theorem 5 for $L \neq 1$ for Setup 1.

Compactly supported shearlets. In this case, it is sufficient to consider the following two cases:

Case 1. The shearlet ψ_λ intersects a corner point, in which two C^2 curves ∂B_0 and ∂B_1, say, meet (see Fig. 11a).

Case 2. The shearlet ψ_λ intersects two edge curves ∂B_0 and ∂B_1, say, simultaneously, but it does not intersect a corner point (see Fig. 11b).

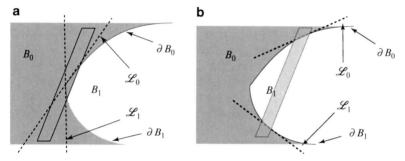

Fig. 11 (**a**) A shearlet ψ_λ intersecting a corner point, in which two edge curves ∂B_0 and ∂B_1 meet. $\mathscr{L}_0$ and $\mathscr{L}_1$ are tangents to the edge curves ∂B_0 and ∂B_1 in this corner point. (**b**) A shearlet ψ_λ intersecting two edge curves ∂B_0 and ∂B_1 which are part of the boundary of sets B_0 and B_1. $\mathscr{L}_0$ and $\mathscr{L}_1$ are tangents to the edge curves ∂B_0 and ∂B_1 in points contained in the support of ψ_λ

We aim to show that $\#|\Lambda(\varepsilon)| \lesssim \varepsilon^{-\frac{2}{3}}$ in both cases. By Lemma 2, this will be sufficient.

Case 1. Since there exist only finitely many corner points with total number not depending on scale $j \geq 0$ and the number of shearlets ψ_λ intersecting each of corner points is bounded by $C2^{j/2}$, we have

$$\#|\Lambda(\varepsilon)| \lesssim \sum_{j=0}^{\frac{4}{3}\log_2(\varepsilon^{-1})} 2^{j/2} \lesssim \varepsilon^{-\frac{2}{3}}.$$

Case 2. As illustrated in Fig. 11b, we can write the function f as

$$f_0\chi_{B_0} + f_1\chi_{B_1} = (f_0 - f_1)\chi_{B_0} + f_1 \quad \text{in } Q,$$

where $f_0, f_1 \in C^2([0,1]^2)$ and B_0, B_1 are two disjoint subsets of $[0,1]^2$. As we indicated before, the rate for optimal sparse approximation is achieved for the smooth function f_1. Thus, it is sufficient to consider $f := g_0 \chi_{B_0}$ with $g_0 = f_0 - f_1 \in C^2([0,1]^2)$. By a *truncated* estimate, we can replace two boundary curves ∂B_0 and ∂B_1 by hyperplanes of the form

$$\mathscr{L}_i = \left\{ x \in \mathbb{R}^2 : \langle x - x_0, (-1, s_i) \rangle = 0 \right\} \quad \text{for } i = 0, 1.$$

In the sequel, we assume $\max_{i=0,1} |s_i| \le 3$ and mention that the other cases can be handled similarly. Next define

$$M_{j,k,Q}^i = \left\{ m \in \mathbb{Z}^2 : |\operatorname{supp} \psi_{j,k,m} \cap \mathscr{L}_i \cap Q| \ne 0 \right\} \quad \text{for } i = 0, 1,$$

for each $Q \in \tilde{\mathscr{Q}}_j^0$, where $\tilde{\mathscr{Q}}_j^0$ denotes the dyadic squares containing the two distinct boundary curves. By an estimate similar to (39), we obtain

$$\# \left| M_{j,k,Q}^0 \cap M_{j,k,Q}^1 \right| \lesssim \min_{i=0,1} \left(|k + 2^{j/2} s_i| + 1 \right). \tag{41}$$

Applying Theorem 9(i) to each of the hyperplanes $\mathscr{L}_0$ and $\mathscr{L}_1$, we also have

$$|\langle f, \psi_{j,k,m} \rangle| \le C \cdot \max_{i=0,1} \left\{ \frac{2^{-\frac{3}{4}j}}{|2^{j/2} s_i + k|^3} \right\}. \tag{42}$$

Let $\hat{k}_i = k + 2^{j/2} s_i$ for $i = 0, 1$. Without loss of generality, we may assume that $\hat{k}_0 \le \hat{k}_1$. Then, (41) and (42) imply that

$$\# \left| M_{j,Q}^0 \cap M_{j,Q}^1 \right| \lesssim |\hat{k}_0| + 1 \tag{43}$$

and

$$|\langle f, \psi_{j,k,m} \rangle| \lesssim \frac{2^{-\frac{3}{4}j}}{|\hat{k}_0|^3}. \tag{44}$$

Using (43) and (44), we now estimate $\# |\Lambda(\varepsilon)|$ as follows:

$$\# |\Lambda(\varepsilon)| \lesssim \sum_{j=0}^{\frac{4}{3} \log_2(\varepsilon^{-1}) + C} \sum_{Q \in \tilde{\mathscr{Q}}_j^0} \sum_{\hat{k}_0} (1 + |\hat{k}_0|)$$

$$\lesssim \sum_{j=0}^{\frac{4}{3} \log_2(\varepsilon^{-1}) + C} \# \left| \tilde{\mathscr{Q}}_j^0 \right| (\varepsilon^{-2/3} 2^{-j/2}) \lesssim \varepsilon^{-2/3}.$$

Note that $\# |\tilde{\mathscr{Q}}_j^0| \le C$ since the number of $Q \in \mathscr{Q}_j$ containing two distinct boundary curves ∂B_0 and ∂B_1 is bounded by a constant independent of j. The result is proved.   □

5.2 Optimal Sparse Approximations in 3D

When passing from 2D to 3D, the complexity of anisotropic structures changes significantly. In particular, as opposed to the two-dimensional setting, geometric structures of discontinuities for piecewise smooth 3D functions consist of two morphologically different types of structure, namely surfaces and curves. Moreover, as we saw in Sect. 5.1, the analysis of sparse approximations in 2D heavily depends on reducing the analysis to affine subspaces of $\mathbb{R}^2$. Clearly, these subspaces always have dimension one in 2D. In dimension three, however, we have subspaces of dimensions one and two, and therefore the analysis needs to be performed on subspaces of the "correct" dimension.

This issue manifests itself when performing the analysis for band-limited shearlets, since one needs to replace the Radon transform used in 2D with a so-called X-ray transform. For compactly supported shearlets, one needs to perform the analysis on carefully chosen hyperplanes of dimension two. This will allow for using estimates from the two-dimensional setting in a slice by slice manner.

As in the two-dimensional setting, analyzing the decay of individual shearlet coefficients $\langle f, \mathring{\psi}_\lambda \rangle$ can be used to show optimal sparsity for compactly supported shearlets while the sparsity of the sequence of shearlet coefficients with respect to the weak ℓ^p quasi norm should be analyzed for band-limited shearlets. Furthermore, we will only consider the case $\mathring{\psi} = \psi$ since the two other cases can be handled similarly.

5.2.1 A heuristic analysis

As in the heuristic analysis for the 2D situation debated in Sect. 5.1.1, we can again split the proof into similar three cases as shown in Fig. 12.

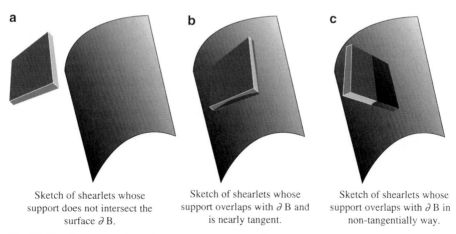

a

Sketch of shearlets whose support does not intersect the surface ∂ B.

b

Sketch of shearlets whose support overlaps with ∂ B and is nearly tangent.

c

Sketch of shearlets whose support overlaps with ∂ B in non-tangentially way.

Fig. 12 The three types of shearlets $\psi_{j,k,m}$ and boundary ∂B interactions considered in the heuristic 3D analysis. Note that only a section of ∂B is shown

Only case (b) differs significantly from the 2D setting, so we restrict out attention to that case.

For case (b), there are at most $O(2^j)$ coefficients at scale $j > 0$ since the plate-like elements are of size $2^{-j/2}$ times $2^{-j/2}$ (and "thickness" 2^{-j}). By Hölder's inequality, we see that

$$\left| \langle f, \psi_{j,k,m} \rangle \right| \leq \|f\|_{L^\infty} \|\psi_{j,k,m}\|_{L^1} \leq C_1 \, 2^{-j} \|\psi\|_{L^1} \leq C_2 \cdot 2^{-j}$$

for some constants $C_1, C_2 > 0$. Hence, we have $O(2^j)$ coefficients bounded by $C_2 \cdot 2^{-j}$.

Assuming the coefficients in case (a) and (c) to be negligible, the nth largest shearlet coefficient c_n^* is therefore bounded by

$$|c_n^*| \leq C \cdot n^{-1},$$

which in turn implies

$$\sum_{n>N} |c_n^*|^2 \leq \sum_{n>N} C \cdot n^{-2} \leq C \cdot \int_N^\infty x^{-2} \mathrm{d}x \leq C \cdot N^{-1}.$$

Hence, we meet the optimal rates (6) and (7) from Definition 1. This, at least heuristically, shows that shearlets provide optimally sparse approximations of 3D cartoon-like images.

5.2.2 Main result

The hypotheses needed for the band-limited case, stated in Setup 3, are a straightforward generalization of Setup 1 in the 2D setting.

Setup 3. The generators $\phi, \psi, \tilde{\psi}, \check{\psi} \in L^2(\mathbb{R}^3)$ are band-limited and C^∞ in the frequency domain. Furthermore, the shearlet system $\mathrm{SH}(\phi, \psi, \tilde{\psi}, \check{\psi}; c)$ forms a frame for $L^2(\mathbb{R}^3)$ (cf. the construction in Sect. 4.2).

For the compactly supported generators, we will also use hypotheses in the spirit of Setup 2, but with slightly stronger and more sophisticated assumption on vanishing moment property of the generators i.e., $\delta > 8$ and $\gamma \geq 4$.

Setup 4. The generators $\phi, \psi, \tilde{\psi}, \check{\psi} \in L^2(\mathbb{R}^3)$ are compactly supported, and the shearlet system $\mathrm{SH}(\phi, \psi, \tilde{\psi}, \check{\psi}; c)$ forms a frame for $L^2(\mathbb{R}^3)$. Furthermore, the function ψ satisfies, for all $\xi = (\xi_1, \xi_2, \xi_3) \in \mathbb{R}^3$,

(i) $|\hat{\psi}(\xi)| \leq C \cdot \min\{1, |\xi_1|^\delta\} \min\{1, |\xi_1|^{-\gamma}\} \min\{1, |\xi_2|^{-\gamma}\} \min\{1, |\xi_3|^{-\gamma}\}$,
   and
(ii) $\left|\frac{\partial}{\partial \xi_i} \hat{\psi}(\xi)\right| \leq |h(\xi_1)| \left(1 + \frac{|\xi_2|}{|\xi_1|}\right)^{-\gamma} \left(1 + \frac{|\xi_3|}{|\xi_1|}\right)^{-\gamma}$,

for $i = 2, 3$, where $\delta > 8$, $\gamma \geq 4$, $h \in L^1(\mathbb{R})$, and C a constant, and $\tilde{\psi}$ and $\check{\psi}$ satisfy analogous conditions with the obvious change of coordinates (cf. the construction in Sect. 4.3).

The main result can now be stated as follows.

Theorem 11 ([8, 12]). *Assume Setup 3 or 4. Let $L = 1$. For any $\nu > 0$ and $\mu > 0$, the shearlet frame $\mathrm{SH}(\phi, \psi, \tilde{\psi}, \check{\psi}; c)$ provides optimally sparse approximations of functions $f \in \mathcal{E}_L^2(\mathbb{R}^3)$ in the sense of Definition 1, i.e.,*

$$\|f - f_N\|_{L^2}^2 \lesssim N^{-1}(\log N)^2), \quad as\ N \to \infty,$$

and

$$|c_n^*| \lesssim n^{-1}(\log n), \quad as\ n \to \infty,$$

where $c = \{\langle f, \check{\psi}_\lambda \rangle : \lambda \in \Lambda, \check{\psi} = \psi, \check{\psi} = \tilde{\psi},$ or $\check{\psi} = \check{\psi}\}$ and $c^* = (c_n^*)_{n \in \mathbb{N}}$ is a decreasing (in modulus) rearrangement of c.

We now give a sketch of proof for this theorem, and refer to [8, 12] for detailed proofs.

5.2.3 Sketch of proof of Theorem 11

Band-limited shearlets. The proof of Theorem 11 for band-limited shearlets follows the same steps as discussed in Sect. 5.1.5 for the 2D case. To indicate the main steps, we will use the same notation as for the 2D proof with the straightforward extension to 3D.

Similar to Theorems 6 and 7, one can prove the following results on the sparsity of the shearlets coefficients for each dyadic square $Q \in \mathcal{Q}_j$.

Theorem 12 ([8]). *Let $f \in \mathcal{E}^2(\mathbb{R}^3)$. $Q \in \mathcal{Q}_j^0$, with $j \geq 0$ fixed, the sequence of shearlet coefficients $\{d_\lambda := \langle f_Q, \check{\psi}_\lambda \rangle : \lambda \in \Lambda_j\}$ obeys*

$$\|(d_\lambda)_{\lambda \in \Lambda_j}\|_{w\ell^1} \lesssim 2^{-2j}.$$

Theorem 13 ([8]). *Let $f \in \mathcal{E}^2(\mathbb{R}^3)$. For $Q \in \mathcal{Q}_j^1$, with $j \geq 0$ fixed, the sequence of shearlet coefficients $\{d_\lambda := \langle f_Q, \check{\psi}_\lambda \rangle : \lambda \in \Lambda_j\}$ obeys*

$$\|(d_\lambda)_{\lambda \in \Lambda_j}\|_{\ell^1} \lesssim 2^{-4j}.$$

The proofs of Theorems 12 and 13 follow the same principles as the proofs of the analog results in 2D, Theorems 6 and 7, with one important difference: In the proof of Theorems 6 and 7, the Radon transform (cf. (27)) is used to deduce estimates for the integral of edge-curve fragments. In 3D one needs to use a different transform, namely the so-called X-ray transform, which maps a function on $\mathbb{R}^3$ into the sets of its line integrals. The X-ray transform is then used to deduce estimates for the integral of the *surface* fragments. We refer to [8] for a detailed exposition.

As a consequence of Theorems 12 and 13, we have the following result.

Theorem 14 ([8]). *Suppose $f \in \mathscr{E}^2(\mathbb{R}^3)$. Then, for $j \geq 0$, the sequence of the shear-let coefficients $\{c_\lambda := \langle f, \mathring{\psi}_\lambda \rangle : \lambda \in \Lambda_j\}$ obeys*

$$\|(c_\lambda)_{\lambda \in \Lambda_j}\|_{w\ell^1} \lesssim 1.$$

Proof. The result follows by the same arguments used in the proof of Theorem 8.
□

By Theorem 14, we can now prove Theorem 11 for the band-limited setup and for $f \in \mathscr{E}_L^2(\mathbb{R}^3)$ with $L = 1$. The proof is very similar to the proof of Theorem 5 in Sect. 5.1.5, wherefore we will not repeat it.

Compactly supported shearlets. In this section we will consider the key estimates for the linearized term for compactly supported shearlets in 3D. This is an extension of Theorem 9 to the 3D setting. Hence, we will assume that the discontinuity surface is a plane, and consider the decay of the shearlet coefficients of shearlets interacting with such a discontinuity.

Theorem 15 ([12]). *Let $\psi \in L^2(\mathbb{R}^3)$ be compactly supported, and assume that ψ satisfies the conditions in Setup 4. Further, let λ be such that $\operatorname{supp} \psi_\lambda \cap \partial B \neq \emptyset$. Suppose that $f \in \mathscr{E}^2(\mathbb{R}^3)$ and that ∂B is linear on the support of ψ_λ in the sense that*

$$\operatorname{supp} \psi_\lambda \cap \partial B \subset \mathscr{H}$$

for some affine hyperplane $\mathscr{H}$ of $\mathbb{R}^3$. Then,

(i) if $\mathscr{H}$ has normal vector $(-1, s_1, s_2)$ with $s_1 \leq 3$ and $s_2 \leq 3$,

$$|\langle f, \psi_\lambda \rangle| \lesssim \min_{i=1,2} \left\{ \frac{2^{-j}}{|k_i + 2^{j/2} s_i|^3} \right\},$$

(ii) if $\mathscr{H}$ has normal vector $(-1, s_1, s_2)$ with $s_1 \geq 3/2$ or $s_2 \geq 3/2$,

$$|\langle f, \psi_\lambda \rangle| \lesssim 2^{-5j/2},$$

(iii) if $\mathscr{H}$ has normal vector $(0, s_1, s_2)$ with $s_1, s_2 \in \mathbb{R}$, then

$$|\langle f, \psi_\lambda \rangle| \lesssim 2^{-3j},$$

Proof. Fix λ, and let $f \in \mathscr{E}^2(\mathbb{R}^3)$. We first consider the case (ii) and assume $s_1 \geq 3/2$. The hyperplane can be written as

$$\mathscr{H} = \left\{ x \in \mathbb{R}^3 : \langle x - x_0, (-1, s_1, s_2) \rangle = 0 \right\}$$

for some $x_0 \in \mathbb{R}^3$. For $\hat{x}_3 \in \mathbb{R}$, we consider the restriction of $\mathscr{H}$ to the slice $x_3 = \hat{x}_3$. This is clearly a line of the form

$$\mathscr{L} = \left\{ x = (x_1, x_2) \in \mathbb{R}^2 : \langle x - x_0', (-1, s_1) \rangle = 0 \right\}$$

for some $x_0' \in \mathbb{R}^2$, and hence we have reduced the singularity to a line singularity, which was already considered in Theorem 9. We apply now Theorem 9 to each on slice, and we obtain

$$|\langle f, \psi_\lambda \rangle| \lesssim 2^{j/4} 2^{-9j/4} 2^{-j/2} = 2^{-5j/2}.$$

The first term $2^{j/4}$ in the estimate above is due to the different normalization factor used for shearlets in 2D and 3D, the second term is the conclusion from Theorem 9, and the third is the length of the support of ψ_λ in the direction of x_3. The case $s_2 \geq 3/2$ can be handled similarly with restrictions to slices $x_2 = \hat{x}_2$ for $\hat{x}_2 \in \mathbb{R}$. This completes the proof of case (ii).

The other two cases, i.e., case (i) and (ii), are proved using the same slice by slice technique and Theorem 9. □

Neglecting truncated estimates, Theorem 15 can be used to prove the optimal sparsity result in Theorem 11. The argument is similar to the one in Sect. 5.1.6 and will not be repeated here. Let us simply argue that the decay rate $|\langle f, \psi_\lambda \rangle| \lesssim 2^{-5j/2}$ from Theorem 15(ii) is what is needed in the case $s_i \geq 3/2$. It is easy to see that in 3D an estimate of the form

$$\#|\Lambda(\varepsilon)| \lesssim \varepsilon^{-1}.$$

will guarantee optimal sparsity. Since we in the estimate $|\langle f, \psi_\lambda \rangle| \lesssim 2^{-5j/2}$ have no control of the shearing parameter $k = (k_1, k_2)$, we have to use a crude counting estimate, where we include all shears at a given scale j, namely $2^{j/2} \cdot 2^{j/2} = 2^j$. Since the number of dyadic boxes Q where ∂B intersects the support of f is of order $2^{3j/2}$, we arrive at

$$\#|\Lambda(\varepsilon)| \lesssim \sum_{j=0}^{\frac{2}{5} \log_2(\varepsilon^{-1})} 2^{5j/2} \asymp \varepsilon^{-1}.$$

5.2.4 Some extensions

Paralleling the 2D setting (see Sect. 5.1.7), we can extend the optimality result in Theorem 11 to the cartoon-like image class $\mathscr{E}_L^2(\mathbb{R}^3)$ for $L \in \mathbb{N}$, in which the discontinuity surface ∂B is allowed to be *piecewise* C^2-smooth.

Moreover, the requirement that the edge ∂B is piecewise C^2 might be too restrictive in some applications. Therefore, in [12], the cartoon-like image model class was enlarged to allow less regular images, where ∂B is piecewise C^α-smooth for $1 < \alpha \leq 2$, and not necessarily a C^2. This class $\mathscr{E}^\beta_{\alpha,L}(\mathbb{R}^3)$ was introduced in Sect. 2 consisting of *generalized* cartoon-like images having C^β-smoothness apart from a piecewise C^α discontinuity curve. The sparsity results presented above in Theorem 11 can be extended to this generalized model class for compactly supported shearlets with a scaling matrix dependent on α. The optimal approximation error rate, as usual measured in $\|f - f_N\|^2_{L^2}$, for this generalized model is $N^{-\alpha/2}$; compare this to N^{-1} for the case $\alpha = 2$ considered throughout this chapter. For brevity, we will not go into detail of this, but mention that the approximation error rate obtained by shearlet frames is slightly worse than in the $\alpha = \beta = 2$ case since the error rate is not only a poly-log factor away from the optimal rate, but a small polynomial factor; and we refer to [12] the precise statement and proof.

5.2.5 Surprising observations

Capturing anisotropic phenomenon in 3D is somewhat different from capturing anisotropic features in 2D as discussed in Sect. 1.3. While in 2D we "only" have to handle curves, in 3D a more complex situation can occur since we find two geometrically very different anisotropic structures: curves and surfaces. Curves are clearly 1D anisotropic features and surfaces 2D features. Since our 3D shearlet elements are plate-like in spatial domain by construction, one could think that these 3D shearlet systems would *only* be able to efficiently capture 2D anisotropic structures, and *not* 1D structures. Nonetheless, surprisingly, as we have discussed in Sect. 5.2.4, these 3D shearlet systems still perform optimally when representing and analyzing 3D data $\mathscr{E}^2_L(\mathbb{R}^3)$ that contain *both* curve and surface singularities (see e.g., Fig. 2).

Acknowledgments The first author acknowledges support by the Einstein Foundation Berlin and by Forschungsgemeinschaft (DFG) Grant KU 1446/14, and the first and third authors acknowledge support by DFG Grant SPP-1324 KU 1446/13.

References

1. E. J. Candés and D. L. Donoho, *New tight frames of curvelets and optimal representations of objects with piecewise C^2 singularities*, Comm. Pure and Appl. Math. **56** (2004), 216–266.
2. I. Daubechies, *Ten Lectures on Wavelets*, SIAM, Philadelphia, 1992.
3. R. A. DeVore, G. G. Lorentz, *Constructive Approximation*, Springer, Berlin, 1993.
4. M. N. Do and M. Vetterli, *The contourlet transform: an efficient directional multiresolution image representation*, IEEE Trans. Image Process. **14** (2005), 2091–2106.
5. D. L. Donoho, *Sparse components of images and optimal atomic decomposition*, Constr. Approx. **17** (2001), 353–382.
6. D. L. Donoho, *Wedgelets: nearly minimax estimation of edges*, Ann. Statist. **27** (1999), 859–897.

7. K. Guo and D. Labate, *Optimally sparse multidimensional representation using shearlets*, SIAM J. Math Anal. **39** (2007), 298–318.

8. K. Guo and D. Labate, *Optimally sparse representations of 3D data with C^2 surface singularities using Parseval frames of shearlets*, preprint.

9. A. C. Kak and Malcolm Slaney, *Principles of Computerized Tomographic Imaging*, IEEE Press, 1988.

10. P. Kittipoom, G. Kutyniok, and W.-Q. Lim, *Construction of Compactly Supported Shearlet Frames*, Constr. Approx. **35** (2012), 21–72.

11. S. Kuratsubo, *On pointwise convergence of Fourier series of the indicator function of D dimensional ball*, J. Fourier Anal. Appl. **16** (2010), 52–59.

12. G. Kutyniok, J. Lemvig, and W.-Q. Lim, *Compactly supported shearlet frames and optimally sparse approximations of functions in $L^2(\mathbb{R}^3)$ with piecewise C^α singularities*, preprint.

13. G. Kutyniok, J. Lemvig, and W.-Q. Lim, *Compactly Supported Shearlets*, Approximation Theory XIII (San Antonio, TX, 2010), Springer Proc. Math. **13** (2012), 163–186.

14. G. Kutyniok and W.-Q. Lim, *Compactly Supported Shearlets are Optimally Sparse*, J. Approx. Theory **163** (2011), 1564–1589.

15. M. A. Pinsky, N. K. Stanton, P. E. Trapa, *Fourier series of radial functions in several variables*, J. Funct. Anal. **116** (1993), 111–132.

16. E. M. Stein, G. Weiss, *Introduction to Fourier analysis on Euclidean spaces*, Princeton University Press, Princeton, N.J., 1971.

Shearlet Multiresolution and Multiple Refinement

Tomas Sauer

Abstract Starting from the concept of filterbanks and subband coding, we present an entirely digital approach to shearlet multiresolution which is not a discretization of the continuous transform but is naturally connected to the filtering of discrete data, the usual procedure in digital signal processing. It will be shown that a full analogy of multiresolution analysis (MRA) can be derived also for shearlets as a special instance of multiple MRA (MMRA) based on cascading a finite number of filterbanks. In this discrete shearlet transform, the MMRA concept goes hand in hand with shear-based scaling matrices of a particularly appealing and simple geometry. Finally, also some application issues of such discrete transformations will be considered briefly.

Key words: Filterbank, Multiresolution analysis, Multiple MRA, Shearlet MRA, Subdivision

1 Introduction

Like the continuous wavelet transform, the continuous shearlet transform, now familiar from all the other chapters of this book, is applicable in practice by "simply" discretizing the transform parameters. There are even definitely lots of interesting applications where such discretizations are of use, theoretically as well as practically.

Nevertheless, the approach pursued here is quite radically different and aims at reproducing fundamental aspects of the discrete wavelet transforms, namely, the *pyramid scheme* and the concept of *multiresolution analysis*. These two ideas form

T. Sauer
Lehrstuhl für Numerische Mathematik, Justus-Liebig-Universität Gießen,
Heinrich-Buff-Ring 44, 35392 Gießen, Germany
e-mail: Tomas.Sauer@math.uni-giessen.de

G. Kutyniok and D. Labate (eds.), *Shearlets: Multiscale Analysis for Multivariate Data*, 199
Applied and Numerical Harmonic Analysis, DOI 10.1007/978-0-8176-8316-0_6,
© Springer Science+Business Media, LLC 2012

the foundation for the practical implementation of (literally) Fast Wavelet Transforms (FWT) with essentially linear complexity entirely by means of *filterbanks*. A very nice and highly recommended introduction to the filterbank wavelet approach for univariate, onedimensional signals is given in [39], also [27] gives some valuable information on filterbanks and subband coding. In this approach, the analytic properties or the explicit form of refinable functions and wavelets do not appear expicitly in the numerical computations which can actually be performed completely without any analysis behind, but take the role of a theoretical, analyticals *justification* for these computations.

The main goal of this presentation here is to provide a sufficiently general framework of filterbanks and the associated discrete manipulations that allows for an extension to shearlets. Indeed, most of the material here will be dedicated to providing exactly that type of background, from which the shearlet multiresolution will then follow like some sort of an extended corollary. Nevertheless, the hope is that this approach will offer some insight into the ideas and intuitions that lead to [22] but had to be sacrificed there to the more important goal of providing details and proofs of this discrete shearlet approach. Here, we can make use of the liberty to provide a more introductory approach and background material.

The story here is told from a very particular if not peculiar perspective, without distraction and with a very narrow focus on these particular aspects and the connection between them. I have no ambition or intention to make this a complete presentation of all more or less relevant facts nor do I want to create any type of survey on filterbanks, subdivision, or multiresolution. As a consequence, everything that does not fit into the narrow focus of providing this particular connection is omitted, not as a judgement of quality, value, or interest, but simply because it has no place within this presentation. An excellent exposition of this general narrative problem can be found in the initial chapter of [34].

The content of this story is very simple, natural, and straightforward. We begin with recalling some basic facts and definitions on filterbanks for fairly general decimations and show how they can be conveniently described and studied by means of symbol calculus, the twin sister of the z-transform that is so popular in engineering. Filterbanks decompose and reassemble signals and the good ones among them are of course those which reconstruct the original signal as long as no changes are applied to the decomposition coefficients. Such filterbanks can be easily characterized and there is a systematic way of building such filterbanks by specifying only a small partition of the filterbank.

This small part from which the full filterbank emerges can be chosen to be the mask of a stationary subdivision scheme which gives us an excuse to have a look at stationary subdivision schemes and their useful consequences: refinability and multiresolution. They connect back to the filterbank and allow for an analytic decomposition of *functions* by means of *discrete* operations on discrete coefficient signals. For the purpose of shearlets, stationary subdivision is not flexible enough, however. Therefore, we pass to a slight generalization called *multiple subdivision schemes*. Also in this situation there exists a notion of refinability; however, in a somewhat generalized sense, as well as a decomposition of functions and signals, but now into a tree of functions and signals. Once all this is settled, shearlets reduce to very particular and even simple choice of the dilation factors in this multiple subdivision filterbank framework. The nice consequence is that many of the things that so far were simply generalizations find their *geometric* meaning as terms describing *directional* components in interpolatory and orthogonal discrete shearlet constructions. In addition, shearlets always allow for canonical tensor product constructions and thus for efficient though nontrivial implementations.

And that's enough of the outline—let's start to have fun with the math ...

2 Filters and Filterbanks

In principle, a *filter* is just an operator that maps a signal class to itself or maybe even another signal class. As such a definition is clearly too general and thus too vague for our purposes, let us become a little bit more specific here. To that end, we denote by $\ell(\mathbb{Z}^s)$ the vector space of all functions from $\mathbb{Z}^s$ to $\mathbb{R}$, i.e., the space of all bi-infinite sequences. Moreover, we will use $\ell_p(\mathbb{Z}^s)$ for all sequences with finite p-norm, $1 \le p \le \infty$, and $\ell_{00}(\mathbb{Z}^s)$ for those with finite support. A filter $F : \ell(\mathbb{Z}^s) \to \ell(\mathbb{Z}^s)$ is called an *LTI filter* (Linear and Time Invariant) provided that it acts as a linear operator and commutes with translation:

$$(Fc)(\cdot + \alpha) = F(c(\cdot + \alpha)), \qquad c \in \ell(\mathbb{Z}^s), \quad \alpha \in \mathbb{Z}^s.$$

With the *translation operator* τ_α, defined as $\tau_\alpha c = c(\cdot + \alpha)$, this is more conveniently written as $F\tau_\alpha = \tau_\alpha F$. In the spirit of Hamming [17], the word "(digital) filter" will always mean an LTI filter. Any filter can be written as a convolution

$$Fc = f * c = \sum_{\alpha \in \mathbb{Z}^s} f(\cdot - \alpha) c(\alpha)$$

with the *impulse response* $f \in \ell(\mathbb{Z}^s)$ which is obtained as $f = F\delta$, where $\delta = (\delta_{0,\alpha} : \alpha \in \mathbb{Z}^s)$ stands for the engineer's good friend, the *pulse signal*. A filter is called an Finite Impulse Response (FIR) if $f \in \ell_{00}(\mathbb{Z}^s)$.

Good and substantial introductions to signal processing and filters can be found for example in [17, 27, 37, 39].

2.1 Filterbanks

Filterbanks add a nice little twist to the convolution concept of filters which is usually referred to as *subband coding*. To that end, let $M \in \mathbb{Z}^{s \times s}$ be an *expanding* integer matrix, that is, it has all eigenvalues greater than one in modulus. The set $M\mathbb{Z}^s \subset \mathbb{Z}^s$ forms a subgrid of $\mathbb{Z}^s$ and even a proper subgrid. One can decompose $\mathbb{Z}^s$ into an overly of shifts of this subgrid, namely as

$$\mathbb{Z}^s = \bigcup_{\varepsilon \in E_M} \varepsilon + M\mathbb{Z}^s, \qquad E_M := M[0,1)^s \cap \mathbb{Z}^s. \tag{1}$$

In particular, $0 \in E_M$ and $\varepsilon = 0$ corresponds to the subgrid $M\mathbb{Z}^s$. It is indeed very easy to see that the quotient space $\mathbb{Z}^s/M\mathbb{Z}^s$ has $|\det M|$ elements and that E_M is a set of representers for this quotient space. To that end, use the well-known *Smith decomposition* $M = PDQ$ of M, cf. [28], where P and Q are *unimodular*, i.e., $|\det P| = |\det Q| = 1$, and D is a nonzero diagonal matrix; it can even be made a *normal form* by requiring successive nonzero diagonal elements to divide their predecessors. In [3], this result is referred to as the *invariant factor theorem*. All this works in general for matrices over a euclidean ring, and the proof of the Smith factorization is a nice variant of Gaussian elimination coupled with euclidean division (with remainder). Now note that $\alpha - \beta \in M\mathbb{Z}^s$ is equivalent to $P^{-1}\alpha - P^{-1}\beta \in D\mathbb{Z}^s$ so that the quotient spaces $\mathbb{Z}^s/M\mathbb{Z}^s$ and $\mathbb{Z}^s/D\mathbb{Z}^s$ are isomorphic. Moreover, the representer for some $\alpha \in \mathbb{Z}^s$ is obtained as

$$\varepsilon := \alpha - M\lfloor M^{-1}\alpha \rfloor = M\left(M^{-1}\alpha - \lfloor M^{-1}\alpha \rfloor\right) \in M[0,1)^s,$$

and the equivalence classes are disjoint since $\varepsilon - \eta = M\beta$ with $\varepsilon, \eta \in M[0,1)^s$ implies directly that $\beta \in (-1,1)^s \cap \mathbb{Z}^s$, hence $\beta = 0$. Having observed that, we only need to substitute (1) into the convolution definition of the filter to obtain that

$$(f * c)(\alpha) = (f * c)(\varepsilon + M\beta) = (f_\varepsilon * c)(M\beta) \tag{2}$$

depending on the "parity" ε of α. Another consequence of the Smith decomposition and the reason why it is often also referred to as the *fundamental theorem for finite groups*, cf. [23] is the useful identity

$$\frac{1}{|\det M|} \sum_{\varepsilon \in E_M} e^{2\pi i \varepsilon^T M^{-T} \varepsilon'} = \delta_{\varepsilon',0}, \qquad \varepsilon' \in E'_M := E_{M^T}, \tag{3}$$

from which it follows that the *Fourier matrix*

$$F_M := \left[e^{2\pi i \varepsilon^T M^{-T} \varepsilon'} \; : \; \begin{array}{c} \varepsilon' \in E'_M \\ \varepsilon \in E_M \end{array} \right]$$

is unitary up to the factor $|\det M|$.

When studying filterbanks, cf. [39], it is common practice to use the *subsampling operator* oder *downsampling* $\downarrow_M$, defined as $\downarrow_M c = c(M\cdot)$, and then to define the *filterbank* as the operator which maps c to the vector

$$(\downarrow_M (f_\varepsilon * c) \; : \; \varepsilon \in E_M), \tag{4}$$

where, at least in principle, f_ε, $\varepsilon \in E_M$, can be impulse responses of more or less arbitrary filters. It is worthwhile, however, to keep in mind that (ideally) the output of the filterbank contains the same amount of information as the original signal. This is due to the fact that the increase of information by the $|\det M|$ different filters is compensated by the subsequent subsampling.

Given the output $(c_\varepsilon \ : \ \varepsilon \in E_M)$, called *subband decomposition* generated by the filterbank, it is a natural task to revert the filterbank process and to hopefully reconstruct the signal from which the vector of subbands was generated. To that end, we invert the downsampling by means of *upsampling*,

$$\uparrow_M c(\alpha) = \begin{cases} c\left(M^{-1}\alpha\right), & \alpha \in M\mathbb{Z}^s, \\ 0, & \alpha \notin M\mathbb{Z}^s, \end{cases}$$

then convolve the result with another family g_ε, $\varepsilon \in E_M$, of *synthesis filters* and sum these components:

$$\sum_{\varepsilon \in E_M} g_\varepsilon * \uparrow_M c_\varepsilon. \tag{5}$$

Schematically, if we write $E_M = \{\varepsilon, \dots, \eta\}$, the filterbank looks as follows:

$$
\begin{array}{ccccccccc}
 & \nearrow & \boxed{F_\varepsilon} & \to \boxed{\downarrow_M} & \to c_\varepsilon \to & \boxed{\uparrow_M} & \to \boxed{G_\varepsilon} & \searrow & \\
c \to & \odot & \vdots & \vdots & \vdots & \vdots & \vdots & & \oplus \to c_* \\
 & \searrow & \boxed{F_\eta} & \to \boxed{\downarrow_M} & \to c_\eta \to & \boxed{\uparrow_M} & \to \boxed{G_\eta} & \nearrow &
\end{array}
$$

$$\text{Analysis} \qquad\qquad\qquad \text{Synthesis}$$

and a natural and reasonable minimal condition on the filterbank is that of *perfect reconstruction*: the *synthesis filterbank* is the inverse of the *analysis filterbank* or simply $c_* = c$. There is a trivial way to build a perfect reconstruction filterbank, namely the "lazy" one obtained by setting $F_\varepsilon = G_\varepsilon = \tau_\varepsilon$. This follows from the simple identity

$$\sum_{\varepsilon \in E_M} \tau_\varepsilon \uparrow_M \downarrow_M \tau_\varepsilon = I. \tag{6}$$

For later use, let us finally rewrite a summand of (5) explicitly as

$$g * \uparrow_M c = \sum_{\alpha \in M\mathbb{Z}^s} g\left(\cdot - \alpha\right) c\left(M^{-1}\alpha\right) = \sum_{\alpha \in \mathbb{Z}^s} g\left(\cdot - M\alpha\right) c(\alpha). \tag{7}$$

We will reencounter this type of convolution-like structures that appears in the synthesis filterbank as *subdivision operators* later.

The way how we introduced filterbanks here implies that the roles of their components are completely symmetric and exchangeable. In practice, however, some of the filters will be *low-pass filters*, i.e. they reproduce constants and thus have an "averaging" nature, and some will be *high-pass filters* which annihilate constant data. In the wavelet context, and also in the shearlet case, the filterbank usually consists of a single low-pass filter and all the others are high-pass ones. In view of this structure, it is convenient to denote the low-pass components by f_0 and g_0, respectively, and to use g_ε and f_ε, $\varepsilon \in E_M \setminus \{0\}$, for the high-pass filters.

In most practical cases all the filter sequences f_ε and g_ε are finitely supported, hence FIR filters, and can easily be implemented. Even if their IIR (Infinite Impulse

Response) analogs exist in the world of rational filters which can be realized by means of delayed feedback, those filters are somewhat more tricky and additional issues like stability have to be considered. Because of that, we make an important assumption that serves the purpose of making our life considerably easier.

> All impulse responses considered from now on are finitely supported, hence belong to $\ell_{00}(\mathbb{Z}^s)$.

2.2 Symbols and Transforms

For a better mathematical treatment of filterbanks, it is useful to switch to algebraic representations that turn sequences into formal Laurent series. This is either the *z-transform*

$$\ell(\mathbb{Z}^s) \ni c \mapsto c^{\flat}(z) = \sum_{\alpha \in \mathbb{Z}^s} c(\alpha) z^{-\alpha}$$

or the *symbol*

$$\ell(\mathbb{Z}^s) \ni c \mapsto c^{\sharp}(z) = \sum_{\alpha \in \mathbb{Z}^s} c(\alpha) z^{\alpha}.$$

Obviously, there is not much of a difference between the z-transform calculus and the symbol calculus, and all results obtained by one of them can be equivalently expressed by the other. While the z-transform is more popular in signal processing, cf. [14, 17], symbol calculus has become the established method in subdivision even before the fundamental monograph [3], which is why we will rely here on symbols rather than on the z-transform.

Symbols and z-transform are *formal* Laurent series, that is, in the spirit of Gauß [15] convergence of these series is not considered. If, however, c is finitely supported, then both become *Laurent polynomials*. Let Λ stand for the ring of Laurent polynomials and $\Pi = \mathbb{C}[z_1, \ldots, z_s]$ for the ring polynomials, then any element of Λ can be written as a polynomial from Π multiplied by a Laurent monomial with non-positive exponent. These monomials, more precisely their nonzero multiples, are the *units* in Λ, denoted by Λ^*. Hence, $\Lambda = \Lambda^* \Pi$ seems not to be too far away from Π. Nevertheless, the two rings differ quite substantially, in particular when ideals in these rings are considered, cf. [30]. In particular, there is no notion of degree for Laurent polynomials while for polynomials there exists a rich variety of degrees, usually all in the framework of *graded rings*, cf. [12]. These different notions of degree can be actively used in the context of Gröbner bases, see [4].

Symbols are also turned into trigonometric polynomials by the substitution $z = e^{i\theta}$, yielding the trigonometric polynomial or series

$$\widehat{c}(\theta) = c^{\sharp}\left(e^{i\theta}\right) = \sum_{\alpha \in \mathbb{Z}^s} c(\alpha) e^{i\langle \alpha, \theta \rangle}, \qquad \theta \in [-\pi, \pi]^s.$$

Since a Laurent polynomial is uniquely determined by its behavior on the *torus* $\mathbb{T}^s = \{z \in \mathbb{C}^s : |z_j| = 1\}$, the two concepts are again mutually exchangeable and the choice between them is once more essentially a matter of taste. Nevertheless, the algebra on (Laurent) polynomials provides a somewhat larger toolbox of methods and techniques which allow at least for some analogies of the univariate factorizations. This is another reason why this presentation here is based entirely on symbols.

The first simple exercise in symbol calculus is to note that

$$(f * c)^{\sharp}(z) = f^{\sharp}(z) g^{\sharp}(z) \tag{8}$$

and that

$$(\uparrow_M c)^{\sharp}(z) = c^{\sharp}\left(z^M\right), \qquad (\downarrow_M c)^{\sharp}\left(z^M\right) = \frac{1}{|\det M|} \sum_{\varepsilon' \in E'_M} c^{\sharp}\left(e^{-2\pi i M^{-T} \varepsilon'} z\right). \tag{9}$$

Here, z^M stands for the vector of Laurent polynomials consisting of z^{m_j}, $j = 1, \ldots, s$, where m_j are the columns of the matrix M and the product $e^{-2\pi i M^{-T} \varepsilon} z$ has to be understood in a componentwise sense. While the formula for $(\uparrow_M c)^{\sharp}$ is really immediate, the proof of the expression for $(\downarrow_M c)^{\sharp}$ makes use of the Fourier character identity (3) to compute

$$\begin{aligned}
(\downarrow_M c)^{\sharp}\left(z^M\right) &= \sum_{\alpha \in \mathbb{Z}^s} c(M\alpha) z^{M\alpha} = \sum_{\alpha \in \mathbb{Z}^s} \sum_{\varepsilon \in E_M} \delta_{\varepsilon,0} c(\varepsilon + M\alpha) z^{\varepsilon + M\alpha} \\
&= \frac{1}{|\det M|} \sum_{\varepsilon' \in E'_M} \sum_{\alpha \in \mathbb{Z}^s} \sum_{\varepsilon \in E_M} e^{2\pi i \varepsilon'^T M^{-1}(\varepsilon + M\alpha)} c(\varepsilon + M\alpha) z^{\varepsilon + M\alpha} \\
&= \frac{1}{|\det M|} \sum_{\varepsilon' \in E'_M} \sum_{\beta \in \mathbb{Z}^s} c(\beta) \left(e^{-2\pi i M^{-T} \varepsilon'} z\right)^{\beta},
\end{aligned}$$

which is the second identity of (9).

By means of symbol calculus, we can now give an algebraic description of the filterbank in terms of matrices of Laurent polynomial or matrix-valued Laurent polynomials, which is the same anyway. To that end, we write the signal c in its *polyphase form*

$$p_c(z) := \left[c^{\sharp}\left(e^{-2\pi i M^{-T} \varepsilon'} z\right) : \varepsilon' \in E'_M \right],$$

which implies that the analysis filterbank has the representation

$$\left[c_{\varepsilon}^{\sharp}\left(z^M\right) : \varepsilon \in E_M \right] = F(z) p_c(z), \tag{10}$$

where

$$F(z) = \frac{1}{|\det M|} \left[f_{\varepsilon}^{\sharp}\left(e^{-2\pi i M^{-T} \varepsilon'} z\right) : \begin{array}{l} \varepsilon \in E_M \\ \varepsilon' \in E'_M \end{array} \right], \tag{11}$$

is called the *modulation matrix* associated to the filterbank given by f_{ε}, $\varepsilon \in E_M$.

The polyphase form of a signal is in 1–1 correspondence with the signal itself. Clearly, the polyphase vector is read off directly from the symbol, while, conversely, (6) and (9) yield for any signal c that

$$
c^\sharp(z) = \frac{1}{|\det M|} \sum_{\varepsilon \in E_M} z^{-\varepsilon} \sum_{\varepsilon' \in E'_M} (\tau_\varepsilon c)^\sharp \left(e^{-2\pi i M^{-T} \varepsilon'} z \right)
$$

$$
= \frac{1}{|\det M|} \sum_{\varepsilon \in E_M} z^{-\varepsilon} \sum_{\varepsilon' \in E'_M} \sum_{\beta \in \mathbb{Z}^s} c(\varepsilon + \beta) \left(e^{-2\pi i M^{-T} \varepsilon'} z \right)^\beta,
$$

and consequently

$$
c^\sharp(z) = \sum_{\varepsilon' \in E'_M} c^\sharp \left(e^{-2\pi i M^{-T} \varepsilon'} z \right) \frac{1}{|\det M|} \sum_{\varepsilon \in E_M} \left(e^{2\pi i M^{-T} \varepsilon'} z \right)^\varepsilon = \langle p_c(z), F_M u(z) \rangle,
$$
(12)

where $u \in \Lambda^{E_M}$ such that $u_\varepsilon(z) = z^\varepsilon$, $\varepsilon \in E_M$, allows us to reconstruct the signal from *all* its polyphase components.

The synthesis filterbank, on the other hand, is more easy to write in symbol form, namely

$$
c^\sharp_*(z) = \sum_{\varepsilon \in E_M} g^\sharp_\varepsilon(z) c^\sharp_\varepsilon \left(z^M \right).
$$

For comparison, however, it is more convenient though unnecessary to consider the polyphase form

$$
p_{c_*}(z) = \left[c^\sharp_* \left(e^{-2\pi i M^{-T} \varepsilon'} z \right) : \varepsilon' \in E'_M \right]
$$

$$
= \left[\sum_{\varepsilon \in E_M} g^\sharp_\varepsilon \left(e^{-2\pi i M^{-T} \varepsilon'} z \right) c^\sharp_\varepsilon \left(z^M \right) : \varepsilon' \in E'_M \right]
$$

$$
= \left[g^\sharp_\varepsilon \left(e^{-2\pi i M^{-T} \varepsilon'} z \right) : \begin{matrix} \varepsilon' \in E'_M \\ \varepsilon \in E_M \end{matrix} \right] \left[c^\sharp_\varepsilon \left(z^M \right) : \varepsilon \in E_M \right]
$$

$$
=: G(z) \left[c^\sharp_\varepsilon \left(z^M \right) : \varepsilon \in E_M \right],
$$

with

$$
G(z) = \left[g^\sharp_\varepsilon \left(e^{-2\pi i M^{-T} \varepsilon'} z \right) : \begin{matrix} \varepsilon' \in E'_M \\ \varepsilon \in E_M \end{matrix} \right]
$$
(13)

being the modulation matrix associated to the synthesis filterbank based on g_ε, $\varepsilon \in E_M$. A simple consequence of all this formalism is now that $p_{c_*}(z) = G(z) F(z) p_c(z)$ and we have the following fundamental result for filterbanks.

Theorem 1. *The filterbank provides perfect reconstruction if and only if* $G(z)$ $F(z) = I$.

This of course somewhat restricts the possible choice of filterbanks as perfect reconstruction requires the existence of an inverse that again has Laurent polynomials as components. But the question whether a square matrix over a ring is invertible or not can be easily answered by looking at its determinant.

Theorem 2. *Let R be a ring and $A \in R^{n \times n}$. Then A is invertible with an inverse in $R^{n \times n}$ if and only if A is* unimodular, *that is $\det A \in R^*$, the set of* units *of R.*

The proof of this result is as simple as well known, cf. again [28]. Indeed, if A is invertible, then $1 = \det I = \det A (\det A)^{-1}$ implies that $\det A \in R$, while the converse is obtained from Cramer's rule which says that all components of A^{-1} lie in $(\det A)^{-1} R$.

In order to be part of a perfect reconstruction filterbank, the modulation matrices of both the analysis and the synthesis part have to be invertible, hence, unimodular. If the matrices are considered in Λ, this means that after proper normalization $\det G(z) = z^\alpha$ for some $\alpha \in \mathbb{Z}^s$, if we restrict ourselves to Π, then only $\det G(z) = 1$ is allowed up to normalization. Since the factor z^α corresponds (only?) to a shift, it is often neglected except in the case when causality of the filters plays a role. Recall that *causal* filter is one that is only supported on a subset of $\mathbb{Z}_+^s$, which means that its symbol is a polynomial. Practically, a causal filter for time signals only uses information from the past which has some advantages in terms of realistic implementation.

Suppose that $F(z)$ and $G(z)$ are modulation matrices of a perfect reconstruction filterbank. Clearly, since $G(z)F(z) = I$, this implies that $F(z)$ and $G(z)$ are both unimodular. But what about the converse? Obviously, if, for example, $F(z)$ is unimodular, then there exists $F^{-1}(z) \in \Lambda$, which, however, does not automatically imply that this matrix-valued Laurent polynomial has the structure (13) of a synthesis filterbank. But fortunately it has.

Lemma 1. *If $F(z)$ is the modulation matrix of an analysis filterbank, then there exist $g_\varepsilon^\sharp \in \Lambda$, $\varepsilon \in E_M$, such that*

$$F^{-1}(z) = \left[g_\varepsilon^\sharp \left(e^{-2\pi i M^{-T} \varepsilon'} z \right) \; : \; \begin{matrix} \varepsilon' \in E_M' \\ \varepsilon \in E_M \end{matrix} \right]. \tag{14}$$

Even if this result is probably well known, a simple proof of it might be illustrative.

Proof. We set $g_\varepsilon^\sharp := \left(F^{-1}(z) \right)_{0,\varepsilon}$, $\varepsilon \in E_M$, and show that the right-hand side of (14), defined by these Laurent polynomials indeed gives an inverse of $F(z)$. First we note that

$$|\det M| \, \delta_{0,\varepsilon'} = \left(F^{-1}(z) F(z) \right)_{0,\varepsilon'} = \sum_{\varepsilon \in E_M} g_\varepsilon^\sharp(z) f_\varepsilon^\sharp \left(e^{-2\pi i M^{-T} \varepsilon'} z \right). \tag{15}$$

Setting $\varepsilon' = 0$ in (15) and then replacing z by $e^{-2\pi i M^{-T} \eta'} z$, $\eta' \in E_M'$, we immediately see that also

$$\sum_{\varepsilon \in E_M} g_\varepsilon^\sharp \left(e^{-2\pi i M^{-T} \eta'} z \right) f_\varepsilon^\sharp \left(e^{-2\pi i M^{-T} \eta'} z \right) = |\det M|, \qquad \eta' \in E_M'. \tag{16}$$

On the other hand, with $\varepsilon' \neq 0$ and again the substitution $z \to e^{-2\pi i M^{-T}\eta'}z$, $\eta' \neq \varepsilon'$, we get that

$$\sum_{\varepsilon \in E_M} g_\varepsilon^\sharp \left(e^{-2\pi i M^{-T}\eta'}z\right) f_\varepsilon^\sharp \left(e^{-2\pi i M^{-T}(\varepsilon'+\eta')}z\right) = 0,$$

and since the set $\{\eta' + \varepsilon' : \varepsilon' \neq 0\}$ is isomorphic to $E_M' \setminus \{\eta'\}$ modulo M, we have that

$$\sum_{\varepsilon \in E_M} g_\varepsilon^\sharp \left(e^{-2\pi i M^{-T}\eta'}z\right) f_\varepsilon^\sharp \left(e^{-2\pi i M^{-T}\varepsilon'}z\right) = 0, \qquad \eta' \neq \varepsilon'. \qquad (17)$$

But (16) and (17) are precisely expressing the fact that the matrix $G(z)$ defined by the right-hand side of (14) satisfies $G(z)F(z) = I$, which is what was claimed in the lemma. □

An identical argument also shows that the inverse of the modulation matrix of any synthesis filterbank is the modulation matrix of an analysis filterbank. Consequently, any "partial," that is, analysis or synthesis, filterbank with a unimodular modulation matrix can be completed to a perfect reconstruction filterbank. We can summarize these observations in the following theorem.

Theorem 3. *An analysis or synthesis filterbank F or G can be completed to a perfect reconstruction filterbank if and only if $F(z)$ or $G(z)$ is unimodular, respectively.*

This theorem can be considered a general extension principle of filterbanks in the sense of the *unitary extension principles* introduced in [33], see [2] for a survey on such techniques. These results are usually stated and proved in terms of Fourier transforms instead of symbol calculus. In the shearlet context, such extension principles were considered in [19].

Another reasonable and natural idea is to use essentially *the same* filterbank for synthesis and reconstruction. Of course, "essentially" is the keyword here since (11) and (13) show that they clearly have a different, transposed, structure. So the question is whether $G(z) = F^H(z)$ is possible, at least up to constant. This means that $F(z)$ or $G(z)$ has to be not only unimodular but *unitary*, or, if the equivalent restriction $F\left(e^{2\pi i \theta}\right)$, $\theta \in \mathbb{R}^s$, to the unit circle is considered, then F and G are called *paraunitary*. Paraunitary matrix-valued Laurent polynomials, especially for $s = 1$, are well-studied objects in signal processing and routines for their treatment can be found in many programs and toolboxes.

2.3 Filterbanks by Matrix Completion

The symbol calculus of the preceding section suggests a way to define and construct filterbanks. As Theorem 1 tells us, a filterbank gives perfect reconstruction if $F(z)$ or $G(z)$ are mutual inverses of each other which by Theorem 2 means that they both are unimodular matrices. Moreover, $F(z)$ already determines $G(z) = F^{-1}(z)$ and vice versa, so it suffices to know one of these matrices, for example $F(z)$.

Therefore, a general method for the construction of filterbanks is as follows: start with one filter, for convenience the one indexed by $\varepsilon = 0$ to obtain the row

$$g_0(z) := \left[g_0^{\sharp} \left(e^{-2\pi i M^{-T} \varepsilon'} z \right) \ : \ \varepsilon' \in E_M' \right] \tag{18}$$

of the synthesis modulation matrix $G(z)$. That we use $G(z)$ here is owed to the fact that subdivision schemes naturally play the role of low-pass synthesis filters and thus make a good and natural starting point for synthesis filterbanks.

Now, "simply" choose the other rows, $\left[g_\varepsilon^{\sharp} \left(e^{-2\pi i M^{-T} \varepsilon'} z \right) \ : \ \varepsilon' \in E_M' \right]$, $\varepsilon \in E_M \setminus \{0\}$ such that the resulting matrix $G(z)$ is unimodular. This requirement is easy to formulate but very hard to achieve, in particular, in several variables. In [3], an ingenious general method is given that works under quite general circumstances, however, only in one variable where $\Pi = \mathbb{C}[z]$, the ring of univariate polynomials, is a *euclidean* and *principal ideal* ring which allows for a Smith factorization of (Laurent) polynomial matrices, too.

Nevertheless, we can get, by fairly elementary techniques, an idea how and under which circumstances such a matrix completion works in the general case too, at least in the case of polynomial symbols.

There is an immediate necessary condition on the row vector (18) that we want to complete. Since the matrix $G(z)$ is desired to be nonsingular, there has to be, for any vector $f(z) = (f_\varepsilon(z) \ : \ \varepsilon \in E_M) \in \Pi^{E_M}$ a vector $a(z) \in \Pi^{E_M'}$ such that $f(z) = G(z)a(z)$. Choosing $f_0(z) = 1$, this especially implies that

$$1 = \sum_{\varepsilon' \in E_M'} a_{\varepsilon'}(z) g_0^{\sharp} \left(e^{-2\pi i M^{-T} \varepsilon'} z \right),$$

that is, $1 \in \left\langle g_0^{\sharp} \left(e^{-2\pi i M^{-T} \varepsilon'} z \right) \ : \ \varepsilon' \in E_M' \right\rangle$, the ideal generated by the entries of the zero row of $G(z)$. In the language of [31], it means that the row $g_0(z)$ of $G(z)$ is *unimodular*. Recall that

$$\langle F \rangle := \sum_{f \in F} a_f(z) f(z)$$

denotes, as usually, the *ideal generated* by the (finite) set F of (Laurent) polynomials. Information about ideals and how to handle them computationally can be found, for example in [1, 4, 5, 6]. Geometrically, unimodular vectors have the property that they have no common zeros, or, in the wording of algebraic geometry, that the variety associated to the ideal they generate is empty. This is the geometric equivalence of the fact that 1 is contained in the ideal and can be checked efficiently by testing whether a reduced Gröbner basis of the ideal consists of the constant polynomial, cf. [4].

The really fascinating and deep fact about unimodular vectors, however, is that even the converse of our necessary condition "unimodularity" holds true.

Theorem 4. *A row vector $g(z) \in \Pi^n$ can be completed to a unimodular matrix $G(z) \in \Pi^{n \times n}$ if and only if it is unimodular.*

Algorithmic proofs of this result can be found in [25] as well as in [31], and Park's Ph.D. thesis [32] even provides the explicit connection to filterbanks, even if "only" the case $M = 2I$ is treated there. Note, however, that Theorem 4 itself does not yet imply that the completion is the modulation matrix of a filterbank, i.e., that all its completed rows are of the form $\left[g^\sharp_\varepsilon \left(e^{-2\pi i M^{-T} \varepsilon'} z \right) \; : \; \varepsilon' \in E'_M \right]$, $\varepsilon \in E_M \setminus \{0\}$. Nevertheless, in principle all this can be done and it can be done *algorithmically* even if these algorithms will have to rely on some time-consuming techniques and methods from Computer Algebra. The general and fundamental backbone of matrix completions is the following algebraic principle that can be verified, for example, by Gröbner basis or H-basis computations.

> A low-pass-analysis or synthesis filter f_0 or g_0 can be extended into a full per-fect reconstruction filterbank if and only if $1 \in \left\langle f^\sharp_0 \left(e^{-2\pi i M^{-T} \varepsilon'} z \right) \; : \; \varepsilon' \in E'_M \right\rangle$
> or $1 \in \left\langle g^\sharp_0 \left(e^{-2\pi i M^{-T} \varepsilon'} z \right) \; : \; \varepsilon' \in E'_M \right\rangle$, respectively.

In yet other words: to obtain a perfect reconstruction filterbank, all we have to do is to provide a suitable low-pass synthesis filter, the rest can be obtained almost automatically from it. Low-pass synthesis filters, on the other hand, are subdivision schemes which we will consider in more detail in Sect. 3. This may be a good motivation to continue reading after that brief algebraic excursion.

2.4 Subbands and Multiresolution

Filterbanks have provided us with a method to decompose a signal into subbands and, at least when the filterbank is a perfect reconstruction one, to reconstruct the signal again from these subbands. As mentioned before, the appearance of subsam-pling operators $\downarrow_M$ after the application of the filters f_ε, $\varepsilon \in E_M$, takes care that all the subbands *together* contain "the same amount" of information as the original signal. For example, when the signal has finite support, then any subband contains (up to overlap effects of the filters which depend on the support size of the filters) essentially $(1/\#E_M)$th of the nonzero values. In other words, subbands do what is expected of a reasonable decomposition.

The usual model of a filterbank is that F_0 is a low-pass filter and that all the other F_ε are high-pass filters which can be expressed as

$$f_\varepsilon * 1 = \delta_{\varepsilon,0} 1, \qquad \varepsilon \in E_M. \tag{19}$$

In this situation, the result $c_0 = \downarrow_M f_0 * c$ of low-pass filtering yields a smoothed, averaged signal, while the other subbands $c_\varepsilon = \downarrow_M f_\varepsilon * c$, $\varepsilon \in E_M \setminus \{0\}$, cover the oscillatory part of the signal which corresponds to high frequency content. Nev-ertheless, this is only a *model* and the *motivation* for what is going to follow, but formally there is no need to have any distinction between low-pass and high-pass filters—even if in applications this is definitely useful and to some extent responsible for the good performance of wavelet multiresolution.

Keeping this in mind, we divide E_M into two disjoint subsets $E_\searrow$ and $E_\nearrow$ where, according to our model, $E_\searrow$ indexes the "low-pass" components of the filterbank and $E_\nearrow$ the "high-pass" ones. Once more: formally it is not necessary that these filters satisfy (19) but it is no disadvantage if they do. Now we perform one step of the analysis filterbank on an initial signal c and obtain "low-pass" and "high-pass" components

$$c^1 = \left(c_\varepsilon^1 := F_\varepsilon c \ : \ \varepsilon \in E_\searrow \right) \qquad \text{and} \qquad d^1 = \left(d_\varepsilon^1 := F_\varepsilon c \ : \ \varepsilon \in E_\nearrow \right).$$

The multiresolution approach now cascades the filterbank by applying it once more to the "low frequency" part c^1 giving

$$c^2 = \left(F_\eta c^1 = F_\eta F_\varepsilon c \ : \ \varepsilon, \eta \in E_\searrow \right), \qquad d^2 = \left(F_\eta c^1 = F_\eta F_\varepsilon c \ : \ \eta \in E_\nearrow, \varepsilon \in E_\nearrow \right),$$

from which the general decomposition rule

$$c^n = \left(F_{\varepsilon_n} \cdots F_{\varepsilon_1} c \ : \ (\varepsilon_1, \ldots, \varepsilon_n) \in E_\searrow^n \right), \tag{20}$$

$$d^n = \left(F_\eta F_{\varepsilon_{n-1}} \cdots F_{\varepsilon_1} c \ : \ (\varepsilon_1, \ldots, \varepsilon_{n-1}) \in E_\searrow^{n-1}, \eta \in E_\nearrow \right), \tag{21}$$

$n \in \mathbb{N}$, follows immediately. This gives a decomposition of the signal c into the components $d^1, \ldots, d^n$ and c^n. And as long as $E_\searrow$ and $E_\nearrow$ really index low-pass and high-pass filters, c^n describes a highly smoothed, averaged, and severely subsampled "coarse" version of c while $d^1, \ldots, d^n$ contain the detail information lost in this process. Schematically this means

$$c \longrightarrow c^1 \longrightarrow c^2 \longrightarrow \cdots \longrightarrow c^n$$
$$\searrow \qquad \searrow \qquad \searrow \qquad \searrow$$
$$d^1 \qquad d^2 \qquad \cdots \qquad d^n$$

where the total amount of information in $d^1, \ldots, d^n$ and c^n *together* is again roughly the same as in the original signal c.

Reconstruction of c from these data makes use of the perfect reconstruction property and cascades of the synthesis part G of the filterbank. Indeed, we combine c^n and d^n into a vector

$$b^n := (c^n, d^n) = (b_\varepsilon \ : \ \varepsilon \in E_M)$$

and pass it through the synthesis part to reconstruct c^{n-1}. After all, we simply invert the decomposition process that generated c^n and d^n from c^{n-1}. Having obtained c^{n-1}, we can combine that with d^{n-1} to generate c^{n-2}, and so on. Hence, the reconstruction process is

$$c^n \longrightarrow c^{n-1} \longrightarrow \cdots c^1 \longrightarrow c,$$
$$\nearrow \qquad \nearrow \qquad \nearrow$$
$$d^n \qquad d^{n-1} \qquad \cdots d^1$$

now as a repeated application of the synthesis filterbank.

The subband information $d^1,\dots,d^n$ and c^n can now be exposed to the usual operations like, for example, thresholding in the case of denoising, before the signal is reconstructed again. Especially for denoising it is of course a good idea to really collect high-pass filters in $E_\nearrow$ as otherwise denoising should not be expected to work.

There is also the special case that $E_\searrow = E_M$ and $E_\nearrow = \emptyset$ which generates only a long vector of many small parts or *packets* $F_{\varepsilon_n}\cdots F_{\varepsilon_1}c$. This is precisely the idea behind the concept of *wavelet packages* and can be found, for example, in [27].

3 Subdivision and Refinability

Originally, subdivision is a simple way to iteratively generate functions on finer and finer discrete grids. When these grids become dense in $\mathbb{R}^s$, the limit of the discrete functions, provided that it exists, will be a function defined on all of $\mathbb{R}^s$. The underlying procedure is indeed very simple: starting with discrete data $c \in \ell(\mathbb{Z}^s)$ defined on the integer grid $\mathbb{Z}^s$, an expanding scaling matrix M and a finitely supported *mask* $a \in \ell_{00}(\mathbb{Z}^s)$, one computes a new sequence

$$c^1 := \mathscr{S}_a c = \sum_{\alpha \in \mathbb{Z}^s} a(\cdot - M\alpha)\, c(\alpha) = a * \uparrow_M c, \qquad (22)$$

which is now considered as a function on the finer grid $M^{-1}\mathbb{Z}^s$, i.e., $c^1(\alpha) \sim f(M^{-1}\alpha)$. It is important to note that the *subdivision operator* $\mathscr{S}_a$ from (22) perfectly fits into the filterbank framework as subdivision operators are the blocks of the synthesis filterbank.

3.1 Convergence and Basic Properties

The subdivision operator $\mathscr{S}_a$ from (22) can be iterated, leading to sequences $c^n = \mathscr{S}_a a^n c$, $n \in \mathbb{N}$, associated to the grids $M^{-n}\mathbb{Z}^s$. Since $M^{-n}\mathbb{Z}^s \to \mathbb{R}^s$, one can try to associate a limit function to this process which is, in the filterbank language, obtained by cascading the synthesis block of the filterbank. To recall the notion of convergence of a stationary subdivision scheme, we define, for a nonsingular matrix $A \in \mathbb{R}^{s \times s}$, the vector of averages

$$\mu(f, A) = \left(|\det A| \int_{A(\alpha + [0,1]^s)} f(t)\, dt \; : \; \alpha \in \mathbb{Z}^s \right),$$

which leads to the following definition.

Definition 1. The subdivision scheme associated to a and M is said to be p-*convergent*, $1 \le p < \infty$, if for any initial data c there exists a limit function $f_c \in L_p(\mathbb{R}^s)$ such that

$$\lim_{n \to \infty} |\det M|^{-n/p} \left\| \mathscr{S}_a^n c - \mu \left(f_c, M^{-n} \right) \right\|_p = 0 \tag{23}$$

and if $f_c \ne 0$ for at least one choice of c.

For $p = \infty$, convergence of subdivision schemes is slightly more tricky. The problem is not (23) but the fact that the limit function has to belong to $C_u(\mathbb{R}^s)$, the space of uniformly continuous, uniformly bounded functions on $\mathbb{R}^s$ for which the norm

$$\|f\|_\infty := \sup_{x \in \mathbb{R}^s} |f(x)|$$

is finite. With this space instead of the usually troublesome $L_\infty(\mathbb{R}^s)$, the notion of convergence given in (23) then covers the full range $1 \le p \le \infty$.

By linearity of the subdivision operator, we can restrict ourselves to the convergence of a particular initial sequence, namely the peak sequence $\delta : \alpha \mapsto \delta_{0\alpha}$ which we already met in the definition of the impulse response and which satisfies the quite trivial identity

$$c = \sum_{\alpha \in \mathbb{Z}^s} c(\alpha) \tau_{-\alpha} \delta = c * \delta.$$

Indeed, the subdivision scheme converges for any $c \in \ell_p(\mathbb{Z}^s)$ if and only if it converges for δ and so there only must exist one *(basic) limit function f* defined by

$$\lim_{n \to \infty} |\det M|^{-n/p} \left\| \mathscr{S}_a^n \delta - \mu \left(f, M^{-n} \right) \right\|_p = 0; \tag{24}$$

limit functions for arbitrary initial data then have the form

$$f_c = f * c := \sum_{\alpha \in \mathbb{Z}^s} f(\cdot - \alpha) c(\alpha). \tag{25}$$

The limit functions of convergent subdivision schemes satisfy an important property which makes them the fundamental building blocks for any multiresolution analysis:

> The limit function f of a convergent subdivision scheme is *refinable*, i.e.,
>
> $$f = \uparrow_M (f * a) = (f * a)(M \cdot) = \sum_{\alpha \in \mathbb{Z}^s} a(\alpha) f(M \cdot - \alpha). \tag{26}$$

So any convergent subdivision scheme gives a refinable function. But also the converse is (partially or essentially—which to choose is a question of point of view) true. To that end, we need another fundamental definition.

Definition 2. A function $f \in L_p(\mathbb{R}^s)$, $1 \le p < \infty$, or $f \in C_u(\mathbb{R}^s)$ for $p = \infty$ is called p-*stable* if there exist constants $0 < A \le B < \infty$ such that for any $c \in \ell_p(\mathbb{Z}^s)$

$$A \, \|c\|_p \le \|f * c\|_p \le B \, \|c\|_p \tag{27}$$

holds, that is, the sequence norm $\|c\|_p$ and the function norm $\|f * c\|_p$ are equivalent.

Stability as well as linear independence of the translates of a refinable function f can be described in terms of the symbol $a^\sharp(z)$ and in both cases this is surprisingly independent of the actual value of p, cf. [20]. What makes stability interesting here is that it is a sufficient condition for the aforementioned converse: Whenever there exists a *stable* solution of the refinement equation (26), the associated subdivision scheme converges.

There are two simple but important special cases of stable functions. One is $p = 2$ and a function with orthonormal integer translates,

$$\int_{\mathbb{R}^s} f(x - \alpha) \, f(x - \beta) \, \mathrm{d}x = \delta_{\alpha,\beta}$$

as then $\|f * c\|_2 = \|c\|_2$ by Parseval's identity, the other is $p = \infty$ and a bounded *cardinal* function, $f(\alpha) = \delta_{\alpha,0}$, hence $f * c(\alpha) = c(\alpha)$, $\alpha \in \mathbb{Z}^s$, and thus

$$\|c\|_\infty \le \|f * c\|_\infty \le \|c\|_\infty \|f\|_\infty.$$

It is also well known that in general the existence of a solution for the refinement equation is not sufficient to yield convergence of the subdivision scheme. One of the simplest examples and the "mother" of many counterexamples is the univariate refinement equation with symbol $a^\sharp(z) = z^2 + 1$ with the nonstable solution $f = \chi_{[0,2]}$ which belongs to any L_p, $1 \le p < \infty$, but nevertheless the subdivision scheme is not convergent as it does not satisfy the necessary condition $a^\sharp(-1) = 0$, cf. [3].

This is one reason why interpolatory subdivision schemes are somewhat special: if they converge, their limit function is cardinal, $f(\alpha) = \delta_{\alpha,0}$, and if a refinable function is cardinal then the associated subdivision automatically converges. But another good, even better reason to consider interpolatory schemes is that they can always be used as building blocks for a perfect reconstruction filterbank. We will detail this in the next section.

3.2 Interpolatory Subdivision and Filterbanks

Subdivision can be considered by its restrictions on the natural decomposition of the grid $M^{-1}\mathbb{Z}^s$ due to (1). Indeed, since

$$M^{-1}\mathbb{Z}^s = \bigcup_{\varepsilon \in E_M} M^{-1}\varepsilon + \mathbb{Z}^s,$$

the integer grid $\mathbb{Z}^s$ is the subgrid of $M^{-1}\mathbb{Z}^s$ in the above decomposition that corresponds to the index $\varepsilon = 0$. The values on this grid are defined by c already and $\mathscr{S}_a$ can either modify them or can leave them untouched. In the latter case, which can be characterized as $a(M\cdot) = \delta$, the subdivision scheme is called *interpolatory* since any convergent interpolatory subdivision scheme interpolates the data: $f_c(\alpha) = c(\alpha)$, $\alpha \in \mathbb{Z}^s$, or, in other words, the limit function f is a *cardinal* function, $f|_{\mathbb{Z}^s} = \delta$.

The great advantage of interpolatory subdivision schemes is that they can quite easily be used to define a perfect reconstruction filterbank, allowing for an easily implemented interpolatory multiresolution whose principal idea dates back at least to Faber's classical paper [13] from 1909 where, in modern terminology, interpolatory wavelets are used to describe continuity and differentiability of functions.

The filterbank construction starts by choosing the synthesis low-pass as $g_0 = a$ where $\mathscr{S}_a$ is such an interpolatory subdivision scheme, i.e., $a(M\cdot) = \delta$. The low-pass filter for the analysis filterbank is even simpler: we set $f_0 = \delta$ so that the low-pass consists of subsampling only. This somewhat trivial filter is often referred to as the *lazy* filter, cf. [38]. Geometrically, we can interpret the low-pass synthesis filter $g_0 = a$ as a *prediction* from the subsampled values by means of

$$\mathscr{S}_a(c(M\cdot)) = a * \uparrow_M \downarrow_M \delta * c = a * \uparrow_M \downarrow_M c = \mathscr{S}_a \downarrow_M c. \tag{28}$$

Recall that only $\downarrow_M \uparrow_M$ is an identity while the $\uparrow_M \downarrow_M$-operator from (28) always leads to a loss of information due to decimation, giving

$$\uparrow_M \downarrow_M c(\alpha) = \begin{cases} c(\alpha), & \alpha \in M\mathbb{Z}^s, \\ 0, & \alpha \notin \mathbb{Z}^s, \end{cases}$$

Since the prediction from decimated data cannot be expected to be sufficient for the reconstruction of the complete information, we have to compensate the error $(I - \mathscr{S}_a \downarrow_M)c$ by a *correction* process which is performed by the "high-pass" filters

$$F_\varepsilon = \tau_{-\varepsilon}(I - \mathscr{S}_a \downarrow_M), \qquad \varepsilon \in E_M \setminus \{0\}, \tag{29}$$

where each f_ε takes care of the correction on the subgrid $\varepsilon + M\mathbb{Z}^s$. The high-pass reconstruction filters, on the other hand, are simple again since their only task is to shift the correction values to their appropriate location:

$$G_\varepsilon = \tau_\varepsilon. \tag{30}$$

Thus, the action of the filterbank can be expressed as

$$\sum_{\varepsilon \in E_M} G_\varepsilon \uparrow_M \downarrow_M F_\varepsilon = G_0 \uparrow_M \downarrow_M F_0 + \sum_{\varepsilon \in E_M \setminus \{0\}} G_\varepsilon \uparrow_M \downarrow_M F_\varepsilon$$

$$= \mathscr{S}_a \downarrow_M + \sum_{\varepsilon \in E_M \setminus \{0\}} \tau_\varepsilon \uparrow_M \downarrow_M \tau_{-\varepsilon}(I - \mathscr{S}_a \downarrow_M)$$

$$= \left(I - \sum_{\varepsilon \in E_M \setminus \{0\}} \tau_\varepsilon \uparrow_M \downarrow_M \tau_{-\varepsilon} \right) \mathscr{S}_a \downarrow_M + \sum_{\varepsilon \in E_M \setminus \{0\}} \tau_\varepsilon \uparrow_M \downarrow_M \tau_{-\varepsilon}$$

$$=\uparrow_M\downarrow_M + \sum_{\varepsilon\in E_M\setminus\{0\}} \tau_\varepsilon \uparrow_M\downarrow_M \tau_{-\varepsilon} = \sum_{\varepsilon\in E_M} \tau_\varepsilon \uparrow_M\downarrow_M \tau_{-\varepsilon} = I$$

which verifies that it is a perfect reconstruction one indeed.

Also in symbol calculus, this filterbank is described quite easily. We have that $f_0^\sharp(z)=1$, $g_0(z)=a^\sharp(z)$, as well as $f_\varepsilon^\sharp(z)=z^\varepsilon\left(1-a^\sharp\left(z^M\right)\right)$, and $g_\varepsilon^\sharp=z^{-\varepsilon}$, $\varepsilon\in E_M\setminus\{0\}$.

3.3 Multiresolution

The concept of *multiresolution analysis (MRA)*, as introduced by Mallat in [26], has become a fundamental theoretical basis in wavelet signal processing. We will recall this idea from [27], but adapt it to our setting here.

Definition 3 (Multiresolution Analysis). A sequence of nested closed subspaces $V_j\subset V_{j+1}\subset V$, $j\in\mathbb{Z}$, in a space V is called a *multiresolution analysis* if the following conditions hold:

1. *(Translation invariance)* $f\in V_j$ if and only if $f\left(\cdot-M^{-j}\alpha\right)\in V_j$ for all $\alpha\in\mathbb{Z}^s$.
2. *(Scaling property)* $f\in V_j$ if and only if $f(M\cdot)\in V_{j+1}$.
3. *(Limits)* The spaces are exhaustive and nonredundant:

$$\lim_{j\to\infty} V_j = V, \qquad \lim_{j\to-\infty} V_j = \{0\}.$$

4. *(Basis)* There exists a *scaling function* $\varphi\in V$ such that

$$S(\varphi)=\{\varphi(\cdot-\alpha) : \alpha\in\mathbb{Z}^s\}$$

is a stable basis of V_0.

The scaling property and the nestedness of the spaces require that the scaling function φ is refinable, that is, that there exist coefficients $a\in\ell(\mathbb{Z}^s)$ such that

$$\varphi = \sum_{\alpha\in\mathbb{Z}^s} a(\alpha)\,\varphi(M\cdot-\alpha).$$

Once an MRA is present, it gives a new view on the filterbank mechanism by interpreting data $c\in\ell(\mathbb{Z}^s)$ as the function

$$\varphi_c^n := (\varphi*c)(M^n\cdot) = \sum_{\alpha\in\mathbb{Z}^s} c(\alpha)\varphi(M^n\cdot-\alpha). \tag{31}$$

The choice of the *detail level n* is arbitrary and unrelated to the sequence c from a formal point of view, but normally and in practice the values of c are related to some grid $M^{-n}\mathbb{Z}^s$, and are obtained from discretizing a function g relative to that grid: $c(\alpha) \sim g(M^{-n}\alpha)$.

Now suppose that (F,G) is a *perfect reconstruction* filterbank with the property that $g_0 = a$, where a is the refinement mask of the multiresolution analysis. We take a signal c and decompose this signal with the analysis filterbank based on $E_{\searrow} = \{0\}$ into $c^1 = (c_0^1 = \downarrow_M F_0 c)$ and $d^1 = (d_\varepsilon^1 = \downarrow_M F_\varepsilon c : \varepsilon \in E_M \setminus \{0\})$. Since we have a perfect reconstruction filterbank, we have that

$$c = \sum_{\alpha \in \mathbb{Z}^s} g_0(\cdot - M\alpha) c_0^1(\alpha) + \sum_{\varepsilon \in E_M \setminus \{0\}} \sum_{\alpha \in \mathbb{Z}^s} g_\varepsilon(\cdot - M\alpha) d_\varepsilon^1(\alpha)$$

and therefore

$$
\begin{aligned}
\varphi_c^1 &= \sum_{\alpha \in \mathbb{Z}^s} c(\alpha) \varphi(M \cdot - \alpha) \\
&= \sum_{\alpha \in \mathbb{Z}^s} \sum_{\beta \in \mathbb{Z}^s} a(\alpha - M\beta) c_0^1(\beta) \varphi(M \cdot - \alpha) \\
&\quad + \sum_{\alpha \in \mathbb{Z}^s} \sum_{\varepsilon \in E_M \setminus \{0\}} \sum_{\beta \in \mathbb{Z}^s} g_\varepsilon(\alpha - M\beta) d_\varepsilon^1(\beta) \varphi(M \cdot - \alpha) \\
&= \varphi * c_0^1 + \sum_{\varepsilon \in E_M \setminus \{0\}} \sum_{\beta \in \mathbb{Z}^s} d_\varepsilon^1(\beta) \left(\sum_{\alpha \in \mathbb{Z}^s} g_\varepsilon(\alpha) \varphi(M(\cdot - \beta) - \alpha) \right)
\end{aligned}
$$

from which we obtain the decomposition

$$\varphi_c^1 = \varphi * c_0^1 + \sum_{\varepsilon \in E_M \setminus \{0\}} \psi_\varepsilon * d_\varepsilon^1 \tag{32}$$

with the *wavelet function*

$$\psi_\varepsilon = (g_\varepsilon * \varphi)(M \cdot) = \sum_{\alpha \in \mathbb{Z}^s} g_\varepsilon(\alpha) \varphi(M \cdot - \alpha). \tag{33}$$

Now we only need to iterate (32) to obtain the wavelet decomposition for our initial function from (31):

$$\varphi_c^n = \varphi * c^n + \sum_{k=1}^{n} \sum_{\varepsilon \in E_M \setminus \{0\}} \left(\psi_\varepsilon * d_\varepsilon^k \right) \left(M^{k-1} \cdot \right). \tag{34}$$

So far, there is nothing new or spectacular with these computations, they just are the usual formalism in defining wavelets. However, we should keep in mind which were our assumptions here. The only requirement was that we demanded (F,G) to be a perfect reconstruction filterbank whose synthesis low-pass filter we used for the subdivision scheme. Then the subband signals generated by the cascaded analysis filterbank *automatically* give us the coefficients in the *wavelet decomposition* (34) of the function φ_c^n. And as long as we are working only digitally, we do not even have to know any of these functions φ and ψ_ε explicitly.

But also the converse has its charming side: Whenever we have a subdivision scheme whose mask can be completed to a unimodular modulation matrix, then we have a filterbank with which all computations on a coefficient level can be performed.

> Any perfect reconstruction filterbank whose low-pass synthesis filter defines a convergent subdivision scheme, admits an analytical *interpretation* as coefficients in a wavelet decomposition (34). All *computations*, on the other hand, are done entirely within the filterbank that can be digitally realized.
>
> There always exist such perfect reconstruction filterbanks, namely those based on *interpolatory* subdivision schemes, where the filterbank can be given explicitly.

This was a fairly long introduction. But it hopefully explains why a subdivision approach to shearlets makes sense. Once a subdivision and multiresolution concept applicable to shear geometries is settled, we can obtain an analogy of the above interpretation and thus a completely digital and numerical way to compute also shearlet decompositions.

4 Multiple Subdivision and Multiple Refinability

The general framework to construct subdivision schemes which allow for adapted multiresolution is that of *multiple subdivision*, see [36]. The idea behind multiple subdivision is very simple, intuitive, and natural: Instead of using a single subdivision scheme in any step, there is a finite family of subdivision schemes *and dilation matrices* from which an arbitrary one can be chosen in each step of the subdivision process. To be concrete: let $a = (a_j : j \in \mathbb{Z}_m)$ and $M = (M_j : j \in \mathbb{Z}_m)$ be collections of $m \geq 1$ masks and dilation matrices, respectively; from now on we will use the convenient abbreviation $\mathbb{Z}_m = \{0, \ldots, m-1\}$. For any $j \in \mathbb{Z}_m$, we have an individual subdivision operator $\mathscr{S}_j = \mathscr{S}_{a_j} = a_j * \uparrow_{M_j}$ with associated dilation matrix M_j. The subdivision process is then ruled by an infinite vector $e \in \mathbb{Z}_m^\infty$ of "digits" from $\mathbb{Z}_m$. The nth iteration, $n \in \mathbb{N}$, of the subdivision operators according to e is therefore

$$\mathscr{S}_{e^n} = \mathscr{S}_e^n := \mathscr{S}_{e_n} \cdots \mathscr{S}_{e_1}. \tag{35}$$

We will denote the initial segment of length n of some $e \in \mathbb{Z}_m^\infty$ by $e^n := (e_1, \ldots, e_n)$. The values of $\mathscr{S}_e^n c$ are related to the grid $M_{e^n}^{-1} \mathbb{Z}^s$ where

$$M_{e^n} := M_{e_n} \cdots M_{e_1}. \tag{36}$$

To ensure that this is really a contractive process with grids that converge to $\mathbb{R}^s$, we have to require that

$$\lim_{n \to \infty} \sup_{e^n \in \mathbb{Z}_m^n} \left\| M_{e^n}^{-1} \right\| = 0,$$

which requires that not only the spectral radii of all individual matrix inverses M_j^{-1}, $j \in \mathbb{Z}_m$, are less than one but also that their *joint spectral radius* is less than one.

4.1 Basic Properties

Once again, the immediate and obvious question for this subdivision process is the "usual" one: how to describe convergence and what does it imply, especially in terms of refinable functions. In fact, convergence is described in almost in the same way as it was done for a solitary subdivision scheme in Definition 1—and of course the special case $m = 1$ is exactly the usual stationary subdivision. The following definition is given in [36].

Definition 4. The multiple subdivision scheme based on (a, M) is said to be *p-convergent* if for any $e \in \mathbb{Z}_M^\infty$ there exists a *limit function* f_e such that

$$\lim_{n \to \infty} (\det M_{e^n})^{-1/p} \left\| \mathscr{S}_e^n \delta - \mu \left(f_e, M_{e^n}^{-1} \right) \right\|_p = 0. \tag{37}$$

The choice $e = (j, j, \dots)$ immediately shows that convergence of all the individual subdivision schemes based on (a_j, M_j) is a necessary condition for the convergence of the multiple subdivision scheme; hence, all the classical restrictions have to hold, like the *sum rule of order* 0

$$\sum_{\alpha \in \mathbb{Z}^s} a_j (\varepsilon + M_j \alpha) = 1, \qquad \varepsilon \in E_{M_j}, \quad j \in \mathbb{Z}_m.$$

If all the individual subdivision schemes are *interpolatory* ones, then, of course, so is the multiple one and the limit functions $f_e, e \in \mathbb{Z}_m^\infty$, are all cardinal: $f_e(\alpha) = \delta_{\alpha,0}$, $\alpha \in \mathbb{Z}^s$.

Convergence analysis for multiple subdivision schemes can be performed and this has been done in [22, 36]. Since it is of no real importance for what I have in mind here, I will skip these results which are natural extensions of the results for single subdivision schemes. It is more important to consider the resulting refinement properties of the limit functions as this will lead to the concept of *joint refinement* which will be crucial for the multiresolution analysis we want to construct. Since this type of refinement is not describing one function in terms of its scaled and shifted copies but relates *all* the limit functions f_e, $e \in \mathbb{Z}_m^\infty$, to each other, the name *multiple refinement equation* seems appropriate for (38).

For $e = (e_1, e^\infty) \in \mathbb{Z}_m^\infty$, the *multiple refinement equation* takes the form

$$f_e = \uparrow_{M_{e_1}} (f_{e^\infty} * a_{e_1}) = \sum_{\alpha \in \mathbb{Z}^s} a_{e_1}(\alpha) f_{e^\infty} (M_{e_1} \cdot - \alpha). \tag{38}$$

In other words, The *first* digit e_1 of e selects *which mask* and therefore also *which scaling matrix* has to be used in the refinement equation, the remaining, infinitely many, digits determine *which function* is used for the refinement.

There is an immediate generalization of (38) to an arbitrary decomposition of $e = (e^n, e^\infty)$ into a finite part e^n and an infinite trailing part e^∞, giving

$$f_e = \sum_{\alpha \in \mathbb{Z}^s} a_{e^n}(\alpha) f_{e^\infty} (M_{e^n} \cdot - \alpha), \qquad a_{e^n} := \mathscr{S}_{e^n} \delta. \tag{39}$$

Again, the initial digits select mask and scaling matrix while the infinite rest determines the function used for the refinement. Moreover, it is evident from (38) that the functions $f_j = f_{(j,j,\dots)}$ are refinable in the classical sense with respect to a_j and M_j.

A simple but fundamental observation in [22, 36] was that any convergent multiple subdivision scheme leads to multiply refinable limit functions.

Theorem 5. *The limit functions f_e, $e \in \mathbb{Z}_m^\infty$, of a convergent multiple subdivision scheme are multiply refinable in the sense of (38).*

Like in single stationary subdivision, stability of refinable functions leads to convergence of the subdivision scheme. The following result was proved in [36].

Theorem 6. *If f_e, $e \in \mathbb{Z}_m^\infty$, is a system of stable multiple refinable functions with masks $a = (a_j : j \in \mathbb{Z}_m)$ and dilation matrices $M = (M_j : j \in \mathbb{Z}_m)$, then the multiple subdivision scheme associated to (a, M) converges.*

Convergence of the multiple subdivision scheme implies that the limit functions are similar if the indices are similar. More precisely, we have the following result, again from [36].

Theorem 7. *If the multiple subdivision scheme based on (a, M) converges, then the mapping $e \mapsto f_e$ is continuous with respect to the distance function*

$$|e - e'| := \sum_{k=1}^\infty m^{-k} |e_k - e_k'|$$

on $\mathbb{Z}_m^\infty$.

4.2 The Multiple MRA

Once we have at hand the system of refinable functions, we can mimic the MRA approach, where V_j, $j \in \mathbb{Z}$, were and still are the spaces generated by $\varphi \left(M^j \cdot - \alpha \right)$, $\alpha \in \mathbb{Z}^s$. Now, having a slightly more complex refinement structure, we request that all the functions f_e are stable and could set

$$V_j := \text{span} \left\{ f_e \left(M_d \cdot - \alpha \right) \ : \ e \in \mathbb{Z}_m^{\infty}, d \in \mathbb{Z}_m^j, \alpha \in \mathbb{Z}^s \right\}, \tag{40}$$

hence, in particular,

$$V_0 = \text{span} \left\{ f_e \left(\cdot - \alpha \right) \ : \ e \in \mathbb{Z}_m^{\infty}, \alpha \in \mathbb{Z}^s \right\}. \tag{41}$$

Because of the refinement equation (38), these spaces directly satisfy the crucial property of an MRA, the *scaling property*. However, even V_0 would be a space spanned by a generating set that is not even countable any more. To obtain a somewhat simpler generating system that even works with weaker assumptions, we have to give up the symmetry between the symbols to some extent and introduce some more notation.

Definition 5. By

$$\mathbb{Z}_m^* = \left\{ e = (e^n, 0, \ldots) \ : \ e^n \in \mathbb{Z}_m^n, n \in \mathbb{N} \right\} \subset \mathbb{Z}_m^{\infty} \tag{42}$$

we denote the set of all digit sequences that consist only of *finitely many* nonzero digits. This set is countable and allows for a canonical embedding of all finite sequences by appending zeros. This canonical embedding of $e^n \in \mathbb{Z}_m^n$ into $\mathbb{Z}_m^*$ we will denote by $e_*^n = (e^n, 0, \ldots) \in \mathbb{Z}_m^*$.

Before continuing, let us make some remarks about this definition:

1. The choice of making 0 an "outstanding" digit is completely deliberate. But so is, on the other hand, the numbering of masks and dilation matrices as a_j, M_j anyway.
2. The intuition behind that choice, however, is that M_0 is chosen to be the "simplest" matrix in the family. Often, and especially in the shearlet case, there is one matrix which is "scaling only," hence a diagonal one. That one would clearly be a good choice for M_0. Other matrices may combine scaling with further geometric operations like shears, for example.
3. When considering the functions f_e, $e \in \mathbb{Z}_m^*$, only, the question of existence becomes significantly simpler. These functions are limits of subdivision processes where only $\mathscr{S}_0 = \mathscr{S}_{a_0}$ is repeated infinitely many times; hence only, (a_0, M_0) must yield a convergent subdivision scheme while the other ones can be chosen quite arbitrarily. And this concerns mask *and* dilation!

With this slight modification, we set

$$V_j^* := \text{span} \left\{ f_e \left(M_d \cdot - \alpha \right) \ : \ e \in \mathbb{Z}_m^*, d \in \mathbb{Z}_m^j, \alpha \in \mathbb{Z}^s \right\}, \tag{43}$$

and observe that this set generates a reasonable extension of an MRA. Indeed, translation invariance and the scaling property of the V_j^* are obvious from the construction (43), while nestedness follows from (38): for any $f \in V_0^*$, say $f = f_e(\cdot - \alpha)$, we get that

$$f = f_e(\cdot - \alpha) = \sum_{\beta \in \mathbb{Z}^s} a_{e_1} (\beta - M_{e_1}\alpha) \, f_{e^\infty} (M_{e^1} \cdot -\beta), \qquad e = (e_1, e^\infty),$$

which is a linear combination of functions from V_1^* since $e^\infty \in \mathbb{Z}_M^*$ again.

Definition 6. The generalized multiresolution based on the refinable family f_e, $e \in \mathbb{Z}_m^*$ and the spaces V_j^* is called a *multiple multiresolution analysis (MMRA)*.

Remark 1. The MMRA shows striking similarities to the concept of *AB multiresolution* defined in [16].

The next natural question in extending our analogy obviously is: "what are the wavelets then?" To answer that one, let us recall how the MRA was initialized, namely by considering a quasi interpolant projection to V_n in (31). Here, a priori we have plenty of possible initial choices on some level n, namely

$$\varphi_c^{e^n} := (f_{e_*^n} * c)(M_{e^n}\cdot) = \sum_{\alpha \in \mathbb{Z}^s} c(\alpha) f_{e_*^n} (M_{e^n} \cdot -\alpha), \qquad e^n \in \mathbb{Z}_m^n,$$

where the values of c are related to the (different) grids $M_{e^n}^{-1}\mathbb{Z}^s$, respectively. Thus, we have m^n different samples of an underlying function, say g, at least in principle when the M_j are not sharing some underlying relationship. After all, generality has its value and therefore its price ...

Now, we assume again that $a_j = g_{j,0}$ for some finite family of perfect reconstruction filterbanks (F_j, G_j), $j \in \mathbb{Z}_m$, with associated filters $f_{j,\varepsilon}$ and $g_{j,\varepsilon}$, $\varepsilon \in E_{M_j}$, $j \in \mathbb{Z}_m$, where, in the true spirit of multiresolution, only $f_{j,0}$ and $g_{j,0}$, are low-pass filters. Further generalizations are easy to work out formally but useless in the practical context.

With these filterbanks, we decompose c *for each* $j \in \mathbb{Z}_m$ into $c_j^1 = \left(c_{j,0}^1\right)$ and $d^1 = \left(d_{j,\varepsilon}^1 : \varepsilon \in E_{M_j}\right)$ by means of the analysis filterbank F_j and get for any $e^n = \left(e_1, \widehat{e}^{n-1}\right)$ with *first digit* e_1 that

$$\varphi_c^{e^n} = f_{\widehat{e}_*^{n-1}} * c_{e^1}^1 + \sum_{\varepsilon \in E_{M_{e_1}} \setminus \{0\}} \left(\psi_{e^n,\varepsilon} * d_{e^1,\varepsilon}^1\right)(M_{\widehat{e}^{n-1}}\cdot) \tag{44}$$

with the (generalized) *wavelets*

$$\psi_{e^n,\varepsilon} = \sum_{\alpha \in \mathbb{Z}^s} g_{e_1,\varepsilon}(\alpha) f_{\widehat{e}^{n-1}} (M_{e_1} \cdot -\alpha), \qquad e^n \in \mathbb{Z}_m^n. \tag{45}$$

The general case follows now by iteration of (44), uses general decompositions $e^n = \left(e^k, \widehat{e}^{n-k}\right)$, $k = 1, \ldots, n-1$, and looks as follows:

$$\varphi_c^{e^n} = f_0 * c_{e^n}^n + \sum_{k=1}^{n} \sum_{\varepsilon \in E_{M_k} \setminus \{0\}} \left(\psi_{\widehat{e}^{n-k+1}} * d_{\left(e^k, \varepsilon\right)}^k \right) \left(M_{\widehat{e}^{n-k}} \cdot\right). \tag{46}$$

This expansion relates to any data c a totality of m^n different interpretations of level n, indexed by e^n, and thus also a totality of m^n possibly wavelet decompositions. It need not be said that these decompositions will definitely be highly redundant. Nevertheless, as we will see in the next chapter, the *coefficients* in this representation, $c_{e^n}^n$ and $d_{e^k, \varepsilon}^k$, will be accessible in a quite simple and natural computation as they can be arranged most conveniently and quite efficiently into a tree. The simple point of *shearlets* will then be to give geometrical meaning to this decomposition by relating e^n to a directional component.

This is a good time to recall the philosophy of the filterbank and MRA where wavelet expansions like the one in (46) are only the analytic explanation and justification of the purely digital and discrete filterbank operations. Exactly the latter will be our main focus here.

One brief word on such explanations. An important property of the wavelets in (45) and (46) is that of *vanishing moments* as this can be used to ensure some description of (local) regularity of functions by means of the decay of their wavelet coefficients. Such properties are considered to some extent in [36], but I do not want to go into more detail here, even if the temptation is high since quotient ideals are always a nice thing to look at, cf. [4, 35].

4.3 Filterbanks, Cascades, Trees

The computation of the coefficients in the wavelet decomposition (46) can now be performed quite simply by cascading the filterbanks (F_j, G_j), for decomposition as well as for reconstruction. As mentioned before, we assume that like in Sect. 3.3 the filterbanks consist of a single low-pass filter $f_{j,0}$ and $g_{j,0} = a_j$, respectively, and that the rest of the filters $f_{j,\varepsilon}$ and $g_{j,\varepsilon}$, $\varepsilon \in E_{M_j}$ are the high-pass completions. Using these assumptions, let us give the action of the filterbanks a closer look.

We begin with the data $c \in \ell(\mathbb{Z}^s)$ and feed it into each of the filterbanks, giving, for each $j \in \mathbb{Z}_m$,

$$c_{j,0} = \downarrow_{M_j} F_{j,0} c \qquad \text{and} \qquad d_{j,\varepsilon} = \downarrow_{M_j} F_{j,\varepsilon} c, \quad \varepsilon \in E_{M_j}.$$

These sequences have appeared before, namely as

$$c_{e^1}^1 \qquad \text{and} \qquad d_{e^1, \varepsilon}^1, \quad \varepsilon \in E_{M_{e_1}}$$

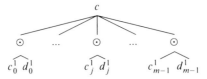

Fig. 1 One level of decomposition with the multiple filterbanks

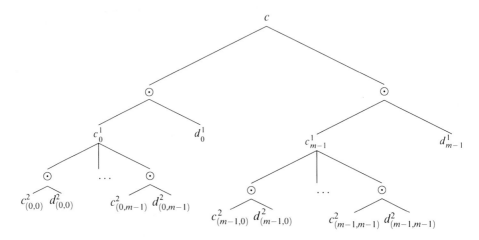

Fig. 2 Two levels of decomposition, to fit the page margins only the first and last branch are shown, which follow c_0^1 and c_{m-1}^1, respectively. The leaves are $c_{e^2}^2$ and $d_{e^k}^k$, $k = 1, 2$

where now e^1 varies over $\mathbb{Z}_m^1$. Schematically, this decomposition can be arranged in the tree form shown in Fig. 1. Next, we apply all the decompositions to all the c^1-signals that we have obtained so far, which leads to $c_{e^2}^2$ and $d_{e^2, \varepsilon}^2$, $\varepsilon \in E_{M_{e_2}}$, $e^2 \in \mathbb{Z}_m^2$, with a tree structure as in Fig. 2. Although the trees become more complex with increasing level of decomposition, their construction is really plain and straightforward and in the end, we will arrive from c to a tree whose leaves are

$$c_{e^n}^n, \quad e^n \in \mathbb{Z}_m^n \quad \text{and} \quad d_{e^k}^k, \quad e^k \in \mathbb{Z}_m^k, k = 1, \ldots, n,$$

and these are precisely the coefficients that show up in the wavelet decomposition (46). The complexity of this computation as well as its storage requirements appear fairly overwhelming at first and clearly the computational complexity rises exponentially as m^n—which is unavoidable. If, on the other hand, we store the coefficient vectors in a suitable way and make use of the fact that $c_{e^{k-1}}^{k-1}$ is not needed any more once the coefficients on level k, that is, $c_{e^k}^k$ and $d_{e^k}^k$, are computed, at least the storage requirement turns out to be reasonably modest, namely only a constant multiple of the storage required for c. This was pointed out in [21] for the shearlet case and reads as follows in the general situation.

Lemma 2. *If the dilation matrices are "sufficiently expansive," the storage requirement for an MMRA tree of depth n for a finitely supported signal c is $O(\#c)$, where $\#c = \#\{\alpha \in \mathbb{Z}^s : c(\alpha) \neq 0\}$.*

Proof. Filtering of c by one of the analysis filters requires a storage of $\#c + C$, where

$$C \geq \max_{j \in \mathbb{Z}_m} \max_{\varepsilon \in E_{M_j}} \#f_{j,\varepsilon}$$

is a constant that covers the "overlap" of the filters. Downsampling by M_j reduces the storage by a factor of $|\det M_j|$, so that the total storage requirement is bounded by

$$\sigma(1) := \sum_{j \in \mathbb{Z}_m} \left(\frac{\#c + C}{|\det M_j|} + \sum_{\varepsilon \in E_{M_j}} \frac{\#c + C}{|\det M_j|} \right),$$

where the first term describes the storage for c^1, the other ones the storage for d^1. Since only the c-part is decomposed in the next level, we have that then the storage consumption is bounded by

$$\sigma(2) = \sum_{e^2 \in \mathbb{Z}_m^2} \left(\frac{\#c + 2C}{|\det M_{e^2}|} + \sum_{\varepsilon \in E_{M_{e^2}}} \frac{\#c + 2C}{|\det M_{e^2}|} \right) + \sum_{e^1 \in \mathbb{Z}_m^1} \sum_{\varepsilon \in E_{M_{e^1}}} \frac{\#c + C}{|\det M_{e^1}|},$$

and, in general, we get the estimate

$$\sigma(n) = \sum_{e^n \in \mathbb{Z}_m^n} \frac{\#c + nC}{|\det M_{e^n}|} + \sum_{k=1}^{n} \sum_{e^k \in \mathbb{Z}_m^k} \sum_{\varepsilon \in E_{M_{e^k}}} \frac{\#c + kC}{|\det M_{e^k}|}. \tag{47}$$

Now let $\rho < 1$ be the joint spectral radius of $\left(M_j^{-1} : j \in \mathbb{Z}_m \right)$ and suppose that

$$m\rho < 1; \tag{48}$$

this is the "sufficiently expansive" property mentioned above. Then, it is easy to see that

$$\sigma(n) \leq (\#c + nC)(m\rho)^n + \left(\max_{j \in \mathbb{Z}_m} |\det M_j| \right) \sum_{k=1}^{n} (\#c + kC)(m\rho)^k$$

is bounded independently of n. $\square$

Provided the joint spectral condition on the matrices is satisfied, we thus have a memory efficient encoding of the original data c into a tree of coefficients by using the different filterbanks. This tree can then be processed further, for example, by thresholding of the wavelet coefficients. Such operations are described in other chapters of this book. For thresholding, things can be done more efficiently if the filterbank has a certain property.

Definition 7. Let $\|\cdot\|$ denote a norm on $\ell(\mathbb{Z}^s)$. The filterbank is called *norm preserving* if there exists a constant $C > 0$ such that

$$\sum_{\varepsilon \in E_M} \|F_\varepsilon c\| \leq C \|c\|. \tag{49}$$

This can be seen as a stability property of the wavelet decomposition (46).

The most prominent example of a norm preserving filterbank are the *orthogonal filterbanks*, known from classical wavelets, cf. [8, 27, 39], where the 2-norm and orthogonal wavelet decompositions are used and where even $C = 1$.

Norm preserving filterbanks can be used for *tree pruning*: whenever some coefficient $c^k_{e^k}$ is small, the whole branch emerging from this coefficient can only contain small values and thus does not have to be computed, for example, when thresholding or basis pursuit methods, cf. [27], are applied to the expansion. This can extensively reduce computation time as well as storage.

The reconstruction of c from the tree is pretty obvious now and simply uses the synthesis filterbanks to determine $c^k_{e^k}$ from its tree successors $c^{k+1}_{e^{k+1}}$ and $d^{k+1}_{e^{k+1},\varepsilon}$, where e^k is the initial segment of e^{k+1}. Thus, we just read the tree backward and reconstruct the value at each node from the values at its successor via the synthesis filterbank. Should the tree have a pruned structure, also some of the reconstructions can be spared.

> It is worthwhile to mention that the whole computational, digital part, does not depend at all on the "preference" for 0 that lead to the definition of $\mathbb{Z}^*_m$ and V^*_j. The only difference is the *interpretation* of the values at the leaves of the tree in (46). If we replace 0 there by an arbitrary $e \in \mathbb{Z}^\infty_m$, then we give a different continuous meaning to the decomposition, the values, however, remain unchanged.

4.4 Things Work Along Trees

We have seen in the preceding section that trees are the natural structure to describe an MMRA. Of course, any MRA is also an MMRA and possesses a tree expansion that consists of a single branch only. But this means that any analogy of wavelet expansions has to be considered for any individual branch of the tree separately and that with whatever we do, always a lot of functions are involved.

For a more detailed consideration of these branches, we will have to distinguish three different cases:

1. The *pragmatic* one as described above, where we start from a function sampled at level n and only proceed down the *finite* tree with branches indexed by e^k,

$k \leq n$. This is the entirely discrete concept where we start with values on the fine grids $M_{e^n}^{-1} \mathbb{Z}^s$ and iterate until we arrive at $\mathbb{Z}^s$ which is then the maximal level of "coarsity".

2. The *function space* one, where we start with a sampling

$$\varphi_c^e = f_e * c \in V_0^*, \qquad e \in \mathbb{Z}_m^*, \tag{50}$$

of level 0 and can then perform an even infinite "wavelet" decomposition

$$\varphi_c^e = \sum_{k=1}^{\infty} \sum_{\varepsilon \in E_M \setminus \{0\}} \left(\psi_{e^{k-1}, \infty, \varepsilon} * d_{e^k, \varepsilon}^k \right) \left(M_{e^k}^{-1} \cdot \right), \qquad e = \left(e^k, e^{k, \infty} \right) \tag{51}$$

into coarser and coarser information parts. This corresponds to the classical "full wavelet" decomposition but can be obtained separately for any branch in the now infinite tree. The use of $e \in \mathbb{Z}_m^*$ has the fundamental advantage that such decomposition exists as soon as all the individual filterbanks consist of FIR filters, allow for perfect reconstruction, and only $g_{0,0}$ has to define a convergent subdivision scheme.

3. The *fully infinite function space approach*, with $\varphi_c^e \in V_0$ and $e \in \mathbb{Z}_M^*$. Here the requirements are slightly stronger, as now the multiple subdivision scheme based on $g_{j,0}$, $j \in \mathbb{Z}_m$, has to be convergent, requiring some more interaction properties between the filterbanks. Such convergence analysis can be found in [36].

Nevertheless, independent of the "model," all relevant properties of the decomposition, even refinability, proceed *vertically*, i.e., along the branches of the tree. This leads to definitions like the following one which lives in the context of the "intermediate" function space decompositions.

Definition 8. A system $\psi_{e,\varepsilon}$ is called a *wavelet frame* if for any $e \in \mathbb{Z}_m^*$ there exist constants A_e and B_e such that

$$A_e \left(\sum_{k=1}^{\infty} \sum_{\varepsilon \in E_M \setminus \{0\}} \left\| d_{e^k, \varepsilon}^k \right\| \right) \leq \| \varphi_c^e \| \leq B_e \left(\sum_{k=1}^{\infty} \sum_{\varepsilon \in E_M \setminus \{0\}} \left\| d_{e^k, \varepsilon}^k \right\| \right).$$

In a *tight frame* we have $A_e = B_e$ and if all the individual MRAs are orthogonal, we can even obtain $A_e = B_e = 1$.

4.5 A Canonical Interpolatory Construction

We have seen before, in Sect. 3.2, that interpolatory subdivision is a simple and natural way to obtain perfect reconstruction filterbanks. Hence, interpolation can be used to generate an MMRA in a very straightforward manner that we are going to describe in some more explicit detail here. The classical paper by Dubuc and Deslauriers [10] introduces a family of *univariate* interpolatory subdivision schemes by means of local polynomial interpolation—the resulting refinable functions are, by the way, the auto-correlations of the Daubechies scaling functions, cf. [8, 29]. The extension of this method to general univariate integer scaling factors can be found, for example, in [24]. Interpolatory schemes for scaling factors which are a power of 2 are actually more easy to obtain: just iterate a *binary* scheme (i.e., a scheme with scaling factor of 2) sufficiently often.

Anyway, let us assume now that for any integer $r \geq 1$ we have available a favorite *univariate* compactly supported, interpolatory r-adic subdivision scheme with mask b_r, that is,

$$c'(r\cdot) = c \qquad \text{where} \qquad c' = \sum_{k \in \mathbb{Z}} b_r(\cdot - rk)\, c(k), \qquad c \in \ell(\mathbb{Z}).$$

These favorite schemes can have further desirable properties like smoothness of the limit function of polynomial preservation or whatever one may look for, cf. [11].

The first step is to build *convergent* interpolatory subdivision schemes for arbitrary M from these univariate schemes. This is easy if M is diagonal matrix with integer diagonal elements ≥ 2 as then we can use tensor product schemes which converge to tensor product limit functions. To be specific, if $d_j \geq 2$, $j = 1, \ldots, s$, denote the diagonal elements of D, then

$$b_D = \bigotimes_{j=1}^{s} b_{d_j} \in \ell_{00}(\mathbb{Z}^s)$$

is a mask for a convergent s-variate interpolatory subdivision scheme that inherits all the properties of its univariate components b_{d_j}. If M is a general expanding matrix with Smith decomposition $M = PDQ$, we begin with $b = b_D(P^{-1}\cdot)$. Then

$$\mathscr{S}_b c = \sum_{\alpha \in \mathbb{Z}^s} b_D(P^{-1}\cdot - DQ\alpha)\, c(\alpha) = \sum_{\alpha \in \mathbb{Z}^s} b_D(P^{-1}\cdot - D\alpha)\, c(Q^{-1}\alpha),$$

hence

$$\mathscr{S}_b c(P\beta) = \sum_{\alpha \in \mathbb{Z}^s} b_D(\beta - D\alpha)\, c(Q^{-1}\alpha) = \left(\mathscr{S}_{b_D} c(Q^{-1}\cdot)\right)(\beta)$$

or, since b_D is an interpolatory mask,

$$\mathscr{S}_b c(PD\beta) = c(Q^{-1}\beta), \qquad \beta \in \mathbb{Z}^s, \tag{52}$$

and since Q is unimodular, we can replace β by $Q\beta$ in (52) and get that $\mathscr{S}_b c(M\cdot) = c$, hence $\mathscr{S}_b$ is indeed interpolatory. The symbol for b is

$$b^{\sharp}(z) = \sum_{\alpha \in \mathbb{Z}^s} b_D(P^{-1}\alpha)\, z^{\alpha} = \sum_{\alpha \in \mathbb{Z}^s} b_D(\alpha)\, z^{P\alpha} =: b_D^{\sharp}(y)$$

with the *invertible* change of variables $y := z^P = (z^{p_1}, \ldots, z^{p_s})$. Since convergence can be expressed in terms of the symbol and since we have just seen a unimodular operation on a mask does only result in a change of variables of the symbol, we just have given a very simple constructive proof of the following fact that is originally due to Derado [9] and Han [18].

Theorem 8. *For any expanding scaling matrix M, there exists a convergent interpolatory subdivision scheme and hence a refinable cardinal function.*

Thus, for our family of dilation matrix M_j, $j \in \mathbb{Z}_m$, we can find associated interpolatory subdivision schemes such that each of them, in particular a_0, is convergent. Thus, f_0 is well defined and thus also all the other components of the MMRA which we are heading for. In particular, the analysis and the synthesis filterbanks are determined by (29) and (30). This leads to the following explicit formulas for decomposition:

$$c_{e^n}^n = c_{e^{n-1}}^{n-1}(M_{e_n}\cdot), \tag{53}$$

$$d_{e^n,\varepsilon}^n = \left(c_{e^{n-1}}^{n-1} - \mathscr{S}_{e_n}c_{e^n}^n\right)(M_{e_n}(\cdot - \varepsilon)), \qquad \varepsilon \in E_{M_{e_n}}, \tag{54}$$

and reconstruction:

$$c_{e^{n-1}}^{n-1}(M_{e_n}\cdot+\varepsilon) = \mathscr{S}_{e_n}c_e^n(M_{e_n}\cdot+\varepsilon) + d_{e^n,\varepsilon}^n, \qquad \varepsilon \in E_{M_{e_n}}, \tag{55}$$

where $d_{e^n,0}^n = 0$.

This shows that *for any* family of dilation matrices there always exists an interpolatory MMRA which, in addition, can be constructed by a straightforward and canonical process entirely from given univariate interpolatory schemes.

5 Shearlet Subdivision and Multiresolution

So far, in the definition of MMRAs, things obviously were fairly general. Actually, it was even possible that low-pass filters were not even low-pass filters and that the scaling matrices could be almost completely unrelated, only the joint spectral radius of their inverses had to be less than one. Of course, in such a generality a geometric meaning of the subdivision operations and the decomposition can and should not be expected. On the other hand, a proper selection of the M_j could be useful to extract geometric information about the underlying function from the MMRA decomposition (34).

Indeed, thanks to the context we set up so far, discrete shearlets can now be considered as just a very special case of multiple subdivision schemes whose scaling matrices also provide geometric information. The main idea of shearlets is to provide rotation-like geometric transformations which give directional information and nevertheless leave the integer grid $\mathbb{Z}^s$ invariant. Since the latter can hardly be achieved by rotations, shears form a useful replacement.

5.1 Shears and Scaling

The geometry of shearlets is a consequence of a very nice interaction between shear and appropriate scaling such that together they operate on planes, i.e., on linear spaces in quite a rotation-like way.

Definition 9. A *shear matrix* $S \in \mathbb{R}^{s \times s}$ is defined in block form as

$$S = S(W) = \begin{pmatrix} I_p & W \\ 0 & I_q \end{pmatrix}, \qquad W \in \mathbb{R}^{p \times q}, \quad p, q \leq s. \tag{56}$$

Here I_p and I_q stand for the $p \times p$ and $q \times q$ identity matrices, respectively. However, we will mostly drop the subscripts of the identity blocks since their dimension follows from the size of W.

Shears on $\mathbb{Z}^s$ are unimodular matrices with $S(W)^{-1} = S(-W) \in \mathbb{Z}^{s \times s}$. Geometrically, a shear maps a vector $\begin{pmatrix} x \\ y \end{pmatrix}$ to $\begin{pmatrix} x + Wy \\ y \end{pmatrix}$, that is, the last q variables are fixed while the first p coordinates are shifted by Wy.

> That the last q coordinates are fixed is clearly a very deliberate choice among the coordinates without any real justification. A *complete* shearlet decomposition would have to consider *all* possible combinations of q invariant variables, cf. [21, 22].

The next step is to combine S with a *parabolic scaling* matrix

$$D = \begin{pmatrix} 4I_p & 0 \\ 0 & 2I_q \end{pmatrix} \tag{57}$$

which scales more substantially in the sheared coordinates than in the invariant ones. Since

$$DS = \begin{pmatrix} 4I & 0 \\ 0 & 2I \end{pmatrix} \begin{pmatrix} I & W \\ 0 & I \end{pmatrix} = \begin{pmatrix} 4I & 4W \\ 0 & 2I \end{pmatrix} = \begin{pmatrix} I & 2W \\ 0 & I \end{pmatrix} \begin{pmatrix} 4I & 0 \\ 0 & 2I \end{pmatrix},$$

we have the simple but useful identities

$$DS(W) = S(2W)D \qquad \text{and} \qquad S(W)S(W') = S(W + W') \tag{58}$$

Definition 10. A *shearlet subdivision* associated to matrices $W_j \in \mathbb{Z}^{p \times q}$, $j \in \mathbb{Z}_m$, $W_0 = 0$, is a multiple subdivision scheme with scaling matrices $M_j = DS(W_j)$, $j \in \mathbb{Z}_m$.

The choice $W_0 = 0$ guarantees that "pure scaling" is included among the family of dilation matrices. Besides the obvious geometric meaning of this it always enables the construction of shearlet filterbanks as a convergent subdivision scheme a_0 for M_0 can be easily and immediately obtained by tensor product constructions. Thus, shearlet MMRAs can also be constructed in the noninterpolatory case.

Example 1. The "classical" shearlet subdivision for $s = 2$ uses $q = 1$ and $W_j = j$, $j \in \mathbb{Z}_1$. This type of subdivision has been introduced and discussed in [22].

From (58) it is easy to give *explicit* formulas for the iterated scaling matrices M_{e^n} of the multiple subdivision scheme, $e^n = (e_1, \ldots, e_n) \in \mathbb{Z}_m^n$, as the shears

$$M_{e^n} = M_{e_n} \cdots M_{e_1} = DS(W_{e_n}) \cdots DS(W_{e_n}) = S(2W_{e_n}) D^2 S\left(W_{e_{n-1}}\right) \cdots DS(W_{e_1})$$

$$= \prod_{k=1}^{n} S\left(2^k W_{e_{n+1-k}}\right) D^n = S\left(\sum_{k=1}^{n} 2^{n+1-k} W_{e_k}\right) D^n$$

and thus also

$$M_{e^n}^{-1} = D^{-n} S\left(-\sum_{k=1}^{n} 2^{n+1-k} W_{e_k}\right) = S\left(-\sum_{k=1}^{n} 2^{1-k} W_{e_k}\right) D^{-n}. \tag{59}$$

Note that (59) means that $M_{e^n}^{-1} = D^{-n} U_{e^n} = V_{e^n} D^{-n}$ where *all* the U_{e^n} and also V_0 are even *unimodular* matrices. This simple observation has a very appealing and useful consequence for the filterbank decompositions! Recall that there we "sampled" the function f that we wanted to analyze on the grid $M_{e^n}^{-1} \mathbb{Z}^s = D^{-n} U_{e^n} \mathbb{Z}^s = D^{-n} \mathbb{Z}^s$; hence, it suffices to know f on the simplest grid, $D^{-n} \mathbb{Z}^s$.

Clearly, the choice of the shear parameters W_j, $j \in \mathbb{Z}_m$, determines the yet unfixed *geometry* behind the shearlet system and it definitely makes sense to determine the W_j such that they are linearly independent. Since

$$M_{e^n}^{-1} = \begin{pmatrix} I - \sum_{k=1}^{n} 2^{1-k} W_{e_k} \\ 0 & I \end{pmatrix} \begin{pmatrix} 4^{-n} I & 0 \\ 0 & 2^{-n} I \end{pmatrix} = \begin{pmatrix} 4^{-n} I & -2^{1-2n} \sum_{k=1}^{n} 2^{-k} W_{e_k} \\ 0 & 2^{-n} I, \end{pmatrix}$$

we have that

$$M_{e^n}^{-1} \begin{pmatrix} x \\ y \end{pmatrix} = 2^{-n} \begin{pmatrix} 2^{-n} \left(x - 2 \sum_{k=1}^{n} 2^{-k} W_{e_k} y\right) \\ y \end{pmatrix} \tag{60}$$

which shows that like in the bivariate case in [22] the digits in e define dyadic numbers which yield to what extent a certain shear W_j has to be applied. Nevertheless, in this generality, we cannot obtain much geometric intuition about the action of $M_{e^n}^{-1}$.

5.2 Shears of Codimension 1: Hyperplane Shearlets

Probably the most important application of shearlets is the detection of local wavefronts or of tangent planes in a certain point. These tangent planes are hyperplanes, so the scaling should only "cancel" one dimension which means that we have to choose $q = 1$ and thus $p = s - 1$. Then any W_j, $j \in \mathbb{Z}_m$, is a *vector* in $\mathbb{Z}^{s-1}$, and the most natural choice leads to $m = s - 1$ and the use of unit vectors: $W_j \in \mathbb{R}^{s-1}$, $j \in \mathbb{Z}_{s-1} \setminus \{0\}$. In three dimensions, for example, this approach leads to

$$M_0 = \begin{pmatrix} 4 & 0 & 0 \\ 0 & 4 & 0 \\ 0 & 0 & 2 \end{pmatrix}, \quad M_1 = \begin{pmatrix} 4 & 0 & 4 \\ 0 & 4 & 0 \\ 0 & 0 & 2 \end{pmatrix}, \quad M_2 = \begin{pmatrix} 4 & 0 & 0 \\ 0 & 4 & 4 \\ 0 & 0 & 2 \end{pmatrix}.$$

A vector $\begin{pmatrix} x \\ 1 \end{pmatrix} \in \mathbb{R}^s$ is then mapped to

$$M_{e^n}^{-1} \begin{pmatrix} x \\ 1 \end{pmatrix} = 4^{-n} \begin{pmatrix} x_1 - \xi_1 \\ \vdots \\ x_{s-1} - \xi_{s-1} \\ 2^n \end{pmatrix}, \qquad \xi_j = \sum_{k=1}^{n} 2^{1-k} \delta_{e_k, j},$$

hence

$$\xi \in \frac{2^n - 1}{2^{n-1}} \Delta_{s-1} \subset 2\Delta_{s-1}, \qquad \Delta_{s-1} = \left\{ x \in \mathbb{R}^{s-1} : s_j \geq 0, \sum s_j = 1 \right\},$$

where Δ_k denotes the $k - 1$-dimensional unit simplex in $\mathbb{R}^k$. In the limit, when $n \to \infty$, the vectors ξ cover the dyadic numbers in $2\Delta_{s-1}$.

Now suppose that $z = \begin{pmatrix} x \\ 1 \end{pmatrix}$ with $x \in [0,1]^n$; hence, $\|z\|_\infty = 1$ is normal to the hyperplane $H \subset \mathbb{R}^s$ whose slope in direction j is given by x_j. After renormalization, we have that

$$z' = 2^n M_{e^n}^{-1} \begin{pmatrix} x \\ 1 \end{pmatrix} = \begin{pmatrix} 2^n (x - \xi) \\ 1 \end{pmatrix} \in \mathbb{Z}^s \tag{61}$$

so that the integer slopes of the hyperplane after the transformation vary between -2^n and 2^n, extending [22, Lemma 2.3] naturally to higher dimensions. Small values in z' are obtained by setting the components of ξ equal to the dyadic expansions of x, but of course this only works if $x \in \Delta_{s-1}$, the set of normalized vectors from the first octant. This is the discrete analog of the *cone-adapted shearlets*: the geometry of the shears can only cover one part of all possible slopes, the other one has to be obtained by considering all combinations of W_j being positive or negative unit vectors and also picking the "fixed" coordinate appropriately. The generalization of this fact, already mentioned in [21, 22], thus looks as follows:

A full discrete shearlet analysis that can handle all slopes requires $s 2^{s-1}$ different implementations of the shearlet transform.

In other words, besides the increase of complexity due to the appearance of trees in MMRAs, we have to consider not one, but $s 2^{s-1}$ of these MMRAs and thus trees in order to get a full coverage of the directional parameter. In addition, even if for any digit sequence e^n there exists an associated slope, hence a directional parameter, this relationship is not uniform and the "directions" obtained that way are distributed in a somewhat peculiar and nonuniform way. But nevertheless:

Decomposing a signal with respect to a shearlet MMRA gives a representation where the location of a certain coefficient $d_{e^n}^n$ within the tree encodes *local directional* information as long as all filters are finitely supported.

Shearlet subdivision of codimension 1 is also "sufficiently expansive" in the sense of Lemma 2, since $\det M_j = 2^{2s-1} > s = m$ for any $s \geq 2$. Therefore, the directional tree decomposition can be computed with the same order of storage space as needed for the original sampling.

5.3 Orthogonal Shearlets by Tensor Product

Another nice property of any "shear MMRA" is the fact that there is a straightforward construction of *discrete orthogonal shearlets*, even of compact support. Choose again a to be the mask of your favorite univariate subdivision scheme whose limit function φ, however, has orthonormal integer translates this time. For example, the masks for the Daubechies scaling functions, cf. [8, 27], shall do well. Moreover, let $a^2 = \mathscr{S}_a a$ be the quaternary scheme (with a scaling factor of 4) obtained by applying $\mathscr{S}_a$ on the mask a. The limit function φ^2 of this process is easily seen to have orthonormal integer translates, too. The tensor product

$$\varphi(x) = \bigotimes_{j=1}^{p} \varphi^2(x_j) \otimes \bigotimes_{j=p+1}^{s} \varphi(x_j)$$

has orthonormal multi-integer translates and is M_0-refinable with respect to the tensor product mask

$$a(\alpha) = \bigotimes_{j=1}^{p} a^2(\alpha_j) \otimes \bigotimes_{j=p+1}^{s} a(\alpha_j).$$

This settles the first building block and directly defines the function f_0 in the MMRA. The remaining masks are obtained from a by the unimodular change of variables

$$a_j := a(U_j \cdot), \qquad U_j = U_{(j)} = DM_j^{-1} = \begin{pmatrix} I & -2W_j \\ 0 & I \end{pmatrix}, \tag{62}$$

as defined implicitly in (59). The functions $f_{e_*^n}$ are then be defined inductively in n by the refinement equation (26), that is,

$$f_{(j,e_*^n)} = |\det M|^{1/2} \sum_{\alpha \in \mathbb{Z}^s} a_j(\alpha) f_e(M_j \cdot - \alpha), \qquad j \in \mathbb{Z}_m, \quad e \in \mathbb{Z}_m^*. \tag{63}$$

This gives a sequence of functions that is orthogonal along branches.

Lemma 3. *For $e \in \mathbb{Z}_m^*$ we have that*

$$\int_{\mathbb{R}^s} f_e(x-\alpha) f_e(x) \, dx = \delta_{\alpha,0}, \qquad \alpha \in \mathbb{Z}^s, \tag{64}$$

and

$$\int_{\mathbb{R}^s} \psi_{e,\varepsilon}(\cdot - \alpha) f_e(x) \, dx = 0, \qquad \alpha \in \mathbb{Z}^s, \quad \varepsilon \in E_M \setminus \{0\}. \tag{65}$$

Proof. We write $e = e_*^n = (e_1, \widehat{e}^{n-1}, 0) = (e_1, \widehat{e})$ and perform induction on n where the case $n = 0$ is trivial.

Applying (63), we get that

$$\int_{\mathbb{R}^s} f_e(x-\alpha) f_e(x) \, dx$$

$$= \sum_{\beta,\gamma \in \mathbb{Z}^s} a_{e_1}(\beta) a_{e_1}(\gamma) \int_{\mathbb{R}^s} f_{\widehat{e}}(M_{e_1}(x-\alpha) - \beta) f_{\widehat{e}}(M_{e_1}x - \gamma) \, dx$$

$$= \sum_{\beta,\gamma \in \mathbb{Z}^s} a_{e_1}(\beta) a_{e_1}(\gamma) \delta_{\beta + M_{e_1}\alpha,\gamma} = \sum_{\beta \in \mathbb{Z}^s} a(U_{e_1}\beta) a\left(DM_{e_1}^{-1}M_{e_1}\alpha + U_{e_1}\beta\right)$$

$$= \sum_{\beta \in \mathbb{Z}^s} a(\beta) a(D\alpha + \beta) = \delta_{\alpha,0};$$

the last identity is the symbol version of orthogonality for the associated refinable function, cf. [3, 7].

Equation (65) is proved in a similar way: one applies the refinement equation to f_e, substitutes (45) for the wavelet, and again uses (64) for $\widehat{e}$ as well as the biorthogonality of the filters and the unimodularity of U_{e_1}. $\quad\square$

This is already the story of the simple and straightforward construction for compactly supported orthogonal shearlets on $\mathbb{R}^s$. All the usual properties for orthogonal wavelets now follow immediately. However, all such properties hold only "vertically" along branches of the tree, not "horizontally" within the levels of the tree. There is no reason to assume that f_e and $f_{e'}$, e, e' should be orthogonal or related in some way, except maybe that they should be similar (hence, have almost orthogonal translates) provided that (a, M) defines a convergent subdivision schemes, see Theorem 7.

5.4 Implementation

In [21], a first "proof of concept" implementation of an interpolatory discrete shearlet transform has been realized and considered. The numerical experiments show that indeed the discrete shearlet transform behaves as it should be expected and that wavelet coefficients are large where position and tangent direction coincide

Fig. 3 Two examples of reconstructions after thresholding. It can be seen (or not) that the artifacts point somewhat perpendicular to the edges

with the localization within the wavelet coefficients and the slope determined by the respective branch of the tree. A test application was compression, that is decomposition, thresholding, and reconstruction, and some results for a very small threshold value can be seen in Fig. 3. It should be mentioned that thresholding was trickier than originally expected as there were many coefficients of about the same absolute value within the tree, so thresholding also became a somewhat randomized process. So far, it is not clear how optimal selection strategies for the huge dictionary formed by the functions f_0 and

$$\left\{ \psi_{e^k,\varepsilon} \ : \ k = 1, \ldots, n, \ \varepsilon \in E_M \setminus \{0\} \right\}$$

should look like. However, it is likely that the tree structure has to be taken into account.

Applying filters of finite, nonzero, length to *finite* data leads to the usual boundary problems and each of the known strategies to overcome this problem, be it data enlargement, periodization or zero padding, can also be applied here with the usual caveats, problems, and side effects. However, there is a slight additional pitfall as the shears can and strongly enhance the effect of the boundary completion. For example, a combination of excessive shearing and periodization can lead to "wrap around" effects that may create edges perpendicular to the wrapping direction.

Another problem is the anisotropy of the sampling matrix D which is $\begin{pmatrix} 4 & 0 \\ 0 & 2 \end{pmatrix}$ in the case $s = 2$. This means that the image is sampled with twice the frequency in x-direction than in y-direction. This, however, appears to be the case only because of the brevity of presentation because, as mentioned before, a complete shearlet decomposition has to consider $\pm W_j$ and all choices of the "special" coordinate among the s. This means that sampling by $\begin{pmatrix} 2 & 0 \\ 0 & 4 \end{pmatrix}$ has to be taken into account as well and suggests to begin with a quadratic image, say of size $2^n \times 2^n$. For the first type of shearlet decomposition, the image is subsampled by n binary subdivision steps (and the scheme can be chosen to be interpolatory if the original image values are to be preserved, otherwise an arbitrary scheme can be used) to yield a resolution of $4^n \times 2^n$ to which we can apply the shearlet decomposition. The same procedure is also done in the other direction. If, conversely, images are reconstructed from shearlet decompositions and thus are of size $4^n \times 2^n$ or $2^n \times 4^n$, the final $2^n \times 2^n$ image will be obtained either by downsampling or averaging and from these images the final result can be formed by one more averaging process. This final averaging is in accordance with the linear nature of the filterbank operations.

All implementations are still at a very early stage. The simple reason for this is the tremendous rise of complexity which necessitates the use of efficient implementations from the very beginning and introduced numerous detail problems that have to be solved. So far a set of octave routines exist to perform a few levels of shearlet decomposition for relatively small images. These were part of the Bachelor thesis [21] and are described there.

References

1. Adams, W.W., Loustaunau, P.: An Introduction to Groebner Bases, *Graduate Studies in Mathematics*, vol. 3. AMS (1994)
2. Benedetto, J.J., Treiber, O.M.: Wavelet frames: multiresolution analysis and extension principles. In: Wavelet transforms and time-frequency signal analysis, Appl. Numer. Harmon. Anal., pp. 3–36. Birkhäuser Boston (2001)
3. Cavaretta, A.S., Dahmen, W., Micchelli, C.A.: Stationary Subdivision, *Memoirs of the AMS*, vol. 93 (453). Amer. Math. Soc. (1991)
4. Cox, D., Little, J., O'Shea, D.: Ideals, Varieties and Algorithms, 2. edn. Undergraduate Texts in Mathematics. Springer–Verlag (1996)
5. Cox, D., Little, J., O'Shea, D.: Using Algebraic Geometry, *Graduate Texts in Mathematics*, vol. 185. Springer Verlag (1998)
6. Cox, D.A., Sturmfels, B. (eds.): Applications of Computational Algebraic Geometry. AMS (1998)
7. Dahmen, W., Micchelli, C.A.: Biorthogonal wavelet expansion. Constr. Approx. **13**, 294–328 (1997)
8. Daubechies, I.: Ten Lectures on Wavelets, *CBMS-NSF Regional Conference Series in Applied Mathematics*, vol. 61. SIAM (1992)
9. Derado, J.: Multivariate refinable interpolating functions. Appl. Comput. Harmonic Anal. **7**, 165–183 (1999)
10. Deslauriers, G., Dubuc, S.: Symmetric iterative interpolation processes. Constr. Approx. **5**, 49–68 (1989)
11. Dyn, N., Levin, D.: Subdivision schemes in geometric modelling. Acta Numerica **11**, 73–144 (2002)
12. Eisenbud, D.: Commutative Algebra with a View Toward Algebraic Geometry, *Graduate Texts in Mathematics*, vol. 150. Springer (1994)
13. Faber, G.: Über stetige Funktionen. Math. Ann. **66**, 81–94 (1909)
14. Föllinger, O.: Laplace-, Fourier- und z-Transformation. Hüthig (2000)
15. Gauss, C.F.: Methodus nova integralium valores per approximationem inveniendi. Commentationes societate regiae scientiarum Gottingensis recentiores **III** (1816)

16. Guo, K., Labate, D., Lim, W., Weiss, G., Wilson, E.: Wavelets with composite dilations and their MRA properties. Appl. Comput. Harmon. Anal. **20**, 231–249 (2006)
17. Hamming, R.W.: Digital Filters. Prentice–Hall (1989). Republished by Dover Publications, 1998
18. Han, B.: Compactly supported tight wavelet frames and orthonormal wavelets of exponential decay with a general dilation matrix. J. Comput. Appl. Math. **155**, 43–67 (2003)
19. Han, B., Kutyniok, G., Shen, Z.: A unitary extension principle for shearlet systems. Math. Comp. (2011). Accepted for publication
20. Jia, R.Q., Micchelli, C.A.: On the linear independence for integer translates of a finite number of functions. Proc. Edinburgh Math. Soc. **36**, 69–85 (1992)
21. Kurtz, A.: Die schnelle Shearletzerlegung. Bachelor Thesis, Justus–Liebig–Universität Gießen (2010)
22. Kutyniok, G., Sauer, T.: Adaptive directional subdivision schemes and shearlet multiresolution analysis. SIAM J. Math. Anal. **41**, 1436–1471 (2009)
23. Latour, V., Müller, J., Nickel, W.: Stationary subdivision for general scaling matrices. Math. Z. **227**, 645–661 (1998)
24. Lian, J.: On a-ary subdivision for curve design. III. $2m$-point and $(2m + 1)$-point interpolatory schemes. Appl. Appl. Math. **4**, 434–444 (2009)
25. Logar, A., Sturmfels, B.: Algorithms for the Quillen–Suslin theorem. J. Algebra **145**, 231–239 (1992)
26. Mallat, S.: Multiresolution approximations and wavelet orthonormal bases of $L^2(\mathbb{R})$. Trans. Amer. Math. Soc. **315**, 69–87 (1989)
27. Mallat, S.: A Wavelet Tour of Signal Processing, 2. edn. Academic Press (1999)
28. Marcus, M., Minc, H.: A Survey of Matrix Theory and Matrix Inequalities. Prindle, Weber & Schmidt (1969). Paperback reprint, Dover Publications, 1992
29. Micchelli, C.A.: Interpolatory subdivision schemes and wavelets. J. Approx. Theory **86**, 41–71 (1996)
30. Möller, H.M., Sauer, T.: Multivariate refinable functions of high approximation order via quotient ideals of Laurent polynomials. Adv. Comput. Math. **20**, 205–228 (2004)
31. Park, H., Woodburn, C.: An algorithmic proof of suslin's stability theorem for polynomial rings. J. Algebra **178**, 277–298 (1995)
32. Park, H.J.: A computational theory of Laurent polynomial rings and multidimensional FIR systems. Ph.D. thesis, University of California at Berkeley (1995)
33. Ron, A., Shen, Z.: Affine systems in $L^2(\mathbb{R}^d)$: The analysis of the analysis operator. J. Funct. Anal. **148**, 408–447 (1997)
34. Rushdie, S.: Shame. Jonathan Cape Ltd. (1983)
35. Sauer, T.: Multivariate refinable functions, difference and ideals—a simple tutorial. J. Comput. Appl. Math. **221**, 447–459 (2008)
36. Sauer, T.: Multiple subdivision. In: Curves and Surfaces 2010, *7th International Conference*, Avignon, France, June 24–30, Lecture Notes in Computer Science, Vol. 6920. Springer (2011)
37. Schüßler, H.W.: Digitale Signalverarbeitung, 3. edn. Springer (1992)
38. Sweldens, W.: The lifting scheme: a custom–design construction of biorthogonal wavelets. Appl. Comput. Harmon. Anal. **3**, 186–200 (1996)
39. Vetterli, M., Kovačević, J.: Wavelets and Subband Coding. Prentice Hall (1995)

Digital Shearlet Transforms

Gitta Kutyniok, Wang-Q Lim, and Xiaosheng Zhuang

Abstract Over the past years, various representation systems which sparsely approximate functions governed by anisotropic features such as edges in images have been proposed. We exemplarily mention the systems of contourlets, curvelets, and shearlets. Alongside the theoretical development of these systems, algorithmic realizations of the associated transforms were provided. However, one of the most common shortcomings of these frameworks is the lack of providing a unified treatment of the continuum and digital world, i.e., allowing a digital theory to be a natural digitization of the continuum theory. In fact, shearlet systems are the only systems so far which satisfy this property, yet still deliver optimally sparse approximations of cartoon-like images. In this chapter, we provide an introduction to digital shearlet theory with a particular focus on a unified treatment of the continuum and digital realm. In our survey we will present the implementations of two shearlet transforms, one based on band-limited shearlets and the other based on compactly supported shearlets. We will moreover discuss various quantitative measures, which allow an objective comparison with other directional transforms and an objective tuning of parameters. The codes for both presented transforms as well as the framework for quantifying performance are provided in the Matlab toolbox ShearLab.

Key words: Digital shearlet system, Fast digital shearlet transform, Performance measures, Pseudo-polar Fourier transform, Pseudo-polar grid, ShearLab, Software package, Tight frames

G. Kutyniok
Institut für Mathematik, Technische Universität Berlin, 10623 Berlin, Germany
e-mail: kutyniok@math.tu-berlin.de

W.-Q. Lim
Institut für Mathematik, Technische Universität Berlin, 10623 Berlin, Germany
e-mail: lim@math.tu-berlin.de

X. Zhuang
Institut für Mathematik, Technische Universität Berlin, 10623 Berlin, Germany
e-mail: xzhuang@math.tu-berlin.de

G. Kutyniok and D. Labate (eds.), *Shearlets: Multiscale Analysis for Multivariate Data*, 239
Applied and Numerical Harmonic Analysis, DOI 10.1007/978-0-8176-8316-0_7,
© Springer Science+Business Media, LLC 2012

1 Introduction

One key property of wavelets, which enabled their success as a universal methodology for signal processing, is the unified treatment of the continuum and digital world. In fact, the wavelet transform can be implemented by a natural digitization of the continuum theory, thus providing a theoretical foundation for the digital transform. Lately, it was observed that wavelets are however suboptimal when sparse approximations of 2D functions are sought. The reason is that these functions are typically governed by anisotropic features such as edges in images or evolving shock fronts in solutions of transport equations. However, Besov models—which wavelets optimally encode—are clearly deficient to capture these features. Within the model of cartoon-like images, introduced by Donoho in [9] in 1999, the suboptimal behavior of wavelets for such 2D functions was made mathematically precise; see also the chapter on "Shearlets and Optimally Sparse Approximations."

Among various directional representation systems which have since then been proposed such as contourlets [8], curvelets [5], and shearlets, the shearlet system is in fact the only one which delivers optimally sparse approximations of cartoon-like images and still also allows for a unified treatment of the continuum and digital world. One main reason in comparison to the other two mentioned systems is the fact that shearlets are affine systems, thereby enabling an extensive theoretical framework, but parameterize directions by slope (in contrast to angles) which greatly supports treating the digital setting. As a thought experiment just note that a shear matrix leaves the digital grid $\mathbb{Z}^2$ invariant, which is in general not true for rotation.

This raises the following questions, which we will answer in this chapter:

(P1) What are the main desiderata for a digital shearlet theory?
(P2) Which approaches do exist to derive a natural digitization of the continuum shearlet theory?
(P3) How can we measure the accuracy to which the desiderata from (P1) are matched?
(P4) Can we even introduce a framework within which different directional transforms can be objectively compared?

Before delving into a detailed discussion, let us first contemplate about these questions on a more intuitive level.

1.1 A Unified Framework for the Continuum and Digital World

Several desiderata come to one's mind, which guarantee a unified framework for both the continuum and digital world, and provide an answer to (P1). The following are the choices of desiderata which were considered in [16, 10]:

- *Parseval Frame Property.* The transform shall ideally have the Parseval frame property, which enables taking the adjoint as inverse transform. This property can be broken into the following two parts, which most, but not all, transforms admit:
  - ◇ *Algebraic Exactness.* The transform should be based on a theory for digital data in the sense that the analyzing functions should be an exact digitization of the continuum domain analyzing elements.
  - ◇ *Isometry of Pseudo-Polar Fourier Transform.* If the image is first mapped into a different domain—here the pseudo-polar domain—, then this map should be an isometry.
- *Space-Frequency-Localization.* The analyzing elements of the associated transform should ideally be highly localized in space and frequency—to the extent to which uncertainty principles allow this.
- *Shear Invariance.* Shearing naturally occurs in digital imaging, and it can—in contrast to rotation—be precisely realized in the digital domain. Thus, the transform should be shear invariant, i.e., a shearing of the input image should be mirrored in a simple shift of the transform coefficients.
- *Speed.* The transform should admit an algorithm of order $O(N^2 \log N)$ flops, where N^2 is the number of digital points of the input image.
- *Geometric Exactness.* The transform should preserve geometric properties parallel to those of the continuum theory, for example, edges should be mapped to edges in transform domain.
- *Stability.* The transform should be resilient against impacts such as (hard) thresholding.

1.2 Band-Limited vs. Compactly Supported Shearlet Transforms

In general, two different types of shearlet systems are utilized today: Band-limited shearlet systems and compactly supported shearlet systems (see also the chapters on "Introduction to Shearlets" and "Shearlets and Optimally Sparse Approximations."). Regarding those from an algorithmic viewpoint, both have their particular advantages and disadvantages:

Algorithmic realizations of the *band-limited shearlet transform* have on the one hand typically a higher computational complexity due to the fact that the windowing takes place in frequency domain. However, on the other hand, they do allow a high localization in frequency domain which is important, for instance, for handling seismic data. Even more, band-limited shearlets do admit a precise digitization of the continuum theory.

In contrast to this, algorithmic realizations of the *compactly supported shearlet transform* are much faster and have the advantage of achieving a high accuracy in spatial domain. But for a precise digitization one has to lower one's sights slightly. A more comprehensive answer to (P2) will be provided in this chapter, where we will present the digital transform based on band-limited shearlets introduced in [16] and the digital transform based on compactly supported shearlets from [18, 19].

1.3 Related Work

Since the introduction of directional representation systems by many pioneer researchers ([4, 5, 6, 7, 8]), various numerical implementations of their directional representation systems have been proposed. Let us next briefly survey the main features of the two closest to shearlets: the contourlet and curvelet algorithms.

- *Curvelets* [3]. The discrete curvelet transform is implemented in the software package *CurveLab*, which comprises two different approaches. One is based on unequispaced FFTs, which are used to interpolate the function in the frequency domain on different tiles with respect to different orientations of curvelets. The other is based on frequency wrapping, which wraps each subband indexed by scale and angle into a fixed rectangle around the origin. Both approaches can be realized efficiently in $O(N^2 \log N)$ flops, N being the image size. The disadvantage of this approach is the lack of an associated continuum domain theory.
- *Contourlets* [8]. The implementation of contourlets is based on a directional filter bank, which produces a directional frequency partitioning similar to the one generated by curvelets. The main advantage of this approach is that it allows a tree-structured filter bank implementation, in which aliasing due to subsampling is allowed to exist. Consequently, one can achieve great efficiency in terms of redundancy and good spatial localization. A drawback is that various artifacts are introduced and that an associated continuum domain theory is missing.

Summarizing, all the above implementations of directional representation systems have their own advantages and disadvantages; one of the most common shortcomings is the lack of providing a unified treatment of the continuum and digital world.

Besides the shearlet implementations we will present in this chapter, we would like to refer to the chapter on "Image Processing using Shearlets" for a discussion of the algorithm in [12] based on the Laplacian pyramid scheme and directional filtering. It should be though noted that this implementation is not focussed on a natural digitization of the continuum theory, which is a crucial aspect of the work presented in the sequel. We further would like to draw the reader's attention to the chapter on "Shearlet Multiresolution and Multiple Refinement" which is based on [17] aiming at introducing a shearlet MRA from a subdivision perspective. Finally, we remark that a different approach to a shearlet MRA was recently undertaken in [13].

1.4 Framework for Quantifying Performance

A major problem with many computation-based results in applied mathematics is the nonavailability of an accompanying code, and the lack of a fair and objective comparison with other approaches. The first problem can be overcome by following the philosophy of "reproducible research" [11] and making the code publicly available with sufficient documentation. In this spirit, the shearlet transforms presented in this chapter are all downloadable from http://www.shearlab.org. One approach

to overcome the second obstacle is the provision of a carefully selected set of prescribed performance measures aiming to prohibit a biased comparison on isolated
tasks such as denoising and compression of specific standard images like "Lena,"
"Barbara," etc. It seems far better from an intellectual viewpoint to carefully decompose performance according to a more insightful array of tests, each one motivated
by a particular well-understood property we are trying to obtain. In this chapter we
will present such a framework for quantifying performance specifically of implementations of directional transforms, which was originally introduced in [16, 10].
We would like to emphasize that such a framework does not only provide the possibility of a fair and thorough comparison but also enables the tuning of the parameters
of an algorithm in a rational way, thereby providing an answer to both (P3) and (P4).

1.5 ShearLab

Following the philosophy of the previously detailed thoughts, ShearLab[1] was introduced by Donoho, Shahram, and the authors. This software package contains

- An algorithm based on band-limited shearlets introduced in [16].
- An algorithm based on compactly supported separable shearlets introduced in
  [18].
- An algorithm based on compactly supported non-separable shearlets introduced
  in [19].
- A comprehensive framework for quantifying performance of directional representations in general.

This chapter is also devoted to provide an introduction to and discuss the mathematical foundation of these components.

1.6 Outline

In Sect. 2, we introduce and analyze the fast digital shearlet transform (FDST),
which is based on band-limited shearlets. Section 3 is then devoted to the presentation and discussion of the digital separable shearlet transform (DSST) and the digital
non-separable shearlet transform (DNST). The framework of performance measures
for parabolic scaling-based transforms is provided in Sect. 4. In the same section,
we further discuss these measures for the special cases of the three previously introduced transforms.

---

[1] ShearLab (Version 1.1) is available from http://www.shearlab.org.

2 Digital Shearlet Transform Using Band-Limited Shearlets

The first algorithmic realization of a digital shearlet transform we will present, coined FDST, is based on band-limited shearlets. Let us start by defining the class of shearlet systems we are interested in. Referring to the chapter on "Introduction to Shearlets," we will consider the cone-adapted discrete shearlet system $\mathrm{SH}(\phi,\psi,\tilde{\psi};\Delta,\Lambda,\tilde{\Lambda}) = \Phi(\phi;\Delta) \cup \Psi(\psi;\Lambda) \cup \tilde{\Psi}(\tilde{\psi};\tilde{\Lambda})$ with $\Delta = \mathbb{Z}^2$ and

$$\Lambda = \tilde{\Lambda} = \{(j,k,m) : j \geq 0, |k| \leq 2^j, m \in \mathbb{Z}^2\}.$$

We wish to emphasize that this choice relates to a scaling by 4^j yielding an integer valued parabolic scaling matrix, which is better adapted to the digital setting than a scaling by 2^j. We further let ψ be a classical shearlet ($\tilde{\psi}$ likewise with $\tilde{\psi}(\xi_1,\xi_2) = \psi(\xi_2,\xi_1)$), i.e.,

$$\hat{\psi}(\xi) = \hat{\psi}(\xi_1,\xi_2) = \hat{\psi}_1(\xi_1)\,\hat{\psi}_2\left(\tfrac{\xi_2}{\xi_1}\right), \tag{1}$$

where $\psi_1 \in L^2(\mathbb{R})$ is a wavelet with $\hat{\psi}_1 \in C^\infty(\mathbb{R})$ and supp $\hat{\psi}_1 \subseteq [-4,-\frac{1}{4}] \cup [\frac{1}{4},4]$, and $\psi_2 \in L^2(\mathbb{R})$ a "bump" function satisfying $\hat{\psi}_2 \in C^\infty(\mathbb{R})$ and supp $\hat{\psi}_2 \subseteq [-1,1]$. We remark that the chosen support deviates slightly from the choice in the introduction, which is however just a minor adaption again to prepare for the digitization. Further, recall the definition of the cones $\mathscr{C}_{11}$—$\mathscr{C}_{22}$ from the chapter on "Introduction to Shearlets."

The digitization of the associated discrete shearlet transform will be performed in the frequency domain. Focussing, on the cone $\mathscr{C}_{21}$, say, the discrete shearlet transform is of the form

$$f \mapsto \langle f, \psi_\eta \rangle = \langle \hat{f}, \hat{\psi}_\eta \rangle = \left\langle \hat{f}, 2^{-j\frac{3}{2}}\,\hat{\psi}(S_k^T A_{4^{-j}} \cdot) \mathrm{e}^{2\pi\mathrm{i}\langle A_{4^{-j}} S_k m, \cdot \rangle} \right\rangle, \tag{2}$$

where $\eta = (j,k,m,\iota)$ indexes *scale* j, *orientation* k, *position* m, and *cone* ι. Considering this shearlet transform for continuum domain data (taking all cones into account) implicitly induces a trapezoidal tiling of frequency space which is evidently not cartesian. A digital grid perfectly adapted to this situation is the so-called "pseudo-polar grid," which we will introduce and discuss subsequently in detail. Let us for now mention that this viewpoint enables representation of the discrete shearlet transform as a cascade of three steps:

(1) Classical Fourier transformation and change of variables to pseudo-polar coordinates.
(2) Weighting by a radial "density compensation" factor.
(3) Decomposition into rectangular tiles and inverse Fourier transform of each tiles.

Before discussing these steps in detail, let us give an overview of how these steps will be faithfully digitized. First, it will be shown in Sect. 2.1, that the two operations in Step (1) can be combined to the so-called pseudo-polar Fourier transform. An oversampling in radial direction of the pseudo-polar grid, on which the pseudo-polar Fourier transform is computed, will then enable the design of

"density-compensation-style" weights on those grid points leading to Steps (1) &
(2) being an isometry. This will be discussed in Sect. 2.2. Section 2.3 is then concerned with the digitization of the discrete shearlets to subband windows. Notice
that a digital analog of (2) moreover requires an additional 2D iFFT. Thus, concluding the digitization of the discrete shearlet transform will cascade the following
steps, which is the exact analogy of the continuum domain shearlet transform (2):

(S1) PPFT: Pseudo-polar Fourier transform with oversampling factor in the radial
     direction.
(S2) Weighting: Multiplication by "density-compensation-style" weights.
(S3) Windowing: Decomposing the pseudo-polar grid into rectangular subband
     windows with additional 2D iFFT.

With a careful choice of the weights and subband windows, this transform is an
isometry. Then, the inverse transform can be computed by merely taking the adjoint
in each step. A final discussion on the FDST will be presented in Sect. 2.4.

2.1 Pseudo-Polar Fourier Transform

We start by discussing Step (S1).

2.1.1 Pseudo-polar grids with oversampling

In [1], a fast pseudo-polar Fourier transform (PPFT) which evaluates the discrete
Fourier transform at points on a trapezoidal grid in frequency space, the so-called
pseudo-polar grid, was already developed. However, the direct use of the PPFT is
problematic, since it is—as defined in [1]—not an isometry. The main obstacle is
the highly nonuniform arrangement of the points on the pseudo-polar grid. This intuitively suggests to downweight points in regions of very high density by using
weights which correspond roughly to the density compensation weights underlying the continuous change of variables. This will be enabled by a sufficient radial
oversampling of the pseudo-polar grid.

   This new pseudo-polar grid, which we will denote in the sequel by Ω_R to indicate
the oversampling rate R, is defined by

$$\Omega_R = \Omega_R^1 \cup \Omega_R^2, \tag{3}$$

where

$$\Omega_R^1 = \{(-\tfrac{2n}{R} \cdot \tfrac{2\ell}{N}, \tfrac{2n}{R}) : -\tfrac{N}{2} \le \ell \le \tfrac{N}{2}, -\tfrac{RN}{2} \le n \le \tfrac{RN}{2}\}, \tag{4}$$

$$\Omega_R^2 = \{(\tfrac{2n}{R}, -\tfrac{2n}{R} \cdot \tfrac{2\ell}{N}) : -\tfrac{N}{2} \le \ell \le \tfrac{N}{2}, -\tfrac{RN}{2} \le n \le \tfrac{RN}{2}\}. \tag{5}$$

This grid is illustrated in Fig. 1. We remark that the pseudo-polar grid introduced
in [1] coincides with Ω_R for the particular choice $R = 2$. It should be emphasized

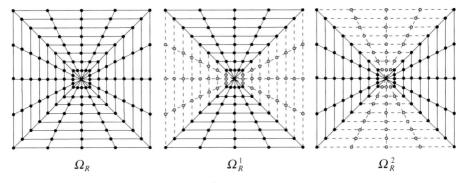

Ω_R Ω_R^1 Ω_R^2

Fig. 1 The pseudo-polar grid $\Omega_R = \Omega_R^1 \cup \Omega_R^2$ for $N = 4$ and $R = 4$

that $\Omega_R = \Omega_R^1 \cup \Omega_R^2$ is neither a disjoint partitioning nor is the mapping $(n,\ell) \mapsto (-\frac{2n}{R} \cdot \frac{2\ell}{N}, \frac{2n}{R})$ or $(\frac{2n}{R}, -\frac{2n}{R} \cdot \frac{2\ell}{N})$ injective. In fact, the center

$$\mathscr{C} = \{(0,0)\} \tag{6}$$

appears $N + 1$ times in Ω_R^1 as well as Ω_R^2, and the points on the seam lines

$$\mathscr{S}_R^1 = \{(-\tfrac{2n}{R}, \tfrac{2n}{R}) : -\tfrac{RN}{2} \le n \le \tfrac{RN}{2}, n \neq 0\},$$
$$\mathscr{S}_R^2 = \{(\tfrac{2n}{R}, -\tfrac{2n}{R}) : -\tfrac{RN}{2} \le n \le \tfrac{RN}{2}, n \neq 0\}.$$

appear in both Ω_R^1 and Ω_R^2.

Definition 1. Let N, R be positive integers, and let Ω_R be the pseudo-polar grid given by (3). For an $N \times N$ image $I := \{I(u,v) : -\frac{N}{2} \le u, v \le \frac{N}{2} - 1\}$, the *pseudo-polar Fourier transform (PPFT)* $\hat{I}$ of I evaluated on Ω_R is then defined to be

$$\hat{I}(\omega_1, \omega_2) = \sum_{u,v=-N/2}^{N/2-1} I(u,v)e^{-\frac{2\pi i}{m_0}(u\omega_1 + v\omega_2)}, \quad (\omega_1, \omega_2) \in \Omega_R,$$

where $m_0 \ge N$ is an integer.

We wish to mention that $m_0 \ge N$ is typically set to be $m_0 = \frac{2}{R}(RN + 1)$ for computational reasons (see also [1]), but we for now allow more generality.

2.1.2 Fast PPFT

It was shown in [1], that the PPFT can be realized in $O(N^2 \log N)$ flops with $N \times N$ being the size of the input image. We will now discuss how the extended pseudo-polar Fourier transform (cf. Definition 1) can be computed with similar complexity.

For this, let I be an image of size $N \times N$. Also, m_0 is set—but not restricted— to be $m_0 = \frac{2}{R}(RN + 1)$; we will elaborate on this choice at the end of this section.

We now focus on Ω_R^1, and mention that the PPFT on the other cone can be computed similarly. Rewriting the pseudo-polar Fourier transform from Definition 1, for $(\omega_1, \omega_2) = (-\frac{2n}{R} \cdot \frac{2\ell}{N}, \frac{2n}{R}) \in \Omega_R^1$, we obtain

$$\hat{I}(\omega_1, \omega_2) = \sum_{u,v=-N/2}^{N/2-1} I(u,v) e^{-\frac{2\pi i}{m_0}(u\omega_1 + v\omega_2)} = \sum_{u=-N/2}^{N/2-1} \sum_{v=-N/2}^{N/2-1} I(u,v) e^{-\frac{2\pi i}{m_0}(u\frac{-4n\ell}{RN} + v\frac{2n}{R})}$$

$$= \sum_{u=-N/2}^{N/2-1} \left(\sum_{v=-N/2}^{N/2-1} I(u,v) e^{-\frac{2\pi i v n}{RN+1}} \right) e^{-2\pi i u \ell \cdot \frac{-2n}{(RN+1)\cdot N}}. \qquad (7)$$

This rewritten form, i.e., (7), suggests that the pseudo-polar Fourier transform $\hat{I}$ of I on Ω_R^1 can be obtained by performing the 1D FFT on the extension of I along direction v and then applying a fractional Fourier transform (frFT) along direction u. To be more specific, we require the following operations:

Fractional Fourier Transform. For $c \in \mathbb{C}^{N+1}$, the *(unaliased) discrete fractional Fourier transform by* $\alpha \in \mathbb{C}$ is defined to be

$$(F_{N+1}^\alpha c)(k) := \sum_{j=-N/2}^{N/2} c(j) e^{-2\pi i \cdot j \cdot k \cdot \alpha}, \quad k = -\frac{N}{2}, \dots, \frac{N}{2}.$$

It was shown in [2], that the fractional Fourier transform $F_{N+1}^\alpha c$ can be computed using $O(N \log N)$ operations. For the special case of $\alpha = 1/(N+1)$, the fractional Fourier transform becomes the (unaliased) 1D discrete Fourier Transform (1D FFT), which in the sequel will be denoted by F_1. Similarly, the 2D discrete Fourier Transform (2D FFT) will be denoted by F_2, and the inverse of the F_2 by F_2^{-1} (2D iFFT).

Padding Operator. For N even, $m > N$ an odd integer, and $c \in \mathbb{C}^N$, the *padding operator* $E_{m,n}$ gives a symmetrically zero padding version of c in the sense that

$$(E_{m,N}c)(k) = \begin{cases} c(k) & k = -\frac{N}{2}, \dots, \frac{N}{2} - 1, \\ 0 & k \in \{-\frac{m}{2}, \dots, \frac{m}{2}\} \setminus \{-\frac{N}{2}, \dots, \frac{N}{2} - 1\}. \end{cases}$$

Using these operators, (7) can be computed by

$$\hat{I}(\omega_1, \omega_2) = \sum_{u=-N/2}^{N/2-1} F_1 \circ E_{RN+1,N} \circ I(u,n) e^{-2\pi i u \ell \cdot \frac{-n}{(RN+1)\cdot N/2}}$$

$$= \sum_{u=-N/2}^{N/2} E_{N+1,N} \circ F_1 \circ E_{RN+1,N} \circ I(u,n) e^{-2\pi i u \ell \cdot \frac{-2n}{(RN+1)\cdot N}} = (F_{N+1}^{\alpha_n} \tilde{I}(\cdot, n))(\ell),$$

where $\tilde{I} = E_{N+1,N} \circ F_1 \circ E_{RN+1,N} \circ I \in \mathbb{C}^{(RN+1)\times(N+1)}$ and $\alpha_n = -\frac{n}{(RN+1)N/2}$. Since the 1D FFT and 1D frFT require only $O(N \log N)$ operations for a vector of size N, the total complexity of this algorithm for computing the pseudo-polar Fourier transform from Definition 1 is indeed $O(N^2 \log N)$ for an image of size $N \times N$.

We would like to also remark that for a different choice of constant m_0, one can compute the pseudo-polar Fourier transform also with complexity $O(N^2 \log N)$ for an image of size $N \times N$. This however requires application of the fractional Fourier transform in both directions u and v of the image, which results in a larger constant for the computational cost; see also [2].

2.2 Density-Compensation Weights

Next we tackle Step (S2), which is more delicate than it might seem, since the weights will not be derivable from simple density compensation arguments.

2.2.1 A Plancherel theorem for the PPFT

For this, we now aim to choose weights $w : \Omega_R \to \mathbb{R}^+$ so that the extended PPFT from Definition 1 becomes an isometry, i.e.,

$$\sum_{u,v=-N/2}^{N/2-1} |I(u,v)|^2 = \sum_{(\omega_1,\omega_2)\in\Omega_R} w(\omega_1,\omega_2) \cdot |\hat{I}(\omega_1,\omega_2)|^2. \tag{8}$$

Observing the symmetry of the pseudo-polar grid, it seems natural to select weight functions w which have full axis symmetry properties, i.e., for all $(\omega_1,\omega_2) \in \Omega_R$, we require

$$w(\omega_1,\omega_2) = w(\omega_2,\omega_1),\ w(\omega_1,\omega_2) = w(-\omega_1,\omega_2),\ w(\omega_1,\omega_2) = w(\omega_1,-\omega_2). \tag{9}$$

Then, the following "Plancherel theorem" for the pseudo-polar Fourier transform on Ω_R—similar to the one for the Fourier transform on the cartesian grid—can be proved.

Theorem 1 ([16]). *Let N be even, and let $w : \Omega_R \to \mathbb{R}^+$ be a weight function satisfying* (9). *Then,* (8) *holds if and only if the weight function w satisfies*

$$\delta(u,v) = w(0,0)$$
$$+ 4 \cdot \sum_{\ell=0,N/2} \sum_{n=1}^{RN/2} w(\tfrac{2n}{R},\tfrac{2n}{R}\cdot\tfrac{-2\ell}{N}) \cdot \cos(2\pi u \cdot \tfrac{2n}{m_0 R}) \cdot \cos(2\pi v \cdot \tfrac{2n}{m_0 R}\cdot\tfrac{2\ell}{N})$$
$$+ 8 \cdot \sum_{\ell=1}^{N/2-1} \sum_{n=1}^{RN/2} w(\tfrac{2n}{R},\tfrac{2n}{R}\cdot\tfrac{-2\ell}{N}) \cdot \cos(2\pi u \cdot \tfrac{2n}{m_0 R}) \cdot \cos(2\pi v \cdot \tfrac{2n}{m_0 R}\cdot\tfrac{2\ell}{N}) \tag{10}$$

for all $-N+1 \le u,v \le N-1$.

Proof. We start by computing the right-hand side of (8):

$$\sum_{(\omega_1,\omega_2)\in\Omega_R} w(\omega_1,\omega_2)\cdot|\hat{I}(\omega_1,\omega_2)|^2$$

$$=\sum_{(\omega_1,\omega_2)\in\Omega_R} w(\omega_1,\omega_2)\cdot\left|\sum_{u,v=-N/2}^{N/2-1} I(u,v)e^{-\frac{2\pi i}{m_0}(u\omega_1+v\omega_2)}\right|^2$$

$$=\sum_{(\omega_1,\omega_2)\in\Omega_R} w(\omega_1,\omega_2)\cdot\left[\sum_{u,v=-N/2}^{N/2-1}\sum_{u',v'=-N/2}^{N/2-1} I(u,v)\overline{I(u',v')}e^{-\frac{2\pi i}{m_0}((u-u')\omega_1+(v-v')\omega_2)}\right]$$

$$=\sum_{(\omega_1,\omega_2)\in\Omega_R} w(\omega_1,\omega_2)\cdot\sum_{u,v=-N/2}^{N/2-1} |I(u,v)|^2$$

$$+\sum_{\substack{u,v,u',v'=-N/2\\(u,v)\neq(u',v')}}^{N/2-1} I(u,v)\overline{I(u',v')}\cdot\left[\sum_{(\omega_1,\omega_2)\in\Omega_R} w(\omega_1,\omega_2)\cdot e^{-\frac{2\pi i}{m_0}((u-u')\omega_1+(v-v')\omega_2)}\right].$$

Choosing $I=c_{u_1,v_1}\delta(u-u_1,v-v_1)+c_{u_2,v_2}\delta(u-u_2,v-v_2)$ for all $-N/2\le u_1,v_1,$
$u_2,v_2\le N/2-1$ and for all $c_{u_1,v_1},c_{u_2,v_2}\in\mathbb{C}$, we can conclude that (8) holds if and only if

$$\sum_{(\omega_1,\omega_2)\in\Omega_R} w(\omega_1,\omega_2)\cdot e^{-\frac{2\pi i}{m_0}(u\omega_1+v\omega_2)}=\delta(u,v),\quad -N+1\le u,v\le N-1.$$

By the symmetry of the weights (9), this is equivalent to

$$\sum_{(\omega_1,\omega_2)\in\Omega_R} w(\omega_1,\omega_2)\cdot[\cos(\tfrac{2\pi}{m_0}u\omega_1)\cos(\tfrac{2\pi}{m_0}v\omega_2)]=\delta(u,v)\qquad(11)$$

for all $-N+1\le u,v\le N-1$. From this, we can deduce that (11) is equivalent to (10), which proves the theorem. $\square$

Notice that (10) is a linear system with $RN^2/4+RN/2+1$ unknowns and $(2N-1)^2$ equations, wherefore, in general, one needs the oversampling factor R to be at least 16 to enforce solvability.

2.2.2 Relaxed form of weight functions

The computation of the weights satisfying Theorem 1 by solving the full linear system of (10) is much too complex. Hence, we relax the requirement for exact isometric weighting, and represent the weights in terms of undercomplete basis functions on the pseudo-polar grid.

More precisely, we first choose a set of basis functions $w_1, \ldots, w_{n_0} : \Omega_R \to \mathbb{R}^+$ such that $\sum_{j=1}^{n_0} w_j(\omega_1, \omega_2) \neq 0$ for all $(\omega_1, \omega_2) \in \Omega_R$. We then represent weight functions $w : \Omega_R \to \mathbb{R}^+$ by

$$w := \sum_{j=1}^{n_0} c_j w_j, \tag{12}$$

with $c_1, \ldots, c_{n_0}$ being nonnegative constants. This approach now enables solving (10) for the constants $c_1, \ldots, c_{n_0}$ using the least squares method, thereby reducing the computational complexity significantly. The "full" weight function w is then given by (12).

We next present two different choices of weights which were derived by this relaxed approach. Notice that (ω_1, ω_2) and (n, ℓ) will be used interchangeably.

Choice 1. The set of basis functions $w_1, \ldots, w_5$ is defined as follows:
Center: $w_1 = 1_{(0,0)}$.
Boundary: $w_2 = 1_{\{(\omega_1, \omega_2): |n| = NR/2,\, \omega_1 = \omega_2\}}$ and $w_3 = 1_{\{(\omega_1, \omega_2): |n| = NR/2,\, \omega_1 \neq \omega_2\}}$.
Seam lines: $w_4 = |n| \cdot 1_{\{(\omega_1, \omega_2): 1 \leq |n| < NR/2,\, \omega_1 = \omega_2\}}$.
Interior: $w_5 = |n| \cdot 1_{\{(\omega_1, \omega_2): 1 \leq |n| < NR/2,\, \omega_1 \neq \omega_2\}}$.

Choice 2. The set of basis functions $w_1, \ldots, w_{N/2+2}$ is defined as follows:
Center: $w_1 = 1_{(0,0)}$.
Radial Lines: $w_{\ell+2} = 1_{\{(\omega_1, \omega_2): 1 < |n| < NR/2,\, \omega_2 = \frac{\ell}{N/2}\omega_1\}}$, $\quad \ell = 0, 1, \ldots, N/2$.

The associated weight functions are displayed in Fig. 2. In general, suitable weight functions usually obey the pattern of linearly increasing values along the radial direction. Thus, this is a natural requirement for the basis functions.

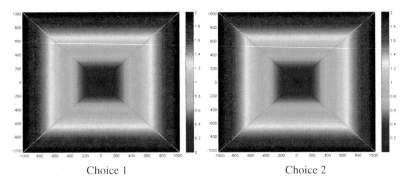

Choice 1   Choice 2

Fig. 2 Weight functions on the pseudo-polar grid for $N = 256$ and $R = 8$

2.2.3 Computation of the weighting

For the FDST—as also in the implementation in ShearLab—the coefficients in the expansion (12) will be computed off-line, and then hardwired in the code.

This enables the weighting of a function on the pseudo-polar grid to simply be a point-wise multiplication in each sampling point. That is, letting $J := \hat{I} : \Omega_R \to \mathbb{C}$ be the pseudo-polar Fourier transform of an $N \times N$ image I and $w : \Omega_R \to \mathbb{R}^+$ be any suitable weight function on Ω_R, the following values need to be computed:

$$J_w(\omega_1, \omega_2) = J(\omega_1, \omega_2) \cdot \sqrt{w(\omega_1, \omega_2)} \quad \text{for all } (\omega_1, \omega_2) \in \Omega_R.$$

Let us comment on why the square root of the weight is utilized. If the weights w satisfy the condition in Theorem 1, we obtain $P^* w P = \text{Id}$ (P is the operator for the PPFT), which can be written in a symmetric form as follows: $(\sqrt{w}P)^* \sqrt{w}P = \text{Id}$. This form shows that the operator $\sqrt{w}P$ can be inverted by taking the adjoint $(\sqrt{w}P)^*$. In other words, each image can be reconstructed from its weighted pseudo-polar Fourier transform by applying the adjoint of the weighted pseudo-polar Fourier transform. This issue will be discussed in further detail in Sect. 2.4.2.

2.3 Digital Shearlets on Pseudo-Polar Grid

We next aim at deriving a faithful digitization of the shearlet transform associated with a band-limited cone-adapted discrete shearlet system to the pseudo-polar grid. This would settle Step (S3).

2.3.1 Preparation for faithful digitization

For this, let us recall the definition of the discrete shearlet transform associated with (2); taking the particular form (1) of the shearlet $\psi \in L^2(\mathbb{R}^2)$ into account. Restricting our attention to the cone $\mathscr{C}_{21}$, we obtain

$$f \mapsto \left\langle \hat{f}, 2^{-j\frac{3}{2}} \hat{\psi}(S_k^T A_{4-j} \cdot) \chi_{\mathscr{C}_{21}} e^{2\pi i \langle A_{4-j} S_k m, \cdot \rangle} \right\rangle$$
$$= \left\langle \hat{f}, 2^{-j\frac{3}{2}} \hat{\psi}_1(4^{-j}\xi_1) \hat{\psi}_2(k + 2^j \tfrac{\xi_2}{\xi_1}) \chi_{\mathscr{C}_{21}} e^{2\pi i \langle A_{4-j} S_k m, \cdot \rangle} \right\rangle,$$

for scale j, orientation k, position m, and cone ι.

To approach a faithful digitization, we first have to partition Ω_R according to the partitioning of the plane into $\mathscr{C}_{11}, \mathscr{C}_{12}, \mathscr{C}_{21}$, and $\mathscr{C}_{22}$, as well as a centered rectangle $\mathscr{R}$. The center $\mathscr{C}$ as defined in (6) will play the role of $\mathscr{R}$. Thus, it remains to partition the set Ω_R beyond the already defined partitioning into Ω_R^1 and Ω_R^2 (cf. (4) and (5)) by setting

$$\Omega_R^1 = \Omega_R^{11} \cup \mathscr{C} \cup \Omega_R^{12} \quad \text{and} \quad \Omega_R^2 = \Omega_R^{21} \cup \mathscr{C} \cup \Omega_R^{22},$$

where

$$\Omega_R^{11} = \{(-\tfrac{2n}{R} \cdot \tfrac{2\ell}{N}, \tfrac{2n}{R}) : -\tfrac{N}{2} \le \ell \le \tfrac{N}{2}, 1 \le n \le \tfrac{RN}{2}\},$$
$$\Omega_R^{12} = \{(-\tfrac{2n}{R} \cdot \tfrac{2\ell}{N}, \tfrac{2n}{R}) : -\tfrac{N}{2} \le \ell \le \tfrac{N}{2}, -\tfrac{RN}{2} \le n \le -1\},$$
$$\Omega_R^{21} = \{(\tfrac{2n}{R}, -\tfrac{2n}{R} \cdot \tfrac{2\ell}{N}) : -\tfrac{N}{2} \le \ell \le \tfrac{N}{2}, 1 \le n \le \tfrac{RN}{2}\},$$
$$\Omega_R^{22} = \{(\tfrac{2n}{R}, -\tfrac{2n}{R} \cdot \tfrac{2\ell}{N}) : -\tfrac{N}{2} \le \ell \le \tfrac{N}{2}, -\tfrac{RN}{2} \le n \le -1\}.$$

When restricting to the cone Ω_R^{21}, say, the exact digitization of the coefficients of the discrete shearlet system is

$$\sum_{\omega := (\omega_1, \omega_2) \in \Omega_R^{21}} J(\omega_1, \omega_2) 2^{-j\frac{3}{2}} \overline{\hat{\psi}(S_k^T A_{4-j} \omega)} e^{-2\pi i \langle A_{4-j} S_k m, \omega \rangle}$$

$$= \sum_{(\omega_1, \omega_2) \in \Omega_R^{21}} J(\omega_1, \omega_2) 2^{-j\frac{3}{2}} \overline{W(4^{-j} \omega_x) V(k + 2^j \tfrac{\omega_2}{\omega_1})} e^{-2\pi i \langle A_{4-j} S_k m, \omega \rangle}$$

$$= \sum_{n=1}^{\frac{RN}{2}} \sum_{\ell=-\frac{N}{2}}^{\frac{N}{2}} J(\omega_1, \omega_2) 2^{-j\frac{3}{2}} \overline{W(4^{-j}\tfrac{2n}{R}) V(k - 2^{j+1} \tfrac{\ell}{N})} e^{-2\pi i \langle m, S_k^T A_{4-j} \omega \rangle}, \quad (13)$$

where V and W as well as the ranges of j, k, and m are to be carefully chosen.

Our main objective will be to achieve a digital shearlet transform, which is an isometry. This— as in the continuum domain situation—is equivalent to requiring the associated shearlet system to form a tight frame for functions $J : \Omega_R \to \mathbb{C}$. For the convenience of the reader let us recall the notion of a Parseval frame in this particular situation. A sequence $(\varphi_\lambda)_{\lambda \in \Lambda}$—$\Lambda$ being some indexing set – is a *tight frame* for all functions $J : \Omega_R \to \mathbb{C}$, if

$$\sum_{\lambda \in \Lambda} \left| \sum_{(\omega_1, \omega_2) \in \Omega_R} J(\omega_1, \omega_2) \overline{\varphi_\lambda(\omega_1, \omega_2)} \right|^2 = \sum_{(\omega_1, \omega_2) \in \Omega_R} |J(\omega_1, \omega_2)|^2.$$

In the sequel we will define digital shearlets on Ω_R^{21} and extend the definition to the other cones by symmetry.

2.3.2 Subband windows on the pseudo-polar grid

We start by defining the scaling function, which will depend on two functions V_0 and W_0, and the generating digital shearlet, which will depend on again two functions V and W. W_0 and W will be chosen to be Fourier transforms of wavelets, and V_0 and V will be chosen to be "bump" functions, paralleling the construction of classical shearlets.

First, let W_0 be the Fourier transform of the Meyer scaling function such that

$$\text{supp} W_0 \subseteq [-1, 1] \quad \text{and} \quad W_0(\pm 1) = 0, \quad (14)$$

and let V_0 be a "bump" function satisfying

$$\mathrm{supp} V_0 \subseteq [-3/2, 3/2] \quad \text{with} \quad V_0(\xi) \equiv 1 \text{ for } |\xi| \leq 1, \xi \in \mathbb{R}.$$

Then, we define the *scaling function* ϕ for the digital shearlet system to be

$$\hat{\phi}(\xi_1, \xi_2) = W_0(4^{-j_L} \xi_1) V_0(4^{-j_L} \xi_2), \quad (\xi_1, \xi_2) \in \mathbb{R}^2.$$

We remark that we will later restrict this function to the pseudo-polar grid.

Let next W be the Fourier transform of the Meyer wavelet function satisfying

$$\mathrm{supp} W \subseteq [-4, 1/4] \cup [1/4, 4] \quad \text{and} \quad W(\pm 1/4) = W(\pm 4) = 0, \tag{15}$$

as well as, choosing the lowest scale j_L to be $j_L := -\lceil \log_4(R/2) \rceil$,

$$|W_0(4^{-j_L} \xi)|^2 + \sum_{j=j_L}^{\lceil \log_4 N \rceil} |W(4^{-j} \xi)|^2 = 1 \quad \text{for all } |\xi| \leq N, \, \xi \in \mathbb{R}. \tag{16}$$

We further choose V to be a "bump" function satisfying

$$\mathrm{supp} V \subseteq [-1, 1] \quad \text{and} \quad V(\pm 1) = 0, \tag{17}$$

as well as

$$|V(\xi - 1)|^2 + |V(\xi)|^2 + |V(\xi + 1)|^2 = 1 \quad \text{for all } |\xi| \leq 1, \, \xi \in \mathbb{R}. \tag{18}$$

Then, the *generating shearlet* ψ for the digital shearlet system on Ω_R^2 is defined as

$$\hat{\psi}(\xi_1, \xi_2) = W(\xi_1) V(\tfrac{\xi_2}{\xi_1}), \quad (\xi_1, \xi_2) \in \mathbb{R}^2. \tag{19}$$

Notice that (18) implies

$$\sum_{k=-2^j}^{2^j} |V(2^j \xi - k)|^2 = 1 \quad \text{for all } |\xi| \leq 1, \, \xi \in \mathbb{R}; j \geq 0, \tag{20}$$

which will become important for the analysis of frame properties. For the particular choice of V_0, W_0, V, and W in `ShearLab`, we refer to Sect. 2.3.7.

2.3.3 Range of parameters

We from now on assume that R and N are both positive even integers and that $N = 2^{n_0}$ for some integer $n_0 \in \mathbb{N}$. This poses no restrictions, since both parameters can be enlarged to satisfy this condition.

We start by analyzing the range of j. Recalling the definition of the shearlet ψ in (19) and the support properties of W and V in (15) and (17), respectively, we observe that the digitized shearlet

$$2^{-j\frac{3}{2}}W(4^{-j}\tfrac{2n}{R})V(k-2^{j+1}\tfrac{\ell}{N})e^{2\pi i\langle m,S_k^T A_{4-j}\omega\rangle} \tag{21}$$

from (13) has radial support

$$n=4^{j-1}\tfrac{R}{2}+t_1,\quad t_1=0,\ldots,4^{j-1}\cdot\tfrac{15R}{2} \tag{22}$$

on the cone Ω_R^{21}. To determine the appropriate range of j, we will analyze the precise support in radial direction. If $j<-\lceil\log(R/2)\rceil$, then $n<1$, which corresponds to only one point—the origin—and is dealt with by the scaling function. If $j>\lceil\log_4 N\rceil$, we have $n\geq\frac{RN}{2}$. Hence, the value $W(1/4)=0$ (cf. (15)) is placed on the boundary, and these scales can be omitted. Therefore, the range of the scaling parameter will be chosen to be

$$j\in\{j_L,\ldots,j_H\},\quad\text{where } j_L:=-\lceil\log(R/2)\rceil\text{ and } j_H:=\lceil\log_4 N\rceil.$$

Next, we determine the appropriate range of k. Again recalling the definition of the shearlet ψ in (19), the digitized shearlet (21) has angular support

$$\ell=2^{-j-1}N(k-1)+t_2,\quad t_2=0,\ldots,2^{-j}N \tag{23}$$

on the cone Ω_R^{21}. To compute the range of k, we start by examining the case $j\geq 0$. If $k>2^j$, we have $\ell\geq N/2$. Hence the value $V(-1)=0$ (cf. (17)) is placed on the seam line, and these parameters can be omitted. By symmetry, we also obtain $k\geq-2^j$. Thus, the shearing parameter will be chosen to be

$$k\in\{-2^j,\ldots,2^j\}.$$

2.3.4 Support size of shearlets

We next compute the support as well as the support size of scaled and sheared version of digital shearlets. This will be used for the normalization of digital shearlets.

As before, we first analyze the radial support. By (22), the radial supports of the windows associated with scales $j_L<j<j_H$ are

$$n=4^{j-1}\tfrac{R}{2}+t_1,\quad t_1=0,\ldots,4^{j-1}\cdot\tfrac{15R}{2}, \tag{24}$$

and the radial support of the windows associated with the scale $j_L=-\lceil\log_4(R/2)\rceil$ and $j_H=\lceil\log_4 N\rceil$ is

$$\begin{aligned}
n=t_1,\qquad\qquad & t_1=1,\ldots,4^{j_L+1}\tfrac{R}{2}, & \text{for } j=j_L,\\
n=4^{j_H-1}\tfrac{R}{2}+t_1,\quad & t_1=0,\ldots,\tfrac{RN}{2}-4^{j_H-1}\tfrac{R}{2}, & \text{for } j=j_H.
\end{aligned} \tag{25}$$

Turning to the angular direction, by (23), the angular support of the windows at scale j associated with shears $-2^j < k < 2^j$ is

$$\ell = 2^{-j-1}N(k-1) + t_2, \quad t_2 = 0,\dots,2^{-j}N, \tag{26}$$

the angular support at scale j associated with the shear parameter $k = -2^j$ is

$$\ell = 2^{-j-1}N(-2^j-1) + t_2, \quad t_2 = 2^{-j}\frac{N}{2},\dots,2^{-j}N,$$

and for $k = 2^j$ it is

$$\ell = 2^{-j-1}N(2^j-1) + t_2, \quad t_2 = 0,\dots,2^{-j}\frac{N}{2}. \tag{27}$$

For the case $j < 0$, we simply let $k = 0$ and $\ell = -N/2 + t_2$ with $t_2 = 0,\dots,N$. Also, for this lower frequency case, the window function $W(4^{-j}\omega_1)V(k+2^j\frac{\omega_2}{\omega_1})$ is slightly modified to be $W(4^{-j}\omega_1)V_0(k+2^j\frac{\omega_2}{\omega_1})$.

These computations now allow us to determine the support size of the function $W(4^{-j}\omega_1)V(k+2^j\frac{\omega_2}{\omega_1})$ in terms of pairs (n,ℓ), which for scale j and shear k, is

$$\mathscr{L}_j^1 = \begin{cases} 4^{j+1}\frac{R}{2} & : j = j_L, \\ 4^{j-1}\cdot\frac{15R}{2} + 1 & : j_L < j < j_H, \\ \frac{RN}{2} - 4^{j-1}\frac{R}{2} + 1 & : j = j_H, \end{cases} \tag{28}$$

and

$$\mathscr{L}_{j,k}^2 = \begin{cases} 2^{-j}N + 1 & : -2^j < k < 2^j \text{ with } j \ge 0, \\ 2^{-j}\frac{N}{2} + 1 & : k \in \{-2^j, 2^j\} \text{ with } j \ge 0, \\ N + 1 & : j < 0. \end{cases} \tag{29}$$

2.3.5 Digitization of the exponential term

We next digitize the exponential term in (13), which can be rewritten as

$$e^{-2\pi i\langle m, S_k^T A_{4-j}\omega\rangle} = e^{-2\pi i\langle m, (4^{-j}\omega_1, 4^{-j}k\omega_1 + 2^{-j}\omega_2)\rangle} = e^{-2\pi i\langle m, (4^{-j}\frac{2n}{R}, 4^{-j}k\frac{2n}{R} - 2^{-j}\frac{4\ell n}{RN})\rangle}.$$

We observe two obstacles:

- The change of variables $\tau := S_k^T A_{4-j}\omega$ possible in (13) cannot be performed similarly in this situation due to the fact that the pseudo-polar grid is *not* invariant under the action of $S_k^T A_{4-j}$. This is however the first step in the continuum domain reasoning for tightness; see the chapter on "Introduction to Shearlets."
- The Fourier transform of a function defined on the pseudo-polar grid does *not* satisfy any Plancherel theorem.

These problems require a slight adjustment of the exponential term, which will be the only adaption we allow us to make when digitizing. This will circumvent the two obstacles and enable us to construct a Parseval frame as well as derive a direct application of the inverse Fast Fourier transform in FDST.

The adjustment will be made by using the mapping $\theta : \mathbb{R} \setminus \{0\} \to \mathbb{R}$ defined by $\theta(x,y) = (x, \frac{y}{x})$. This yields the modified exponential term

$$e^{-2\pi i \left\langle m, (\theta \circ (S_k^T)^{-1})(4^{-j}\frac{2n}{R}, 4^{-j}k\frac{2n}{R} - 2^{-j}\frac{4\ell n}{RN}) \right\rangle} = e^{-2\pi i \left\langle m, (4^{-j}\frac{2n}{R}, -2^{j+1}\frac{\ell}{N}) \right\rangle}, \qquad (30)$$

which can be rewritten as

$$e^{-2\pi i \left\langle m, (4^{-j}\frac{2n}{R}, -2^{j+1}\frac{\ell}{N}) \right\rangle} = e^{-2\pi i (\frac{m_1}{4} + (1-k)m_2)} e^{-2\pi i \left\langle m, (4^{-j}\frac{2t_1}{R}, -2^{j+1}\frac{t_2}{N}) \right\rangle},$$

with t_1 and t_2 ranging over an appropriate set defined by (24)–(27).

Now, taking into account the support size of each $W(4^{-j}\omega_1)V(k + 2^j \frac{\omega_2}{\omega_1})$ as given in (28) and (29), we obtain the following reformulation of (30):

$$\exp \left\{ -2\pi i \left\langle m, \left(\frac{\mathcal{L}_j^1 4^{-j}(2/R)}{\mathcal{L}_j^1} t_1, \frac{-\mathcal{L}_{j,k}^2 2^{j+1}(1/N)}{\mathcal{L}_{j,k}^2} t_2 \right) \right\rangle \right\}, \qquad t_1, t_2. \qquad (31)$$

This version shows that we might regard the exponential terms as characters of a suitable locally compact abelian group (see [14]): with associated annihilator identified with the rectangle

$$\mathcal{R}_{j,k} = \left\{ \left(\frac{4^j \frac{R}{2} \cdot r_1}{\mathcal{L}_j^1}, -\frac{\frac{N}{2^{j+1}} \cdot r_2}{\mathcal{L}_{j,k}^2} \right) : r_1 = 0, \ldots, \mathcal{L}_j^1 - 1, r_2 = 0, \ldots, \mathcal{L}_{j,k}^2 - 1 \right\},$$

where $\mathcal{L}_j^1$ and $\mathcal{L}_{j,k}^2$ were defined in (28) and (29), respectively. This viewpoint will be crucial to guarantee that the digital shearlet system defined in Sect. 2.3.6 provides a Parseval frame on the pseudo-polar grid Ω_R. In practice, (31) also ensures that in Step (S3) on each windowed image on the pseudo-polar grid only a 2D iFFT—in contrast to a fractional Fourier transform—needs to be performed, thereby reducing the computational complexity. For the low-frequency square, we further require the set

$$\mathcal{R} = \left\{ (r_1, r_2) : r_1 = -1, \ldots, 1, r_2 = -\frac{N}{2}, \ldots, \frac{N}{2} \right\}.$$

2.3.6 Digital shearlets

We are now ready to define digital shearlets, which we define as functions on the pseudo-polar grid Ω_R. The spatial domain picture can thus be derived by the inverse pseudo-polar Fourier transform.

Definition 2. Retaining the definitions and notations from Sect. 2.3, for all $(\omega_1, \omega_2) \in \Omega_R^{21}$, we define *digital shearlets* at scale $j \in \{j_L, \ldots, j_H\}$, shear $k = [-2^j, 2^j] \cap \mathbb{Z}$, and spatial position $m \in \mathcal{R}_{j,k}$ by

$$\sigma_{j,k,m}^{21}(\omega_1, \omega_2) = \frac{C(\omega_1, \omega_2)}{\sqrt{|\mathcal{R}_{j,k}|}} W(4^{-j}\omega_1) V^j(k + 2^j \frac{\omega_2}{\omega_1}) \chi_{\Omega_R^{21}}(\omega_1, \omega_2) e^{2\pi i \left\langle m, (4^{-j}\omega_1, 2^j \frac{\omega_2}{\omega_1}) \right\rangle},$$

where $V^j = V$ for $j \geq 0$ and $V^j = V_0$ for $j < 0$, and

$$
C(\omega_1, \omega_2) = \begin{cases} 1 & : (\omega_1, \omega_2) \notin \mathscr{S}_R^1 \cup \mathscr{S}_R^2, \\ \frac{1}{\sqrt{2}} & : (\omega_1, \omega_2) \in (\mathscr{S}_R^1 \cup \mathscr{S}_R^2) \setminus \mathscr{C}, \\ \frac{1}{\sqrt{2(N+1)}} & : (\omega_1, \omega_2) \in \mathscr{C}. \end{cases}
$$

The shearlets $\sigma_{j,k,m}^{11}, \sigma_{j,k,m}^{12}, \sigma_{j,k,m}^{22}$ on the remaining cones are defined accordingly by symmetry with equal indexing sets for scale j, shear k, and spatial location m. For $\iota_0 = 1, 2$, $(\omega_1, \omega_2) \in \Omega_R^{\iota_0}$, and $n_0 \in \mathscr{R}$, we define the *scaling function*

$$
\varphi_{n_0}^{\iota_0}(\omega_1, \omega_2) = \frac{C(\omega_1, \omega_2)}{\sqrt{|\mathscr{R}|}} \hat{\phi}(\omega_1, \omega_2) \chi_{\Omega_R^{\iota_0}}(\omega_1, \omega_2) e^{2\pi i \langle n_0, (\frac{n}{3}, \frac{\ell}{N+1}) \rangle}.
$$

Then, the *digital shearlet system DSH* is defined by

$$
DSH = \{\varphi_{n_0}^{\iota_0} : \iota_0 = 1, 2, n_0 \in \mathscr{R}\} \cup \{\sigma_{j,k,m}^{\iota} : j \in \{j_L, \ldots, j_H\}, k \in \{-2^j, 2^j\},
$$
$$
m \in \mathscr{R}_{j,k}, \iota = 11, 12, 21, 22\}.
$$

As desired, the digital shearlet system *DSH*, which we derived as a faithful digitization of the continuum domain band-limited cone-adapted discrete shearlet system, forms a Parseval frame for $J : \Omega_R \to \mathbb{C}$.

Theorem 2 ([16]). *The digital shearlet system DSH defined in Definition 2 forms a Parseval frame for functions $J : \Omega_R \to \mathbb{C}$.*

Proof. Letting $J : \Omega_R \to \mathbb{C}$, we claim that

$$
\langle J, J \rangle_{\Omega_R} = \sum_{\iota_0, n_0} |\langle J, \varphi_n^{\iota_0} \rangle_{\Omega_R}|^2 + \sum_{\iota, j, k, m} |\langle J, \sigma_{j,k,m}^\iota \rangle_{\Omega_R}|^2 \tag{32}
$$

which proves the result. Here $\langle J_1, J_2 \rangle_{\Omega_R} := \sum_{(\omega_1, \omega_2) \in \Omega_R} J_1(\omega_1, \omega_2) \overline{J_2(\omega_1, \omega_2)}$ for $J_1, J_2 : \Omega_R \to \mathbb{C}$.

We start by analyzing the first term on the RHS of (32). Let $\iota_0 \in \{1, 2\}$ and $J_C : \Omega_R \to \mathbb{C}$ be defined by $J_C(\omega_1, \omega_2) := C(\omega_1, \omega_2) \cdot J(\omega_1, \omega_2)$ for $(\omega_1, \omega_2) \in \Omega_R$. Using the support conditions of $\hat{\phi}$,

$$
\sum_{n_0} |\langle J, \varphi_{n_0}^{\iota_0} \rangle_{\Omega_R}|^2 = \sum_{n_0} \left| \sum_{(\omega_1, \omega_2) \in \Omega_R^{\iota_0}} J(\omega_1, \omega_2) \overline{\varphi_{n_0}^{\iota_0}(\omega_1, \omega_2)} \right|^2
$$

$$
= \frac{1}{|\mathscr{R}|} \sum_{n_0} \left| \sum_{(\omega_1, \omega_2) \in \Omega_R^{\iota_0}} J_C(\omega_1, \omega_2) \cdot \hat{\phi}(\omega_1, \omega_2) \cdot e^{-2\pi i \langle n_0, (\frac{n}{3}, \frac{\ell}{N+1}) \rangle} \right|^2
$$

$$
= \frac{1}{|\mathscr{R}|} \sum_{n_0} \left| \sum_{n=-1}^{1} \sum_{\ell=-N/2}^{N/2} J_C(\omega_1, \omega_2) \cdot \hat{\phi}(\omega_1, \omega_2) \cdot e^{-2\pi i \langle n_0, (\frac{n}{3}, \frac{\ell}{N+1}) \rangle} \right|^2.
$$

The choice of $\mathscr{R}$ now allows us to use the Plancherel formula, see Sect. 2.3.5. Exploiting again support properties (see Sect. 2.3.5), we conclude that

$$\sum_{n_0} |\langle J, \varphi_{n_0}^{\iota_0} \rangle_{\Omega_R}|^2 = \sum_{(\omega_1,\omega_2)\in\Omega_R^{\iota_0}} |C(\omega_1,\omega_2) \cdot J(\omega_1,\omega_2)|^2 \cdot |\hat{\phi}(\omega_1,\omega_2)|^2.$$

Combining $\iota_0 = 1,2$ and using (14), we proved

$$\sum_{\iota_0}\sum_{n_0} |\langle J, \varphi_{n_0}^{\iota_0} \rangle_{\Omega_R}|^2 = \sum_{(\omega_1,\omega_2)\in\Omega_R} |J(\omega_1,\omega_2)|^2 \cdot |W_0(\omega_1)|^2. \tag{33}$$

Next we study the second term on the RHS in (32). By symmetry, it suffices to consider the case $\iota = 21$. By the support conditions on W and V (see (15) and (17)),

$$\sum_{j,k,m} |\langle J, \sigma_{j,k,m}^{21} \rangle_{\Omega_R}|^2 = \sum_{j,k}\sum_{m\in\mathscr{R}_{j,k}} \left| \sum_{(\omega_1,\omega_2)\in\Omega_R^{21}} J(\omega_1,\omega_2)\overline{\sigma_{j,k,m}^{21}(\omega_1,\omega_2)} \right|^2$$

$$= \sum_{j,k}\frac{1}{|\mathscr{R}_{j,k}|}\sum_{m\in\mathscr{R}_{j,k}} \left| \sum_{(\omega_1,\omega_2)\in\Omega_R^{21}} J_C(\omega_1,\omega_2)\cdot\overline{W(4^{-j}\omega_1)} \right.$$
$$\left. \cdot\overline{V^j\left(k+2^j\frac{\omega_2}{\omega_1}\right)}\cdot e^{-2\pi i\left\langle m,(4^{-j}\omega_1,2^j\frac{\omega_2}{\omega_1})\right\rangle} \right|^2$$

$$= \sum_{j,k}\frac{1}{|\mathscr{R}_{j,k}|}\sum_{m\in\mathscr{R}_{j,k}} \left| \sum_{n=4^{j-1}(R/2)}^{4^{j+1}(R/2)}\sum_{\ell=2^{-j-1}N(k-1)}^{2^{-j-1}N(k+1)} J_C(\omega_1,\omega_2) \right.$$
$$\left. \cdot\overline{W(4^{-j}\omega_1)}\cdot\overline{V^j\left(k+2^j\frac{\omega_2}{\omega_1}\right)}\cdot e^{-2\pi i\left\langle m,(4^{-j}\frac{2n}{R},-2^{j+1}\frac{\ell}{N})\right\rangle} \right|^2.$$

Similarly as before, the choice of $\mathscr{R}_{j,k}$ does allow us to use the Plancherel formula, see Sect. 2.3.5. Hence,

$$\sum_{j,k,m} |\langle J, \sigma_{j,k,m}^{21} \rangle_{\Omega_R}|^2 = \sum_{j,k}\sum_{(\omega_1,\omega_2)\in\Omega_R^{21}} \left| J_C(\omega_1,\omega_2)\cdot\overline{W(4^{-j}\omega_1)V^j\left(k+2^j\frac{\omega_2}{\omega_1}\right)} \right|^2.$$

Next we use (20) to obtain

$$\sum_{j,k}\sum_{(\omega_1,\omega_2)\in\Omega_R^{21}} \left| J_C(\omega_1,\omega_2)\cdot\overline{W(4^{-j}\omega_1)}\cdot\overline{V^j\left(k+2^j\frac{\omega_2}{\omega_1}\right)} \right|^2$$

$$= \sum_{(\omega_1,\omega_2)\in\Omega_R^{21}} |J_C(\omega_1,\omega_2)|^2 \sum_{j=j_L}^{j_H}|W(4^{-j}\omega_1)|^2 \cdot \sum_{k=-2^j}^{2^j}|V^j\left(k+2^j\frac{\omega_2}{\omega_1}\right)|^2$$

$$= \sum_{(\omega_1,\omega_2)\in\Omega_R^{21}} |J_C(\omega_1,\omega_2)|^2 \sum_{j=j_L}^{j_H}|W(4^{-j}\omega_1)|^2.$$

Thus the second term on the RHS in (32) equals

$$\sum_{\iota}\sum_{j,k,m} |\langle J, \sigma_{j,k,m}^{\iota} \rangle_{\Omega_R}|^2 = \sum_{(\omega_1,\omega_2)\in\Omega_R} |J(\omega_1,\omega_2)|^2 \cdot \sum_{j=j_L}^{j_H}.|W(4^{-j}\omega_1)|^2. \tag{34}$$

Finally, our claim (32) follows from combining (33), (34), and (16). □

2.3.7 Digital shearlet windowing

The final Step (S3) of the FDST then consists in decomposing the data on the points of the pseudo-polar grid given by the previously—in Steps (S1) and (S2)—computed weighted pseudo-polar image $J_w : \Omega_R \to \mathbb{C}$ into rectangular subband windows according to the digital shearlet system DSH defined in Definition 2, followed by a 2D iFFT. More precisely, given J_w, the set of digital shearlet coefficients

$$c_{n_0}^{t_0} := \left\langle J_w, \varphi_{n_0}^{t_0} \right\rangle_{\Omega_R} \text{ for all } t_0, n_0 \quad \text{and} \quad c_{j,k,m}^{t} := \left\langle J_w, \sigma_{j,k,m}^{t} \right\rangle_{\Omega_R} \text{ for all } j,k,m,t$$

is computed followed by application of the 2D iFFT to each windowed image $J_w \varphi_0^{t_0}$ and $J_w \sigma_{j,k,0}^{t}$ restricted on the support of $\varphi_0^{t_0}$ and $\sigma_{j,k,0}^{t}$, respectively.

The definition of the digital shearlet system DSH in Definition 2 requires appropriate choices of the functions ϕ, V_0, V, W_0, and W, and the required conditions are stated throughout Sect. 2.3.2. We now discuss one particular choice, which is chosen in ShearLab. We start selecting the "wavelets" W_0 and W. In Sect. 2.3.2, these functions were defined to be Fourier transforms of the Meyer scaling function and Meyer wavelet function, respectively, i.e.,

$$W_0(\xi) = \begin{cases} 1 & : |\xi| \leq \frac{1}{4}, \\ \cos\left[\frac{\pi}{2} v(\frac{4}{3}|\xi| - \frac{1}{3})\right] & : \frac{1}{4} \leq |\xi| \leq 1, \\ 0 & : \text{otherwise}, \end{cases}$$

and

$$W(\xi) = \begin{cases} \sin\left[\frac{\pi}{2} v(\frac{4}{3}|\xi| - \frac{1}{3})\right] & : \frac{1}{4} \leq |\xi| \leq 1, \\ \cos\left[\frac{\pi}{2} v(\frac{1}{3}|\xi| - \frac{1}{3})\right] & : 1 \leq |\xi| \leq 4, \\ 0 & : \text{otherwise}, \end{cases}$$

where $v \geq 0$ is a C^k function or C^∞ function such that $v(x) + v(1-x) = 1$ for $0 \leq x \leq 1$. One possible choice for v is the function $v(x) = x^4(35 - 84x + 70x^2 - 20x^3)$, $0 \leq x \leq 1$, which then automatically fixes W_0 and W. Since $|W_0(\xi)|^2 + |W(\xi)|^2 = 1$ for $|\xi| \leq 1$, the required condition (16) is satisfied. The function v can be also used to design the "bump" function V as well, which needs to satisfy (18). One possible choice for V is to define it by $V(\xi) = \sqrt{v(1+\xi) + v(1-\xi)}$, $-1 \leq \xi \leq 1$. V_0 can then simply be chosen as $V_0 \equiv 1$. Let us finally mention that ϕ is defined depending on V_0 and W_0, wherefore fixing these two functions determines ϕ uniquely.

2.4 Algorithmic Realization of the FDST

We have previously discussed all main ingredients of the fast digital shearlet transform (FDST)—Fast PPFT, Weighting, and Digital Shearlet Windowing—, and will now summarize those findings. Depending on the application at hand, a fast inverse

transform is required, which we will also detail in the sequel. In fact, we will present
two possibilities: the Adjoint FDST and the Inverse FDST depending on whether the
weighting allows to use the adjoint for reconstruction or whether an iterative procedure is required for higher accuracy. For a more detailed description of FDST,
Adjoint FDST, and Inverse FDST in form of pseudo-code, we refer to [16].

For the sake of brevity, we now let P, w, and W denote the Fast PPFT from
Sect. 2.1.2, the weighting on the pseudo-polar grid described in Sect. 2.2.3, and
windowing operator consisting of the application of the shearlet windows followed
by 2D iFFT to each array as detailed in Sect. 2.3.7, respectively.

2.4.1 FDST

We can summarize the steps of the algorithm FDST as follows:

- *Step (S1):* For a given image I, apply the Fast PPFT as described in Sect. 2.1.2
  to obtain the function $PI : \Omega_R \to \mathbb{C}$.
- *Step (S2):* Apply the square root of an off-line computed weight function $w :$
  $\Omega_R \to \mathbb{C}$ to PI as described in Sect. 2.2.3, yielding $\sqrt{w}PI : \Omega_R \to \mathbb{C}$.
- *Step (S3):* Apply the shearlet windows to the function $\sqrt{w}PI$, followed by a 2D
  iFFT to each array to obtain the shearlet coefficients $W\sqrt{w}PI$, which we denote
  by $c_{n_0}^{\iota_0}$, ι_0, n_0 and $c_{j,k,m}^{\iota}$, j,k,m,ι.

2.4.2 Adjoint FDST

Assuming that the weight function w used in Step (S2) satisfies the condition in
Theorem 1, and using Theorem 2, we obtain

$$(W\sqrt{w}P)^\star W\sqrt{w}P = P^\star\sqrt{w}(W^\star W)\sqrt{w}P = P^\star wP = Id.$$

Hence in this case, the FDST, which is abbreviated by $W\sqrt{w}P$ can be inverted by
applying the Adjoint FDST, which cascades the following steps:

- *Step 1:* For given shearlet coefficients C, i.e., $c_{n_0}^{\iota_0}$, ι_0, n_0 and $c_{j,k,m}^{\iota}$, j,k,m,ι, compute the linear combination of the shearlet windows with coefficients $c_{n_0}^{\iota_0}$, ι_0, n_0
  and $c_{j,k,m}^{\iota}$, j,k,m,ι. This gives the function $W^\star C : \Omega_R \to \mathbb{C}$.
- *Step 2:* Apply the square root of an off-line computed weight function $w : \Omega_R \to$
  $\mathbb{C}$ to $W^\star C$, yielding the function $\sqrt{w}W^\star C : \Omega_R \to \mathbb{C}$.
- *Step 3:* Apply the Fast Adjoint PPFT by running the Fast PPFT "backward."
  For this, we just notice that the adjoint fractional Fourier transform of a vec-
  tor $c \in \mathbb{C}^{N+1}$ with respect to a constant $\alpha \in \mathbb{C}$ is given by $F_{N+1}^{-\alpha}c$. Also, for
  $m > N$, the adjoint padding operator $E_{m,N}^\star$ applied to a vector $c \in \mathbb{C}^m$ is given by
  $(E_{m,N}^\star c)(k) = c(k), k = -N/2, \ldots, N/2 - 1$. The Adjoint PPFT gives $P^\star\sqrt{w}W^\star C$.

2.4.3 Inverse FDST

Normally—as also with the relaxed form of weights debated in Sect. 2.2.2—the weights will not satisfy the conditions of Theorem 1 precisely. A measure for whether application of the adjoint is still feasible will be discussed in Sect. 4.2. If higher accuracy of the reconstruction is required, one might use iterative methods, such as conjugate gradient methods. Since the digital shearlet system forms a Parseval frame, we always have

$$W^\star W \sqrt{w} P = \sqrt{w} P.$$

Hence, iterative methods need to be "only" applied to reconstruct an image I from knowledge of $J := \sqrt{w} PI$, i.e., to solve the equation

$$P^\star w PI = P^\star w J$$

for I. Since J might not be in the range of P, I is typically computed by solving the weighted least square problem $\min_I \| \sqrt{w} PI - \sqrt{w} J \|_2$. Since the matrix corresponding to $P^\star P$ is symmetric positive definite, iterative methods such as the conjugate gradient methods are applicable. The conjugate gradient method is then applied to the equation $Ax = b$ with $A = P^\star w P$ and $b = P^\star w J$. Its performance can be measured by the condition number of the operator $P^\star w P$: $cond(P^\star w P) = \lambda_{\max}(P^\star w P)/\lambda_{\min}(P^\star w P)$, and it turns out that the weight function serves as a preconditioner. We remark that this measure is more closely studied in Sect. 4.2.

3 Digital Shearlet Transform Using Compactly Supported Shearlets

In this section, we will discuss two implementation strategies for computing shearlet coefficients associated with a cone-adapted discrete shearlet system now based on *compactly supported* shearlets, as introduced in the chapter on "Introduction to Shearlets." Again, one main focus will be on deriving a digitization which is faithful to the continuum setting.

Recall that in the context of wavelet theory, faithful digitization is achieved by the concept of multiresolution analysis, where scaling and translation are digitized by discrete operations: Downsampling, upsampling, and convolution. In the case of directional transforms however, *three* types of operators: Scaling, translation, and direction, need to be digitized. In this section, we will pay particular attention to deriving a framework in which each of the three operators is faithfully interpreted as a digitized operation in digital domain. Both approaches will be based on the following digitization strategies:

- *Scaling and translation*: A multiresolution analysis associated with anisotropic scaling A_{2^j} can be applied for each shear parameter k.

- *Directionality*: A faithful digitization of shear operator $S_{2^{-j/2}k}$ has to be achieved with particular care.

In Sect. 3.1, we present the DSST, which is associated with a shearlet system generated by a separable function alongside with discussions on its properties, e.g., its redundancy. Section 3.2 then presents the digital non-separable shearlet transform (DNST), whose shearlet elements are generated by non-separable shearlet generator.

3.1 Digital Separable Shearlet Transform

We now describe a faithful digitization of the continuum domain shearlet transform based on compactly supported shearlets as introduced in [18], which moreover is highly computationally efficient.

3.1.1 Faithful digitization of the compactly supported shearlet transform

We start by discussing those theoretical aspects which allow a faithful digitization of the shearlet transform associated with the shearlet system generated by the separable shearlet ψ defined by

$$\hat{\psi}(\xi) = m_1(4\xi_1)\hat{\phi}(\xi_1)\hat{\phi}(2\xi_2), \quad \xi = (\xi_1, \xi_2) \in \mathbb{R}^2, \tag{35}$$

where m_1 is a carefully chosen bandpass filter and ϕ an adaptively chosen scaling function, see the chapter on "Introduction to Shearlets." For this, we will only consider shearlets $\psi_{j,k,m}$ for the horizontal cone, i.e., belonging to $\Psi(\psi, c)$. Notice that the same procedure can be applied to compute the shearlet coefficients for the vertical cone, i.e., those belonging to $\tilde{\Psi}(\tilde{\psi}, c)$, except for switching the order of variables.

To construct a separable shearlet generator $\psi \in L^2(\mathbb{R}^2)$ and an associated scaling function $\phi \in L^2(\mathbb{R}^2)$, let $\phi \in L^2(\mathbb{R})$ be a compactly supported 1D scaling function satisfying

$$\phi_1(x_1) = \sum_{n_1 \in \mathbb{Z}} h(n_1)\sqrt{2}\phi_1(2x_1 - n_1) \tag{36}$$

for some "appropriately chosen" filter h—we comment on the required condition below. An associated compactly supported 1D wavelet $\psi_1 \in L^2(\mathbb{R})$ can then be defined by

$$\psi_1(x_1) = \sum_{n_1 \in \mathbb{Z}} g(n_1)\sqrt{2}\phi_1(2x_1 - n_1), \tag{37}$$

where again g is an "appropriately chosen" filter. The selected shearlet generator is then defined to be

$$\psi(x_1, x_2) = \psi_1(x_1)\phi_1(x_2), \tag{38}$$

and the scaling function by $\phi(x_1, x_2) = \phi_1(x_1)\phi_1(x_2)$.

Let us comment on whether this is indeed a special case of the shearlet generators defined in (35). The Fourier transform of ψ defined in (38) takes the form

$$\hat{\psi}(\xi_1,\xi_2) = m_1(\xi_1/2)\hat{\phi}_1(\xi_1/2)\hat{\phi}_1(\xi_2/2),$$

where m_1 is a trigonometric polynomial whose Fourier coefficients are $g(n_1)$. We need to compare this expression with the Fourier transform of the shearlet generator ψ given in (35), which is

$$\hat{\psi}(\xi_1,\xi_2) = m_1(4\xi_1)\hat{\phi}_1(2\xi_1)\hat{\phi}_1(\xi_2),$$

with 1D scaling function ϕ_1 defined in (36). We remark that this later scaling function is slightly different defined as in (35). This small adaption is for the sake of presenting a simpler version of the implementation; essentially the same implementation strategy as the one we will describe can be applied to the shearlet generator given in (35).

The filter coefficients h and g are required to be chosen so that ψ satisfies a certain decay condition (cf. [15] of the chapter on "Introduction to Shearlets") to guarantee a stable reconstruction from the shearlet coefficients.

For the signal $f \in L^2(\mathbb{R}^2)$ to be analyzed, we now assume that, for $J > 0$ fixed, f is of the form

$$f(x) = \sum_{n\in\mathbb{Z}^2} f_J(n)2^J\phi(2^J x_1 - n_1, 2^J x_2 - n_2). \tag{39}$$

Let us mention that this is a very natural assumption for a digital implementation in the sense that the scaling coefficients can be viewed as sample values of f—in fact $f_J(n) = f(2^{-J}n)$ with appropriately chosen ϕ. Aiming toward a faithful digitization of the shearlet coefficients $\langle f, \psi_{j,k,m}\rangle$ for $j = 0,\ldots,J-1$, we first observe that

$$\langle f, \psi_{j,k,m}\rangle = \langle f(S_{2-j/2_k}(\cdot)), \psi_{j,0,m}(\cdot)\rangle, \tag{40}$$

and, WLOG we will from now on assume that $j/2$ is integer; otherwise either $\lceil j/2\rceil$ or $\lfloor j/2\rfloor$ would need to be taken. Our observation (40) shows us in fact precisely how to digitize the shearlet coefficients $\langle f, \psi_{j,k,m}\rangle$: By applying the discrete separable wavelet transform associated with the anisotropic sampling matrix A_{2j} to the sheared version of the data $f(S_{2-j/2_k}(\cdot))$. This however requires—compare the assumed form of f given in (39)—that $f(S_{2-j/2_k}(\cdot))$ is contained in the scaling space

$$V_J = \{2^J\phi(2^J\cdot -n_1, 2^J\cdot -n_2) : (n_1,n_2) \in \mathbb{Z}^2\}.$$

It is easy to see that, for instance, if the shear parameter $2^{-j/2}k$ is non-integer, this is unfortunately not the case. The true reason for this failure is that the shear matrix $S_{2-j/2_k}$ does *not* preserve the regular grid $2^{-J}\mathbb{Z}^2$ in V_J, i.e.,

$$S_{2-j/2_k}(\mathbb{Z}^2) \neq \mathbb{Z}^2.$$

In order to resolve this issue, we consider the new scaling space $V^k_{J+j/2,J}$ defined by

$$V^k_{J+j/2,J} = \{2^{J+4/j}\phi(S_k(2^{J+j/2}\cdot-n_1, 2^J\cdot-n_2)) : (n_1, n_2) \in \mathbb{Z}^2\}.$$

We remark that the scaling space $V^k_{J+j/2,J}$ is obtained by refining the regular grid $2^{-J}\mathbb{Z}^2$ along the x_1-axis by a factor of $2^{j/2}$. With this modification, the new grid $2^{-J-j/2}\mathbb{Z} \times 2^{-J}\mathbb{Z}$ is now invariant under $S_{2^{-j/2}k}$, since with $Q = \mathrm{diag}(2,1)$,

$$2^{-J-j/2}\mathbb{Z} \times 2^{-J}\mathbb{Z} = 2^{-J}Q^{-j/2}(\mathbb{Z}^2) = 2^{-J}Q^{-j/2}(S_k\mathbb{Z}^2) = S_{2^{-j/2}k}(2^{-J-j/2}\mathbb{Z} \times 2^{-J}\mathbb{Z}).$$

This allows us to rewrite $f(S_{2^{-j/2}k}(\cdot))$ in (40) in the following way.

Lemma 1. *Retaining the notations and definitions from this section, letting $\uparrow 2^{j/2}$ and $*_1$ denote the 1D upsampling operator by a factor of $2^{j/2}$ and the 1D convolution operator along the x_1-axis, respectively, and setting $h_{j/2}(n_1)$ to be the Fourier coefficients of the trigonometric polynomial*

$$H_{j/2}(\xi_1) = \prod_{k=0}^{j/2-1} \sum_{n_1 \in \mathbb{Z}} h(n_1)e^{-2\pi i 2^k n_1 \xi_1}, \tag{41}$$

we obtain

$$f(S_{2^{-j/2}k}(x)) = \sum_{n \in \mathbb{Z}^2} \tilde{f}_J(S_k n) 2^{J+j/4}\phi_k(2^{J+j/2}x_1 - n_1, 2^J x_2 - n_2),$$

where $\tilde{f}_J(n) = ((f_J)_{\uparrow 2^{j/2}} *_1 h_{j/2})(n)$.

The proof of this lemma requires the following result, which follows from the cascade algorithm in the theory of wavelet.

Proposition 1 ([18]). *Assume that ϕ_1 and $\psi_1 \in L^2(\mathbb{R})$ satisfy (36) and (37), respectively. For positive integers $j_1 \leq j_2$, we then have*

$$2^{\frac{j_1}{2}}\phi_1(2^{j_1}x_1 - n_1) = \sum_{d_1 \in \mathbb{Z}} h_{j_2-j_1}(d_1 - 2^{j_2-j_1}n_1)2^{\frac{j_2}{2}}\phi_1(2^{j_2}x_1 - d_1) \tag{42}$$

and

$$2^{\frac{j_1}{2}}\psi_1(2^{j_1}x_1 - n_1) = \sum_{d_1 \in \mathbb{Z}} g_{j_2-j_1}(d_1 - 2^{j_2-j_1}n_1)2^{\frac{j_2}{2}}\phi_1(2^{j_2}x_1 - d_1), \tag{43}$$

where h_j and g_j are the Fourier coefficients of the trigonometric polynomials H_j defined in (41) and G_j defined by

$$G_j(\xi_1) = \left(\prod_{k=0}^{j-2}\sum_{n_1 \in \mathbb{Z}} h(n_1)e^{-2\pi i 2^k n_1 \xi_1}\right)\left(\sum_{n_1 \in \mathbb{Z}} g(n_1)e^{-2\pi i 2^{j-1} n_1 \xi_1}\right)$$

for $j > 0$ fixed.

Proof (Proof of Lemma 1). Equation (42) with $j_1 = J$ and $j_2 = J + j/2$ implies that

$$2^{J/2}\phi_1(2^J x_1 - n_1) = \sum_{d_1 \in \mathbb{Z}} h_{J-j/2}(d_1 - 2^{j/2}n_1)2^{J/2+j/4}\phi_1(2^{J+j/2}x_1 - d_1). \quad (44)$$

Also, since ϕ is a 2D separable function of the form $\phi(x_1, x_2) = \phi_1(x_1)\phi_1(x_2)$, we have that

$$f(x) = \sum_{n_2 \in \mathbb{Z}}\Big(\sum_{n_1 \in \mathbb{Z}} f_J(n_1, n_2)2^{J/2}\phi_1(2^J x_1 - n_1)\Big)2^{J/2}\phi_1(2^J x_2 - n_2).$$

By (44), we obtain

$$f(x) = \sum_{n \in \mathbb{Z}^2} \tilde{f}_J(n)2^{J+j/4}\phi(2^J Q^{j/2}x - n),$$

where $Q = \text{diag}(2, 1)$. Using $Q^{j/2}S_{2^{-j/2}k} = S_k Q^{j/2}$, this finally implies

$$f(S_{2^{-j/2}k}(x)) = \sum_{n \in \mathbb{Z}^2} \tilde{f}_J(n)2^{J+j/4}\phi(2^J Q^{j/2}S_{2^{-j/2}k}(x) - n)$$

$$= \sum_{n \in \mathbb{Z}^2} \tilde{f}_J(n)2^{J+j/4}\phi(S_k(2^J Q^{j/2}x - S_{-k}n))$$

$$= \sum_{n \in \mathbb{Z}^2} \tilde{f}_J(S_k n)2^{J+j/4}\phi(S_k(2^J Q^{j/2}x - n)). \quad \square$$

The second term to be digitized in (40) is the shearlet $\psi_{j,k,m}$ itself. A direct corollary from Proposition 1 is the following result.

Lemma 2. *Retaining the notations and definitions from this section, we obtain*

$$\psi_{j,k,m}(x) = \sum_{d \in \mathbb{Z}^2} g_{J-j}(d_1 - 2^{J-j}m_1)h_{J-j/2}(d_2 - 2^{J-j/2}m_2)2^{J+j/4}\phi(2^J x - d).$$

As already indicated before, we will make use of the discrete separable wavelet transform associated with an anisotropic scaling matrix, which, for j_1 and $j_2 > 0$ as well as $c \in \ell(\mathbb{Z}^2)$, we define by

$$W_{j_1, j_2}(c)(n_1, n_2) = \sum_{m \in \mathbb{Z}^2} g_{j_1}(m_1 - 2^{j_1}n_1)h_{j_2}(m_2 - 2^{j_2}n_2)c(m_1, m_2), \quad (n_1, n_2) \in \mathbb{Z}^2.$$

$$(45)$$

Finally, Lemmas 1 and 2 yield the following digitizable form of the shearlet coefficients $\langle f, \psi_{j,k,m} \rangle$.

Theorem 3 ([18]). *Retaining the notations and definitions from this section, and letting$\downarrow 2^{j/2}$ be 1D downsampling by a factor of $2^{j/2}$ along the horizontal axis, we obtain*

$$\langle f, \psi_{j,k,m} \rangle = W_{J-j,J-j/2}\Big(\big((\tilde{f}_J(S_k\cdot) * \Phi_k) *_1 \overline{h}_{j/2}\big)_{\downarrow 2^{j/2}}\Big)(m),$$

where $\Phi_k(n) = \langle \phi(S_k(\cdot)), \phi(\cdot - n)\rangle$ for $n \in \mathbb{Z}^2$, and $\overline{h}_{j/2}(n_1) = h_{j/2}(-n_1)$.

3.1.2 Algorithmic realization

Computing the shearlet coefficients using Theorem 3 now restricts to applying the discrete separable wavelet transform (45) associated with the sampling matrix A_{2^j} to the scaling coefficients

$$S^d_{2^{-j/2}k}(f_J)(n) := \left((\tilde{f}_J(S_k \cdot) * \Phi_k) *_1 \overline{h}_{j/2} \right)_{\downarrow 2^{j/2}}(n) \quad \text{for} \quad f_J \in \ell^2(\mathbb{Z}^2). \tag{46}$$

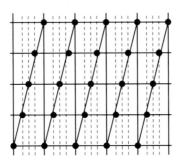

Fig. 3 Illustration of application of the digital shear operator $S^d_{-1/4}$: The *dashed lines* correspond to the refinement of the integer grid. The new sample values lie on the intersections of the *sheared lines* associated with $S_{1/4}$ with this refined grid

Before we state the explicit steps necessary to achieve this, let us take a closer look at the scaling coefficients $S^d_{2^{-j/2}k}(f_J)$, which can be regarded as a new sampling of the data f_J on the integer grid $\mathbb{Z}^2$ by the digital shear operator $S^d_{2^{-j/2}k}$. This procedure is illustrated in Fig. 3 in the case $2^{-j/2}k = -1/4$.

Let us also mention that the filter coefficients $\Phi_k(n)$ in (46) can in fact be easily precomputed for each shear parameter k. For a practical implementation, one may sometimes even skip this additional convolution step assuming that $\Phi_k = \chi_{(0,0)}$.

Concluding, the implementation strategy for DSST cascades the following steps:

- *Step 1:* For given input data f_J, apply the 1D upsampling operator by a factor of $2^{j/2}$ at the finest scale $j = J$.
- *Step 2:* Apply 1D convolution to the upsampled input data f_J with 1D lowpass filter $h_{j/2}$ at the finest scale $j = J$. This gives $\tilde{f}_J$.
- *Step 3:* Resample $\tilde{f}_J$ to obtain $\tilde{f}_J(S_k(n))$ according to the shear sampling matrix S_k at the finest scale $j = J$. Note that this resampling step is straightforward, since the integer grid is invariant under the shear matrix S_k.
- *Step 4:* Apply 1D convolution to $\tilde{f}_J(S_k(n))$ with $\overline{h}_{j/2}$ followed by 1D downsampling by a factor of $2^{j/2}$ at the finest scale $j = J$.
- *Step 5:* Apply the separable wavelet transform $W_{J-j,J-j/2}$ across scales $j = 0, 1, \ldots, J-1$.

3.1.3 Digital realization of directionality

Since the digital realization of a shear matrix $S_{2^{-j/2}k}$ by the digital shear operator $S^d_{2^{-j/2}k}$ is crucial for deriving a faithful digitization of the continuum domain shearlet transform, we will devote this section to a closer analysis.

We start by remarking that in fact in the continuum domain, at least *two* operators exist which naturally provide directionality: Rotation and shearing. Rotation is a very convenient tool to provide directionality in the sense that it preserves important geometric information such as length, angles, and parallelism. However, this operator does not preserve the integer lattice, which causes severe problems for digitization. In contrast to this, a shear matrix S_k does not only provide directionality but also preserves the integer lattice when the shear parameter k is integer. Thus, it is conceivable to assume that directionality can be naturally discretized by using a shear matrix S_k.

To start our analysis of the relation between a shear matrix $S_{2^{-j/2}k}$ and the associated digital shear operator $S^d_{2^{-j/2}k}$, let us consider the following simple example: Set $f_c = \chi_{\{x:x_1=0\}}$. Then digitize f_c to obtain a function f_d defined on $\mathbb{Z}^2$ by setting $f_d(n) = f_c(n)$ for all $n \in \mathbb{Z}^2$. For fixed shear parameter $s \in \mathbb{R}$, apply the shear transform S_s to f_c yielding the sheared function $f_c(S_s(\cdot))$. Next, digitize also this function by considering $f_c(S_s(\cdot))|_{\mathbb{Z}^2}$. The functions f_d and $f_c(S_s(\cdot))|_{\mathbb{Z}^2}$ are illustrated in Fig. 4 for $s = -1/4$. We now focus on the problem that the integer lattice is

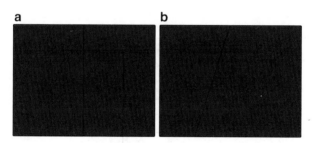

Fig. 4 (**a**) Original image $f_d(n)$. (**b**) Sheared image $f_c(S_{-1/4}n)$

not invariant under the shear matrix $S_{1/4}$. This prevents the sampling points $S_{1/4}(n)$, $n \in \mathbb{Z}^2$ from lying on the integer grid, which causes aliasing of the digitized image $f_c(S_{-1/4}(\cdot))|_{\mathbb{Z}^2}$ as illustrated in Fig. 5a. In order to avoid this aliasing effect, the grid needs to be refined by a factor of 4 along the horizontal axis followed by computing sample values on this refined grid.

More generally, when the shear parameter is given by $s = -2^{-j/2}k$, one can essentially avoid this directional aliasing effect by refining a grid by a factor of $2^{j/2}$ along the horizontal axis followed by computing interpolated sample values on this refined grid. This ensures that the resulting grid contains the sampling points $((2^{-j/2}k)n_2, n_2)$ for any $n_2 \in \mathbb{Z}$ and is preserved by the shear matrix $S_{-2^{-j/2}k}$. This procedure precisely coincides with the application of the digital shear operator

$S^d_{2^{-j/2}k}$, i.e., we just described Steps 1–4 from Sect. 3.1.2 in which the new scaling coefficients $S^d_{2^{-j/2}k}(f_J)(n)$ are computed.

Let us come back to the exemplary situation of $f_c = \chi_{\{x:x_1=0\}}$ and $S_{-1/4}$ we started our excursion with and compare $f_c(S_{-1/4}(\cdot))|_{\mathbb{Z}^2}$ with $S^d_{-1/4}(f_d)|_{\mathbb{Z}^2}$ obtained by applying the digital shear operator $S^d_{-1/4}$ to f_d. And, in fact, the directional aliasing effect on the digitized image $f_c(S_{-1/4}(n))$ in frequency illustrated in Fig. 5a is shown to be avoided in Fig. 5b, c by considering $S^d_{-1/4}(f_d)|_{\mathbb{Z}^2}$. Thus, application of

Fig. 5 (a) Aliased image: DFT of $f_c(S_{-1/4}(n))$. (b) De-aliased image: $S^d_{-1/4}(f_d)(n)$. (c) De-aliased image: DFT of $S^d_{-1/4}(f_d)(n)$

the digital shear operator $S^d_{2^{-j/2}k}$ allows a faithful digitization of the shearing operator associated with the shear matrix $S_{2^{-j/2}k}$.

3.1.4 Redundancy

One of the main issues which practical applicability requires is controllable redundancy. To quantify the redundancy of the discrete shearlet transform, we assume that the input data f is a finite linear combination of translates of a 2D scaling function ϕ at scale J as follows:

$$f(x) = \sum_{n_1=0}^{2^J-1} \sum_{n_2=0}^{2^J-1} d_n \phi(2^J x - n)$$

as it was already the hypothesis in (39). The redundancy—as we view it in our analysis—is then given by the number of shearlet elements necessary to represent f. Furthermore, to state the result in more generality, we allow an arbitrary sampling matrix $M_c = \text{diag}(c_1, c_2)$ for translation, i.e., consider shearlet elements of the form

$$\psi_{j,k,m}(\cdot) = 2^{\frac{3}{4}j} \psi(S_k A_{2^j} \cdot - M_c m).$$

We then have the following result.

Proposition 2 ([18]). *The redundancy of the DSST is $\frac{4}{3c_1c_2}$.*

Proof. For this, we first consider shearlet elements for the horizontal cone for a fixed scale $j \in \{0, \ldots, J-1\}$. We observe that there exist $2^{j/2+1}$ shearing indices k

and $2^j \cdot 2^{j/2} \cdot (c_1 c_2)^{-1}$ translation indices associated with the scaling matrix A_{2^j} and the sampling matrix M_c, respectively. Thus, $2^{2j+1}(c_1 c_2)^{-1}$ shearlet elements from the horizontal cone are required for representing f. Due to symmetry reasons, we require the same number of shearlet elements from the vertical cone. Finally, about c_1^{-2} translates of the scaling function ϕ are necessary at the coarsest scale $j = 0$.

Summarizing, the total number of necessary shearlet elements across all scales is about

$$\left(\frac{4}{c_1 c_2}\right)\left(\sum_{j=0}^{J-1} 2^{2j} + 1\right) = \left(\frac{4}{c_1 c_2}\right)\left(\frac{2^{2J} + 2}{3}\right).$$

The redundancy of each shearlet frame can now be computed as the ratio of the number of coefficients d_n and this number. Letting $J \to \infty$ proves the claim. $\square$

As an example, choose a translation grid with parameters $c_1 = 1$ and $c_2 = 0.4$. Then, the associated DSST has asymptotic redundancy $10/3$.

3.1.5 Computational complexity

A further essential characteristics is the computational complexity (see also Sect. 4.6), which we now formally compute for the digital shearlet transform.

Proposition 3 ([15]). *The computational complexity of DSST is $O(2^{\log_2(\frac{L/2-1}{2})} L \cdot N)$.*

Proof. Analyzing Steps 1–5 from Sect. 3.1.2, we observe that the most time consuming step is the computation of the scaling coefficients in Steps 1–4 for the finest scale $j = J$. This step requires 1D upsampling by a factor of $2^{j/2}$ followed by 1D convolution for each direction associated with the shear parameter k. Letting L denote the total number of directions at the finest scale $j = J$, and N the size of 2D input data, the computational complexity for computing the scaling coefficients in Steps 1–4 is $O(2^{j/2} L \cdot N)$. The complexity of the discrete separable wavelet transform associated with A_{2^j} for Step 5 requires $O(N)$ operations, wherefore it is negligible. The claim follows from the fact that $L = 2(2 \cdot 2^{j/2} + 1)$. $\square$

It should be noted that the total computational cost depends on the number L of shear parameters at the finest scale $j = J$, and it grows approximately by a factor of L^2 as L is increased. It should though be emphasized that L can be chosen so that this shearlet transform is favorably comparable to other redundant directional transforms with respect to running time as well as performance. A reasonable number of directions at the finest scale is 6, in which case the factor $2^{\log_2(1/2(L/2-1))}$ in Proposition 3 equals 1. Hence, in this case the running time of this shearlet transform is only about 6 times slower than the discrete orthogonal wavelet transform, thereby remains in the range of the running time of other directional transforms.

3.1.6 Inverse DSST

Since this transform is not an isometry, the adjoint cannot be used as an inverse transform. However, the "good" ratio of the frame bounds (see the chapters on "Introduction to Shearlets" and "Shearlets and Optimally Sparse Approximations.") leads to fast convergence of iterative methods such as the conjugate gradient method, which requires computing the forward DSST and its adjoint, see [20] and Sect. 2.4.2.

3.2 Digital Non-separable Shearlet Transform

In this section, we describe an alternative approach to derive a faithful digitization of a discrete shearlet transform associated with compactly supported shearlets. This algorithmic realization from [19] resolves the following drawbacks of the DSST:

- Since this transform is not based on a tight frame, an additional computational effort is necessary to approximate the inverse of the shearlet transform by iterative methods.
- Computing the interpolated sampling values in (46) requires additional computational costs.
- This shearlet transform is not shift-variant, even when downsampling associated with A_{2^j} is omitted.

Although this alternative approach resolves these problems, DSST provides a much more faithful digitization in the sense that the shearlet coefficients can be exactly computed in this framework. The main difference between DSST and DNST will be to exploit *non-separable* shearlet generators, which give more flexibility.

3.2.1 Shearlet generators

We start by introducing the *non-separable* shearlet generators utilized in DNST. First, define the shearlet generator ψ^{non} by

$$\hat{\psi}^{non}(\xi) = P(\xi_{1/2}, 2\xi_2)\hat{\psi}(\xi),$$

where ψ is a separable shearlet generator defined in (38) and the trigonometric polynomial P is a 2D fan filter (c.f. [8]). For an illustration of P we refer to Fig. 6a. This in turn defines shearlets $\psi^{non}_{j,k,m}$ generated by non-separable generator function ψ^{non} by setting

$$\psi^{non}_{j,k,m}(x) = 2^{\frac{3}{4}j}\psi^{non}(S_k A_{2^j}x - M_{c_j}m),$$

where M_{c_j} is a sampling matrix given by $M_{c_j} = \text{diag}(c_1^j, c_2^j)$ and c_1^j and c_2^j are sampling constants for translation.

One major advantage of these shearlets $\psi^{non}_{j,k,m}$ is the fact that a fan filter enables refinement of the directional selectivity in frequency domain at each scale.

Figure 6a, b shows the refined essential support of $\hat{\psi}^{non}_{j,k,m}$ as compared to shearlets $\psi_{j,k,m}$ arising from a separable generator as in Sect. 3.1.1.

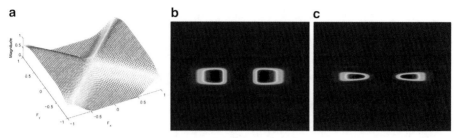

a   b   c

Fig. 6 (**a**) Magnitude response of 2D fan filter. (**b**) Separable shearlet $\psi_{j,k,m}$. (**c**) Non-separable shearlet $\psi^{non}_{j,k,m}$

3.2.2 Algorithmic realization

Next, our aim is to derive a digital formulation of the shearlet coefficients $\langle f, \psi^{non}_{j,k,m} \rangle$ for a function f as given in (39). We will only discuss the case of shearlet coefficients associated with A_{2^j} and S_k; the same procedure can be applied for $\tilde{A}_{2^j}$ and $\tilde{S}_k$ except for switching the order of variables x_1 and x_2.

In Sect. 3.1.3, we discretized a sheared function $f(S_{2^{-j/2}k}\cdot)$ using the digital shear operator $S^d_{2^{-j/2}k}$ as defined in (46). In this implementation, we walk a different path. We digitalize the shearlets $\psi^{non}_{j,k,m}(\cdot) = \psi^{non}_{j,0,m}(S_{2^{-j/2}k}\cdot)$ by combining multiresolution analysis and digital shear operator $S^d_{2^{-j/2}k}$ to digitize the wavelet $\psi^{non}_{j,0,m}$ and the shear operator $S_{2^{-j/2}k}$, respectively. This yields digitized shearlet filters of the form

$$\psi^d_{j,k}(n) = S^d_{2^{-j/2}k}\left(p_{J-j/2} * w_j\right)(n),$$

where w_j is the 2D separable wavelet filter defined by $w_j(n_1,n_2) = g_{J-j}(n_1) \cdot h_{J-j/2}(n_2)$ and $p_{J-j/2}(n)$ are the Fourier coefficients of the 2D fan filter given by $P(2^{J-j}\xi_1, 2^{J-j/2+1}\xi_2)$. The DNST associated with the non-separable shearlet generators ψ^{non}_j is then given by

$$\mathrm{DNST}_{j,k}(f_J)(n) = (f_J * \overline{\psi}^d_{j,k})(2^{J-j}c^j_1 n_1, 2^{J-j/2}c^j_2 n_2), \quad \text{for } f_J \in \ell^2(\mathbb{Z}^2).$$

We remark that the discrete shearlet filters $\psi^d_{j,k}$ are computed by using a similar ideas as in Sect. 3.1.1. As before, those filter coefficients can be precomputed to avoid additional computational effort.

Further notice that by setting $c^j_1 = 2^{j-J}$ and $c^j_2 = 2^{j/2-J}$, the DNST simply becomes a 2D convolution. Thus, in this case, DNST is shift invariant.

3.2.3 Inverse DNST

If $c_1^j = 2^{j-J}$ and $c_2^j = 2^{j/2-J}$, the dual shearlet filters $\tilde{\psi}_{j,k}^d$ can be easily computed by,

$$\hat{\tilde{\psi}}_{j,k}^d(\xi) = \frac{\hat{\psi}_{j,k}^d(\xi)}{\sum_{j,k}|\hat{\psi}_{j,k}^d(\xi)|^2}$$

and we obtain the reconstruction formula

$$f_J = \sum_{j,k}(f_J * \overline{\psi}_{j,k}^d) * \tilde{\psi}_{j,k}^d.$$

This reconstruction is stable due to the existence of positive upper and lower bounds of $\sum_{j,k}|\hat{\psi}_{j,k}^d(\xi)|^2$, which is implied by the frame property of non-separable shearlets $\psi_{j,k,m}^{non}$, see [19]. Thus, no iterative methods are required for the Inverse DNST. The frequency response of a discrete shearlet filter $\psi_{j,k}^d$ and its dual $\tilde{\psi}_{j,k}^d$ is illustrated in Fig. 7. We observe that primal and dual shearlet filters behave similarly in the sense that both the filters are very well localized in frequency.

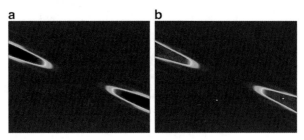

Fig. 7 Magnitude response of shearlet filter $\psi_{j,k}^d$ and its dual filter $\tilde{\psi}_{j,k}^d$

4 Framework for Quantifying Performance

We next present the framework for quantifying performance of implementations of directional transforms, which was originally introduced in [16, 10]. This set of test measures was designed to analyze particular well-understood properties of a given algorithm, which in this case are the desiderata proposed at the beginning of this chapter. This framework was moreover introduced to serve as a tool for tuning the parameters of an algorithm in a rational way and as an objective common ground for comparison of different algorithms. The performance of the three algorithms FDST, DSST, and DNST will then be tested with respect to those measures. This will give us insight into their behavior with respect to the analyzed characteristics, and also allow a comparison. However, the test values of these three algorithms will also show the delicateness of designing such a testing framework in a fair manner, since due to the complexity of algorithmic realizations it is highly difficile to do each aspect of an algorithm justice. It should though be emphasized that—apart from being able to rationally tune parameters—such a framework of quantifying performance is essential for an objective judgement of algorithms. The codes of all measures are available in ShearLab.

In the following, S shall denote the transform under consideration, $S^\star$ its adjoint, and, if iterative reconstruction is tested, $G_A J$ shall stand for the solution of the matrix problem $AI = J$ using the conjugate gradient method with residual error set to be 10^{-6}. Some measures apply specifically to transforms utilizing the pseudo-polar grid, for which purpose we introduce the notation P for the pseudo-polar Fourier transform, w shall denote the weighting applied to the values on the pseudo-polar grid, and W shall be the windowing with additional 2D iFFT.

4.1 Algebraic Exactness

We require the transform to be the precise implementation of a theory for digital data on a pseudo-polar grid. In addition, to ensure numerical accuracy, we provide the following measure, which is designed for transforms using the pseudo-polar grid.

> **Measure 1** *Generate a sequence of five (of course, one can choose any reasonable integer other than 5) random images $I_1, \ldots, I_5$ on a pseudo-polar grid for $N = 512$ and $R = 8$ with standard normally distributed entries. Our quality measure will then be the Monte Carlo estimate for the operator norm $\|W^\star W - \mathrm{Id}\|_{\mathrm{op}}$ given by*
>
> $$M_{alg} = \max_{i=1,\ldots,5} \frac{\|W^\star W I_i - I_i\|_2}{\|I_i\|_2}.$$

This measure applies to the FDST—not to the DSST or DNST—, and we obtain

$$M_{alg} = 6.6E - 16.$$

This confirms that the windowing in the FDST is indeed up to machine precision a Parseval frame, which was already theoretically confirmed by Theorem 2.

4.2 Isometry of Pseudo-Polar Transform

We next test the pseudo-polar Fourier transform itself which might be used in the algorithm under consideration. For this, we will provide three different measures, each being designed to test a different aspect.

Measure 2

- Closeness to isometry. *Generate a sequence of five random images $I_1, \ldots, I_5$ of size 512×512 with standard uniformly distributed entries. Our quality measure will then be the Monte Carlo estimate for the operator norm $\|P^\star wP - \mathrm{Id}\|_{\mathrm{op}}$ given by*

$$M_{\mathrm{isom}_1} = \max_{i=1,\ldots,5} \frac{\|P^\star wPI_i - I_i\|_2}{\|I_i\|_2}.$$

- Quality of preconditioning. *Our quality measure will be the spread of the eigenvalues of the Gram operator $P^\star wP$ given by*

$$M_{\mathrm{isom}_2} = \frac{\lambda_{\max}(P^\star wP)}{\lambda_{\min}(P^\star wP)}.$$

- Invertibility. *Our quality measure will be the Monte Carlo estimate for the invertibility of the operator $\sqrt{w}P$ using conjugate gradient method $G_{\sqrt{w}P}$ (residual error is set to be 10^{-6}, here $G_A J$ means solving matrix problem $AI = J$ using conjugate gradient method) given by*

$$M_{\mathrm{isom}_3} = \max_{i=1,\ldots,5} \frac{\|G_{\sqrt{w}P}\sqrt{w}PI_i - I_i\|_2}{\|I_i\|_2}.$$

This measure applies to the FDST—not to the DSST or DNST—, for which we obtain the following numerical results, see Table 1.

Table 1 The numerical results for the test on isometry of the pseudo-polar transform

	M_{isom_1}	M_{isom_2}	M_{isom_3}
FDST	9.3E-4	1.834	3.3E-7

The slight isometry deficiency of $M_{\mathrm{isom}_1} \approx 9.9\text{E-}4$ mainly results from the isometry deficiency of the weighting. However, for practical purposes this transform can be still considered to be an isometry allowing the utilization of the adjoint as inverse.

4.3 Parseval Frame Property

We now test the overall frame behavior of the system defined by the transform. These measures now apply to more than pseudo-polar based transforms, in particular, to FDST, DSST, and DNST.

Measure 3 *Generate a sequence of five random images $I_1, \ldots, I_5$ of size 512×512 with standard uniformly distributed entries. Our quality measure will then twofold:*

- Adjoint transform. *The measure will be the Monte Carlo estimate for the operator norm $\|S^\star S - \mathrm{Id}\|_{\mathrm{op}}$ given by*

$$M_{\mathrm{tight}_1} = \max_{i=1,\ldots,5} \frac{\|S^\star S I_i - I_i\|_2}{\|I_i\|_2}.$$

- Iterative reconstruction. *Using conjugate gradient method $G_{\sqrt{w}P}$, our measure will be given by*

$$M_{\mathrm{tight}_2} = \max_{i=1,\ldots,5} \frac{\|G_{\sqrt{w}P} W^\star S I_i - I_i\|_2}{\|I_i\|_2}.$$

Table 2 presents the performance of FDST and DNST with respect to these quantitative measures.

Table 2 The numerical results for the test on Parseval property

	M_{tight_1}	M_{tight_2}
FDST	9.9E-4	3.8E-7
DSST	1.9920	1.2E-7
DNST	0.1829	5.8E-16 (with dual filters)

The transform FDST is nearly tight as indicated by the measures $M_{\mathrm{tight}_1} = 9.9\text{E-}4$, i.e., the chosen weights force the PPFT to be sufficiently close to an isometry for most practical purposes. If a higher accurate reconstruction is required, $M_{\mathrm{tight}_2} = 3.8\text{E-}7$ indicates that this can be achieved by the conjugate gradient method. As expected, $M_{\mathrm{tight}_1} = 1.9920$ shows that DSST is not tight. Nevertheless, the conjugate gradient method provides with $M_{\mathrm{tight}_2} = 1.2\text{E-}7$ a highly accurate approximation of its inverse. DNST is much closer to being tight than DSST (see $M_{\mathrm{tight}_1} = 0.1829$). This transform—as discussed—does not require the conjugate gradient method for reconstruction. The value 5.8E-16 was derived by using the dual shearlet filters, which show superior behavior.

4.4 Space-Frequency-Localization

The next measure is designed to test the degree to which the analyzing elements, here phrased in terms of shearlets but can be extended to other analyzing elements, are space-frequency localized.

Measure 4 *Let I be a shearlet in a 512×512 image centered at the origin $(257,257)$ with slope 0 of scale 4, i.e., $\sigma_{4,0,0}^{11} + \sigma_{4,0,0}^{12}$. Our quality measure will be fourfold:*

- Decay in spatial domain. *We compute the decay rates $d_1,\ldots,d_{512}$ along lines parallel to the y-axis starting from the line $[257:512,1]$ and the decay rates $d_{512}, \ldots, d_{1024}$ with x and y interchanged. By decay rate, for instance, for the line $[257:512,1]$, we first compute the smallest monotone majorant $M(x,1)$, $x = 257,\ldots,512$—note that we could also choose an average amplitude here or a different "envelope"—for the curve $|I(x,1)|$, $x = 257,\ldots,512$. Then the decay rate is defined to be the average slope of the line, which is a least square fit to the curve $\log(M(x,1))$, $x = 257,\ldots,512$. Based on these decay rates, we choose our measure to be the average of the decay rates*

$$M_{\text{decay}_1} = \frac{1}{1024} \sum_{i=1,\ldots,1024} d_i.$$

- Decay in frequency domain. *Here we intend to check whether the Fourier transform of I is compactly supported and also the decay. For this, let $\hat{I}$ be the 2D-FFT of I and compute the decay rates d_i, $i = 1,\ldots,1024$ as before. Then we define the following two measures:*

  - Compactly supportedness.

$$M_{\text{supp}} = \frac{\max_{|u|,|v|\leq 3} |\hat{I}(u,v)|}{\max_{u,v} |\hat{I}(u,v)|}.$$

  - Decay rate.

$$M_{\text{decay}_2} = \frac{1}{1024} \sum_{i=1,\ldots,512} d_i.$$

- Smoothness in spatial domain. We will measure smoothness by the average of local Hölder regularity. For each (u_0,v_0), we compute $M(u,v) = |I(u,v) - I(u_0,v_0)|$, $0 < \max\{|u-u_0|,|v-v_0|\} \leq 4$. Then the local Hölder regularity α_{u_0,v_0} is the least square fit to the curve $\log(|M(u,v)|)$. Then our smoothness measure is given by

$$M_{\text{smooth}_1} = \frac{1}{512^2} \sum_{u,v} \alpha_{u,v}.$$

- Smoothness in frequency domain. *We compute the smoothness now for $\hat{I}$, the 2D-FFT of I to obtain the new $\alpha_{u,v}$ and define our measure to be*

$$M_{\text{smooth}_2} = \frac{1}{512^2} \sum_{u,v} \alpha_{u,v}.$$

Let us now analyze the space-frequency localization of the shearlets utilized in FDST, DSST, and DNST by these measures; for the numerical results see Table 3.

The shearlet elements associated with FDST are band-limited and those associated with DSST and DNST are compactly supported, which is clearly indicated

Table 3 The numerical results for the test on space-frequency localization

	M_{decay_1}	M_{supp}	M_{decay_2}	M_{smooth_1}	M_{smooth_2}
FDST	−1.920	5.5E-5	−3.257	1.319	0.734
DSST	$-\infty$	8.6E-3	−1.195	0.012	0.954
DNST	$-\infty$	2.0E-3	−0.716	0.188	0.949

by the values derived for M_{decay_1}, M_{supp}, and M_{decay_2}. It would be expected that $M_{\text{decay}_2} = -\infty$ for FDST due to the band-limitedness of the associated shearlets. The shearlet elements are however defined by their Fourier transform on a pseudo-polar grid, whereas the measure M_{decay_2} is taken after applying the 2D-FFT to the shearlets resulting in data on a cartesian grid, in particular, yielding a non-precisely compactly supported function.

The test values for M_{smooth_1} and M_{smooth_2} show that the associated shearlets are more smooth in spatial domain for FDST than for DSST and DNST, with the reversed situation in frequency domain.

4.5 Shear Invariance

Shearing naturally occurs in digital imaging, and it can—in contrast to rotation—be realized in the digital domain. Moreover, for the shearlet transform, shear invariance can be proven and the theory implies

$$\left\langle 2^{3j/2}\psi(S_k^{-1}A_4^j\cdot -m), f(S_s\cdot)\right\rangle = \left\langle 2^{3j/2}\psi(S_{k+2^js}^{-1}A_4^j\cdot -m), f\right\rangle.$$

We therefore expect to see this or a to the specific directional transform adapted behavior. The degree to which this goal is reached is tested by the following measure.

Measure 5 *Let I be an 256×256 image with an edge through the origin $(129, 129)$ of slope 0. Given $-1 \le s \le 1$, generates an image $I_s := I(S_s\cdot)$ and let S_j be the set of all possible scales j such that $2^j s \in \mathbb{Z}$. Our quality measure will then be the curve*

$$M_{shear,j} = \max_{-2^j<k,k+2^js<2^j} \frac{\|C_{j,k}(SI_s) - C_{j,k+2^js}(SI)\|_2}{\|I\|_2}, \quad scale\ j \in S_j,$$

where $C_{j,k}$ is the shearlet coefficients at scale j and shear k.

We present our results in Table 4.

This table shows that the FDST is indeed almost shear invariant. A closer inspection shows that $M_{shear,1}$ and $M_{shear,2}$ are relatively small compared to the measurements with respect to finer scales $M_{shear,3}$ and $M_{shear,4}$. The reason for this is the

Table 4 The numerical results for the test on shear invariance

	$M_{shear,1}$	$M_{shear,2}$	$M_{shear,3}$	$M_{shear,4}$
FDST	1.6E-5	1.8E-4	0.002	0.003

aliasing effect which shifts some energy to the high frequency part near the boundary away from the edge in the frequency domain.

We did not test DSST and DNST with respect to this measure, since these transforms show a different—not included in this Measure 5—type of shear invariance.

4.6 Speed

Speed is one of the most fundamental properties of each algorithm to analyze. Here, we test the speed up to a size of $N = 512$ which is regarded as sufficient to determining the complexity.

Measure 6 *Generate a sequence of five random images I_i, $i = 5,\ldots,9$ of size $2^i \times 2^i$ with standard normally distributed entries. Let s_i be the speed of the shearlet transform S applied to I_i. Our hypothesis is that the speed behaves like $s_i = c \cdot (2^{2i})^d$; 2^{2i} being the size of the input. Let now $\tilde{d}_a$ be the average slope of the line, which is a least square fit to the curve $i \mapsto \log(s_i)$. Let also f_i be the 2D FFT applied to I_i, $i = 5,\ldots,9$. Our quality measure will then be threefold:*

- Complexity

$$M_{speed_1} = \frac{\tilde{d}_a}{2\log 2}.$$

- The Constant

$$M_{speed_2} = \frac{1}{5}\sum_{i=5}^{9}\frac{s_i}{(2^{2i})^{M_{speed,1}}}.$$

- Comparison with 2D-FFT

$$M_{speed_3} = \frac{1}{5}\sum_{i=5}^{9}\frac{s_i}{f_i}.$$

Table 5 presents the results of testing FDST, DSST and DNST with respect to these speed measures.

To interpret these results correctly, we remark that the DNST was tested only with test images I_i for $i = 7, \ldots, 9$, since it cannot be implemented for small size images.

Table 5 The numerical results for the test on speed

	M_{speed_1}	M_{speed_2}	M_{speed_3}
FDST	1.156	9.3E-6	280.560
DSST	0.821	4.5E-3	88.700
DNST	1.081	9.9E-8	40.519

Interestingly, the results also show that the 2D FFT-based convolution makes DNST comparable to DSST with respect to these speed measures, although it is much more redundant than DSST. Finally, the results show that FDST is comparable with both DSST and DNST with respect to complexity measure M_{speed_1}. From this, it is conceivable to assume that FDST is highly comparable with respect to speed for large scale computations. The larger value $M_{\text{speed}_3} = 280.560$ appears due to the fact that the FDST employs fractional Fourier transforms on an oversampled pseudo-polar grid of size.

4.7 Geometric Exactness

One major advantage of directional transforms is their sensitivity with respect to geometric features alongside with their ability to sparsely approximate those (cf. chapter on "Sparse Approximation"). This measure is designed to analyze this property.

Measure 7 *Let* $I_1, \ldots, I_8$ *be* 256×256 *images of an edge through the origin* $(129, 129)$ *and of slope* $[-1, -0.5, 0, 0.5, 1]$ *and the transpose of the middle three, and let* $c_{i,j}$ *be the associated shearlet coefficients for image* I_i *and scale* j. *Our quality measure will twofold:*

- Decay of significant coefficients. *Consider the curve*

$$\frac{1}{8} \sum_{i=1}^{8} \max |c_{i,j}(\text{of analyzing elements aligned with the line})|, \qquad \text{scale } j,$$

  let d *be the average slope of the line, which is a least square fit to* log *of this curve, and define*
$$M_{\text{geo}_1} = d.$$

- Decay of insignificant coefficients. *Consider the curve*

$$\frac{1}{8}\sum_{i=1}^{8}\max|c_{i,j}(\text{of all other analyzing elements})|, \qquad \text{scale } j,$$

let d be the average slope of the line, which is a least square fit to log *of this curve, and define*

$$M_{\text{geo}_2} = d.$$

Table 6 shows the numerical test results for FDST, DSST, and DNST.

Table 6 The numerical results for the test on geometric exactness

	M_{geo_1}	M_{geo_2}
FDST	-1.358	-2.032
DSST	-0.002	-0.030
DNST	-0.019	-0.342

As expected, the decay rate of the insignificant shearlet coefficients of FDST, i.e., the ones not aligned with the line singularity, measured by $M_{\text{geo}_2} \approx -2.032$ is much larger than the decay rate of the significant shearlet coefficients measured by $M_{\text{geo}_1} \approx -1.358$. Notice that this difference is even more significant in the case of the DSST and DNST.

4.8 Stability

To analyze stability of an algorithm, we choose thresholding as the most common impact on a sequence of transform coefficients.

Measure 8 *Let I be the regular sampling of a Gaussian function with mean 0 and variance 256 on* $\{-128, 127\}^2$ *generating an* 256×256*-image.*

- Thresholding 1. *Our first quality measure will be the curve*

$$M_{\text{thres}_{1,p_1}} = \frac{\|G_{\sqrt{w}P}W^{\star}\,\text{thres}_{1,p_1}\,SI - I\|_2}{\|I\|_2},$$

where thres_{1,p_1} *discards* $100 \cdot (1 - 2^{-p_1})$ *percent of the coefficients* ($p_1 = [2 : 2 : 10]$).

- Thresholding 2. *Our second quality measure will be the curve*

$$M_{thres_{2,p_2}} = \frac{\|G_{\sqrt{w}P}W^\star \, thres_{2,p_2} \, SI - I\|_2}{\|I\|_2},$$

where $thres_{2,p_2}$ *sets all those coefficients to zero with absolute values below the threshold* $m(1 - 2^{-p_2})$ *with* m *being the maximal absolute value of all coefficients.* ($p_2 = [0.001 : 0.01 : 0.041]$)

Table 7 shows that even if we discard $100(1 - 2^{-10}) \sim 99.9\%$ of the FDST coefficients, the original image is still well approximated by the reconstructed image. Thus, the number of the significant coefficients is relatively small compared to the total number of shearlet coefficients. From Table 8, we note that knowledge of the shearlet coefficients with absolute value greater than $m(1 - 1/2^{0.001})(\sim 0.1\%$ of coefficients) is sufficient for precise reconstruction.

DNST shows a similar behavior with worse values for relatively large p_1. It should be however emphasized that firstly, the redundancy of DNST used in this test is 25 and this is lower than the redundancy of FDST, which is about 71. This effect can be more strongly seen by the test results of DSST whose redundancy with 4 even much smaller. Secondly, a significant part of the low-frequency coefficients in both DSST and DNST will be removed by a relatively large threshold, since the ratio between the number of the low-frequency coefficients and the total number of coefficients is much higher than FDST. This prohibits a similarly good reconstruction of a Gaussian function.

This test in particular shows the delicateness of comparing different algorithms by merely looking at the test values without a rational interpretation; in this case, without considering the redundancy and the ratio between the number of the low-frequency coefficients and the total number of coefficients.

Table 7 The numerical results for $M_{thres_{1,p_1}}$

p_1	2	4	6	8	10
FDST	1.5E-08	7.2E-08	2.5E-05	0.001	0.007
DSST	0.02961	0.02961	0.02961	0.0296	0.0331
DNST	5.2E-10	1.2E-04	0.00391	0.0124	0.0396
DNST+DWT	1.2E-09	3.2E-06	3.4E-05	5.5E-05	5.5E-05

Table 8 The numerical results for $M_{thres_{2,p_2}}$

p_2	0.001	0.011	0.021	0.031	0.041
FDST	0.005	0.039	0.078	0.113	0.154
DSST	0.030	0.036	0.046	0.056	0.072
DNST	0.002	0.018	0.035	0.055	0.076
DNST+DWT	0.001	0.013	0.020	0.024	0.031

Acknowledgments

The first author would like to thank David Donoho and Morteza Shahram for many inspiring discussions on topics in this area. She also acknowledges support by the Einstein Foundation Berlin and by Forschungsgemeinschaft (DFG) Grant SPP-1324 KU 1446/13 and DFG Grant KU 1446/14. The second author was supported by DFG Grant SPP-1324 KU 1446/13, and the third author was supported by DFG Grant KU 1446/14.

References

1. A. Averbuch, R. R. Coifman, D. L. Donoho, M. Israeli, and Y. Shkolnisky, *A framework for discrete integral transformations I—the pseudo-polar Fourier transform*, SIAM J. Sci. Comput. **30** (2008), 764–784.
2. D. H. Bailey and P. N. Swarztrauber, *The fractional Fourier transform and applications*, SIAM Review, **33** (1991), 389–404.
3. E. J. Candès, L. Demanet, D. L. Donoho and L. Ying, *Fast discrete curvelet transforms*, Multiscale Model. Simul. **5** (2006), 861–899.
4. E. J. Candès and D. L. Donoho, *Ridgelets: a key to higher-dimensional intermittency?*, Phil. Trans. R. Soc. Lond. A. **357** (1999), 2495–2509.
5. E. J. Candès and D. L. Donoho, *New tight frames of curvelets and optimal representations of objects with C^2 singularities*, Comm. Pure Appl. Math. **56** (2004), 219–266.
6. E. J. Candès and D. L. Donoho, *Continuous curvelet transform: I. Resolution of the wavefront set*, Appl. Comput. Harmon. Anal. **19** (2005), 162–197.
7. E. J. Candès and D. L. Donoho, *Continuous curvelet transform: II. Discretization of frames*, Appl. Comput. Harmon. Anal. **19** (2005), 198–222.
8. M. N. Do and M. Vetterli, *The contourlet transform: an efficient directional multiresolution image representation*, IEEE Trans. Image Process. **14** (2005), 2091–2106.
9. D. L. Donoho, *Wedgelets: nearly minimax estimation of edges*, Ann. Statist. **27** (1999), 859–897.
10. D. L. Donoho, G. Kutyniok, M. Shahram, and X. Zhuang, *A rational design of a digital shearlet transform*, Proceeding of the 9th International Conference on Sampling Theory and Applications, Singapore, 2011.
11. D. L. Donoho, A. Maleki, M. Shahram, V. Stodden, and I. Ur-Rahman, *Fifteen years of Reproducible Research in Computational Harmonic Analysis*, Comput. Sci. Engr. **11** (2009), 8–18.
12. G. Easley, D. Labate, and W.-Q Lim, *Sparse directional image representations using the discrete shearlet transform*, Appl. Comput. Harmon. Anal. **25** (2008), 25–46.
13. B. Han, G. Kutyniok, and Z. Shen, *Adaptive multiresolution analysis structures and shearlet systems*, SIAM J. Numer. Anal. **49** (2011), 1921–1946.
14. E. Hewitt and K.A. Ross, *Abstract Harmonic Analysis I, II*, Springer-Verlag, Berlin/ Heidelberg/New York, 1963.
15. P. Kittipoom, G. Kutyniok, and W.-Q Lim, *Construction of compactly supported shearlet frames*. Constr. Approx. **35** (2012), 21–72.
16. G. Kutyniok, M. Shahram, and X. Zhuang, *ShearLab: A rational design of a digital parabolic scaling algorithm*, preprint.
17. G. Kutyniok and T. Sauer, *Adaptive directional subdivision schemes and shearlet multiresolution analysis*, SIAM J. Math. Anal. **41** (2009), 1436–1471.
18. W.-Q Lim, *The Discrete Shearlet Transform: A new directional transform and compactly supported shearlet frames*, IEEE Trans. Imag. Proc. **19** (2010), 1166–1180.
19. W.-Q Lim, *Nonseparable Shearlet Transforms*, preprint.
20. S. Mallat, *A Wavelet Tour of Signal Processing*, 2nd ed. New York: Academic, 1999.

Image Processing Using Shearlets

Glenn R. Easley and Demetrio Labate

Abstract Since shearlets provide nearly optimally sparse representations for a large class of functions that are useful to model natural images, many image processing methods benefit from their use. In particular, the error rates of data estimation from noise are highly dependent on the sparsity properties of the representation, so that many successful applications of shearlets center around restoration tasks such as denoising and inverse problems. Other imaging problems, where also the application of the shearlet representation turns out to be very beneficial, include image enhancement, image separation, edge detection, and estimation of the geometric features of an object.

Key words: Curvelets, Deconvolution, Denoising, Edge detection, Geometric separation, Image processing, Shearlets, Sparsity, Wavelets, Video denoising

1 Introduction

Shearlets were introduced with the expressed intent to provide a highly efficient representation of images with edges. In fact, the elements of the shearlet representation form a collection of well-localized waveforms, ranging at various locations, scales and orientations, and with highly anisotropic shapes. This makes the shearlet representation particularly well adapted at representing the edges and the other anisotropic objects which are the dominant features in typical images. These properties have been theoretically quantified through the notion of sparse shearlet

G.R. Easley
System Planning Corporation, Arlington, VA 22201, USA
e-mail: geasley@sysplan.com

D. Labate
Department of Mathematics, University of Houston, Houston, TX 77204, USA
e-mail: dlabate@math.uh.edu

G. Kutyniok and D. Labate (eds.), *Shearlets: Multiscale Analysis for Multivariate Data*,
Applied and Numerical Harmonic Analysis, DOI 10.1007/978-0-8176-8316-0_8,
© Springer Science+Business Media, LLC 2012

approximations and the shearlet analysis of singularities (see "Shearlets and Mi-crolocal Analysis" and "Analysis and Identification of Multidimensional Singularities Using the Continuous Shearlet Transform" of this volume). As will be described below, these properties have direct and important implications for the efficient encoding and processing of discrete data. This is demonstrated by an increasing number of very competitive numerical applications of the shearlet transform to the analysis and processing of images and other multidimensional data.

In this chapter, we provide an overview of the most relevant imaging applications of the shearlet approach. For reason of space, we will focus on the general principles used in the development of the algorithms and their significance, rather than on the technical details of the implementations, which can be found in the original papers. Since this is a very active area of investigation, it is understood that improved and newer shearlet imaging applications are currently being developed and, as a consequence, this chapter can only provide a retrospective view on the field. In particular, this chapter will describe the application of the shearlet representation to problems of image denoising, image enhancement, inverse problems, edge analysis and detection, and image separation.

2 Image Denoising

The significance of sparsity for data restoration is well understood and has been addressed in seminal papers such as [20, 27]. Indeed, consider the classical problem of recovering a function $f \in L^2(\mathbb{R})$ from noisy data y, that is, of recovering f from the observation

$$y = f + n,$$

where n is Gaussian white noise[1] with standard deviation σ. We illustrate this problem in dimension $D = 1$, but it generalizes naturally to higher dimensions.

The problem of interest is to optimize the estimation $\tilde{f}$ of f by minimizing the estimation error, usually measured by the L^2 norm $||f - \tilde{f}||$. Hence, the *risk* of the estimator $\tilde{f}$ is given by the Mean Square Error (MSE)

$$E||f - \tilde{f}||^2,$$

where the expectation is calculated with respect to the probability distribution of the noise n. It is clear that the risk depends on f, so that the worst behavior of the estimator is obtained by considering the supremum over all f in a certain class $\mathscr{F}$, that is,

$$\sup_{f \in \mathscr{F}} E||f - \tilde{f}||^2.$$

---

[1] In practice, it is not always accurate to assume that the noise found in applications is Gaussian white noise. However, this assumption is usually needed to make the theory tractable. In particular, it is a standard assumption in the theory of wavelet thresholding and wavelets shrinkage which is discussed below.

The *Minimax MSE* is defined as

$$\inf_{\tilde{f}} \sup_{f \in \mathscr{F}} E||f - \tilde{f}||^2,$$

where all measurable estimation procedures are allowed in the infimum. It is a remarkable result and an important application of the sparsity properties of wavelets that, for uniformly regular and piecewise regular 1D signals, a nearly minimax MSE is achieved using a very simple wavelet estimator known as *wavelet thresholding* [22, 23]. The computation of the wavelet threshold estimator can be briefly described as follows. Let $\{\psi_{j,m}\}$ be a wavelet basis for $L^2(\mathbb{R})$. Then, the noisy function y can be expanded as

$$y = \sum_{j,k} \langle y, \psi_{j,m} \rangle \, \psi_{j,m},$$

with convergence in the L^2 norm. The wavelet *hard thresholding algorithm* consists in setting to zero the wavelet coefficients $\langle y, \psi_{j,m} \rangle$ whose absolute values fall below a certain threshold T. The value of T depends on the standard deviation of the noise σ. Hence, to summarize, the wavelet hard thresholding algorithm consists of the following steps:

1. Compute the wavelet coefficients $\langle y, \psi_{j,m} \rangle$ of y.
2. Determine the threshold value $T(\sigma)$, where σ is estimated from y.
3. Remove (zero out) the wavelet coefficients whose absolute values are smaller than $T(\sigma)$.
4. Compute the estimator $\tilde{f}$ as $\tilde{f} = \sum_{j,k} c_{j,k} \psi_{j,m}$, where

$$c_{j,k} = \langle y, \psi_{j,m} \rangle, \qquad \text{if } |\langle y, \psi_{j,m} \rangle| > T,$$
$$c_{j,k} = 0, \qquad\qquad \text{otherwise.}$$

An alternative thresholding approach is to use a *soft thresholding* algorithm, where the coefficients are modified by the *shrinkage function:* $shr(c) = sgn(c) \max(|c| - T, 0)$. Unlike the hard thresholding that is an "all or nothing procedure" (values above the threshold are kept, values below it are deleted), the soft thresholding function produces a smooth transitions between the original and the deleted values, where values slightly below the threshold are not removed but attenuated. In practice, the main challenge is to find an appropriate value of T, and several strategies have been proposed in the literature [59]. For example, the VisuShrink algorithm [23, 24] uses the universal threshold $T = \sigma\sqrt{2\log M}$, M being the size of the data, and is asymptotically near minimax within the class of Besov spaces. Another classical approach, called BayesShrink [12], uses a different threshold value $T_j = \frac{\sigma^2}{\sigma_j}$ for each resolution level j, where σ_j is the standard deviation of the data at the resolution level j.

As discussed in the previous chapters, wavelets are non-optimal when dealing with piecewise regular multivariable functions, and this implies that wavelet thresholding does not provide a minimax MSE in this situation. Consider, in particular, the class of cartoon-like images $\mathscr{E}^2(\mathbb{R}^2)$. For $f \in \mathscr{E}^2(\mathbb{R}^2)$, define $|c(f)^W|_{(N)}$ to be the Nth largest entry in the sequence of wavelet coefficients of f given by $\{|c(f)_\mu^W| : c(f)_\mu^W = \langle f, \psi_\mu \rangle\}$, where $\{\psi_\mu\}$ is a 2D wavelet basis. Since

$$\sup_{f \in \mathscr{E}^2(\mathbb{R}^2)} |c(f)^W|_{(N)} \leq CN^{-1},$$

it follows that the N-term wavelet estimator $\tilde{f}_N$ satisfies

$$||f - \tilde{f}_N||^2 \leq \sum_{m>N} |c(f)^W|_{(N)}^2 \leq CN^{-1}.$$

This implies that the Mean Square Error (MSE) of the wavelet thresholding estimator satisfies

$$\sup_{f \in \mathscr{E}} E||f - \tilde{f}||^2 \asymp \sigma, \quad \text{as } \sigma \to 0,$$

where σ is the noise level, as indicated above. By contrast, let $|c(f)^S|_{(N)}$ be the Nth largest entry in the sequence of shearlet coefficients of f given by $\{|c(f)_\mu^S| : c(f)_\mu^W = \langle f, s_\mu \rangle\}$ where $\{s_\mu\}$ is a Parseval frame of shearlets. Then, a basic result from [43] shows that

$$\sup_{f \in \mathscr{E}} |c(f)^S|_{(N)} \leq CN^{-3/2}(\log N)^{3/2}.$$

Ignoring the log factor, this gives that the N-term shearlet estimator $\tilde{f}_N$ satisfies

$$||f - \tilde{f}_N||^2 \leq \sum_{m>N} |c(f)^S|_{(N)}^2 \leq CN^{-2}.$$

This implies that a denoising strategy based on the thresholding of the shearlet coefficients yields an estimator $\tilde{f}$ of f whose Mean Square Error (MSE) satisfies (essentially) the minimax MSE

$$\sup_{f \in \mathscr{E}} E||f - \tilde{f}||^2 \asymp \sigma^{4/3}, \quad \text{as } \sigma \to 0,$$

This shows that a denosing estimator based on shearlet thresholding has the ability to achieve a minimax MSE for images with edges. Notice that the same type of theoretical behavior is achieved using curvelets [78, 6].

In the following, we will show some numerical demonstrations to illustrate that indeed a denosing algorithm based on shearlet thresholding outperforms a similar wavelet-based approach. Before presenting these results, let us briefly recall the construction of the Discrete Shearlet Transform, originally introduced in [32].

2.1 Discrete Shearlet Transform

We start by re-writing the cone-based shearlet system, defined in "Introduction to Shearlets" of this volume, in a form more suitable to its digital implementation.

As usual, it is more convenient to work in the Fourier domain. For additional emphasis, in the following we will use the notation $\widehat{\mathbb{R}}^2$ when referring to the plane in the Fourier domain, while we will use the standard symbol $\mathbb{R}^2$ to denote the plane in the space domain.

Set $\mathscr{D}_0 = \{(\xi_1, \xi_2) \in \widehat{\mathbb{R}}^2 : |\xi_1| \geq \frac{1}{8}, |\frac{\xi_2}{\xi_1}| \leq 1\}$ and $\mathscr{D}_1 = \{(\xi_1, \xi_2) \in \widehat{\mathbb{R}}^2 : |\xi_2| \geq \frac{1}{8}, |\frac{\xi_1}{\xi_2}| \leq 1\}$. Given a smooth function $\hat{\psi}_2$, with support in $[-1, 1]$, we define

$$
W_{j,\ell}^{(0)}(\xi) = \begin{cases} \hat{\psi}_2(2^j \frac{\xi_2}{\xi_1} - \ell)\chi_{\mathscr{D}_0}(\xi) + \hat{\psi}_2(2^j \frac{\xi_1}{\xi_2} - \ell + 1)\chi_{\mathscr{D}_1}(\xi) & \text{if } \ell = -2^j \\ \hat{\psi}_2(2^j \frac{\xi_2}{\xi_1} - \ell)\chi_{\mathscr{D}_0}(\xi) + \hat{\psi}_2(2^j \frac{\xi_1}{\xi_2} - \ell - 1)\chi_{\mathscr{D}_1}(\xi) & \text{if } \ell = 2^j - 1 \\ \hat{\psi}_2(2^j \frac{\xi_2}{\xi_1} - \ell) & \text{otherwise} \end{cases}
$$

and

$$
W_{j,\ell}^{(1)}(\xi) = \begin{cases} \hat{\psi}_2(2^j \frac{\xi_2}{\xi_1} - \ell + 1)\chi_{\mathscr{D}_0}(\xi) + \hat{\psi}_2(2^j \frac{\xi_1}{\xi_2} - \ell)\chi_{\mathscr{D}_1}(\xi) & \text{if } \ell = -2^j \\ \hat{\psi}_2(2^j \frac{\xi_2}{\xi_1} - \ell - 1)\chi_{\mathscr{D}_0}(\xi) + \hat{\psi}_2(2^j \frac{\xi_1}{\xi_2} - \ell)\chi_{\mathscr{D}_1}(\xi) & \text{if } \ell = 2^j - 1 \\ \hat{\psi}_2(2^j \frac{\xi_1}{\xi_2} - \ell) & \text{otherwise,} \end{cases}
$$

for $\xi = (\xi_1, \xi_2) \in \widehat{\mathbb{R}}^2$, $j \geq 0$, and $\ell = -2^j, \ldots, 2^j - 1$. This notation allows us to write the elements of the cone-based shearlet system, in the Fourier domain, as

$$
\hat{\psi}_{j,\ell,m}^{(d)}(\xi) = 2^{\frac{3j}{2}} V(2^{-2j}\xi) W_{j,\ell}^{(d)}(\xi) e^{-2\pi i \xi A_d^{-j} B_d^{-\ell} m},
$$

where $d \in \{0, 1\}$, $V(\xi_1, \xi_2) = \hat{\psi}_1(\xi_1)\chi_{\mathscr{D}_0}(\xi_1, \xi_2) + \hat{\psi}_1(\xi_2)\chi_{\mathscr{D}_1}(\xi_1, \xi_2)$ and ψ_1 is the Meyer-type wavelet associated with the classical shearlet. Hence, the *shearlet transform* of $f \in L^2(\mathbb{R}^2)$ can be expressed as

$$
\langle f, \psi_{j,\ell,m}^{(d)} \rangle = 2^{\frac{3j}{2}} \int_{\mathbb{R}^2} \hat{f}(\xi) \overline{V(2^{-2j}\xi) W_{j,\ell}^{(d)}(\xi)} \, e^{2\pi i \xi A_d^{-j} B_d^{-\ell} m} \, d\xi. \tag{1}
$$

To formulate the implementation in the finite domain setting, we consider $\ell^2(\mathbb{Z}_N^2)$ as the discrete analogue of $L^2(\mathbb{R}^2)$. Given an image $f \in \ell^2(\mathbb{Z}_N^2)$, the Discrete Fourier Transform (DFT) is defined as

$$
\hat{f}(k_1, k_2) = \frac{1}{N} \sum_{n_1, n_2 = 0}^{N-1} f(n_1, n_2) e^{-2\pi i(\frac{n_1}{N}k_1 + \frac{n_1}{N}k_2)}, \quad -\frac{N}{2} \leq k_1, k_2 < \frac{N}{2}.
$$

The product $\hat{f}(\xi_1, \xi_2) \overline{V(2^{-2j}\xi_1, 2^{-2j}\xi_2)}$ is found analogously in the DFT domain as the product of the DFT of f with the discretization of the *filter* functions $\overline{V(2^j \cdot)}$, $j \geq 0$. Notice that these functions are associated to specific regions of the frequency plane, roughly near $|\xi_1| \approx 2^{2j}$ or $|\xi_2| \approx 2^{2j}$. In the space domain, this produces a decomposition of f, at various scales j, in terms of elements of the form $f^j(n_1, n_2) = f * v_j(n_1, n_2)$, where v_j corresponds to the function $\overline{V(2^j \cdot)}$ in the Fourier domain. This can be implemented using the Laplacian pyramid filter [5].

In order to obtain the directional localization, $\hat{f}^j$ is resampled onto a pseudo-polar grid, and a one-dimensional band-pass filter is applied. The pseudo-polar grid is parametrized by lines going through the origin and their slopes. Specifically, the pseudo-polar coordinates $(u, p) \in \mathbb{R}^2$ match the following assignments:

$$(u, p) = (\xi_1, \tfrac{\xi_2}{\xi_1}) \quad \text{if } (\xi_1, \xi_2) \in \mathscr{D}_0$$

$$(u, p) = (\xi_2, \tfrac{\xi_1}{\xi_2}) \quad \text{if } (\xi_1, \xi_2) \in \mathscr{D}_1$$

The resampled $\hat{f}^j$ is denoted as F_j, so that

$$\hat{f}(\xi_1, \xi_2) \overline{V(2^{-2j}\xi_1, 2^{-2j}\xi_2) W_{j,\ell}^{(d)}(\xi_1, \xi_2)} = F_j(u, p) \overline{W^{(d)}(2^j p - \ell)}. \qquad (2)$$

This resampling in the DFT can be done by direct re-assignment or by using the Pseudo-polar DFT [2].

Given the one-dimensional DFT defined as

$$\mathscr{F}_1(q)(k_1) = \frac{1}{\sqrt{N}} \sum_{n_1 = -N/2}^{N/2-1} q(n_1) e^{\frac{-2\pi i k_1 n_1}{N}},$$

we denote $\{w_{j,\ell}^{(d)}(n) : n \in \mathbb{Z}_N\}$ to be the sequence of values such that $\mathscr{F}_1\left(w_{j,\ell}^{(d)}(n)\right) = \overline{W^{(d)}(2^j n - \ell)}$. For a fixed $n_1 \in \mathbb{Z}_N$, we then have

$$\mathscr{F}_1\left(\mathscr{F}_1^{-1}\left(F_j(n_1, n_2)\right) * w_{j,\ell}^{(d)}(n_2)\right) = F_j(n_1, n_2)\, \mathscr{F}_1\left(w_{j,\ell}^{(d)}(n_2)\right), \qquad (3)$$

where $*$ denotes the one-dimensional convolution along the n_2 axis. Equation (3) summarizes the algorithmic implementation for computing the discrete samples of $F_j(u, p) \overline{W^{(d)}(2^j p - \ell)}$.

The shearlet coefficients $\langle f, \psi_{j,\ell,m}^{(d)} \rangle$, given by (1), are now formally obtained by inverting the Pseudo-polar Fourier transform of expression (2). This can be either implemented by computing the inverse Pseudo-polar DFT or by directly re-assembling the Cartesian sampled values of (3) and apply the inverse two-dimensional DFT. Using the Fast Fourier Transform (FFT) to implement the DFT, the discrete shearlet transform algorithm runs in $O(N^2 \log N)$ operations. Note that the direct conversion between the Cartesian to pseudo-polar can be pre-conditioned so the operation has a condition number of 1 as explained in [32] and the operation will be L^2 norm preserving (cf. [16] for details). Since no performance improvement was noticed with this adjustment, this pre-conditioning was avoided at the expense of having a faster inversion, which is just a summation, since it was observed to have many advantages for this formulation.

An illustration of the shearlet decomposition produced by this implementation is given in Fig. 1, which shows a 2-level subband decomposition, that is, a decomposition where the scale parameter j is ranging over $j = 0, 1$. Recall that, in the language of image processing, a *subband decomposition* is a decomposition of an image into components associated with different regions of the frequency plane $\widehat{\mathbb{R}}^2$. As the figures shows, there are four subband terms corresponding to $j = 0$ and eight subband terms corresponding to $j = 1$, consistent with the fact that the directional parameter ℓ takes values in $\{-2^j, \ldots, 2^j - 1\}$ for each of the cone regions $d = 0$ and $d = 1$. This corresponds precisely to the illustration of the shearlet decomposition into directional subbands given in "Introduction to Shearlets" of this volume. A Matlab toolbox for this numerical implementation of the Discrete Shearlet Transform is available from http://www.math.uh.edu/~dlabate.

An alternative technique to implement the discrete shearlet transform as an application of M filters was presented in [33]. In this case, filters v_j and $w_{j,\ell}^{(d)}$ are found so that $\langle f, \psi_{j,\ell,m}^{(d)} \rangle$ can be computed as

$$f * (v_j * w_{j,\ell}^{(d)})(m) \triangleq f * g_{j,\ell}^{(d)}[m],$$

where $g_{j,\ell}^{(d)} = v_j * w_{j,\ell}^{(d)}$ are the directionally oriented filters. When the filters $g_{j,\ell}$ are chosen to have significantly smaller support sizes than N as explained in [32], the filter bank implementation is even faster than $O(N^2 \log N)$. It is also a formulation that is easily parallelizable, in the sense that the different directional components of the image can be processed in parallel.

Besides these implementations, it is useful to recall that a reduced-redundancy implementation of the discrete shearlet transform was presented in [42] and a critically sampled version of the discrete shearlet transform was presented [30]. Finally, a compactly supported version of the discrete shearlet transform is discussed in [55]. We refer to "Digital Shearlet Transforms" of this volume for additional detail about the digital implementations of the discrete shearlet transform.

To simplify the notation, in the applications which will be presented below, the superscript (d) will be suppressed and the distinction between $d = 0$ and $d = 1$ will be absorbed by re-indexing the parameter ℓ so that the cardinality is doubled.

2.2 Shearlet Thresholding

In this section, we present the first application of the discrete shearlet transform to the problem of image denoising. This approach can be viewed as a direct adaptation of the classical wavelet thresholding described above.

Suppose that an image f is corrupted by white Gaussian noise with standard deviation σ. Using the discrete shearlet transform, an estimator $\tilde{f}$ of f can be computed by using a thresholding procedure which follows essentially the same ideas of the wavelet thresholding algorithm described in Sect. 2. In particular, for the choice of the threshold parameter it was founds in [32] that an excellent performance is achieved by adapting the BayesShrink algorithm. That is, the threshold is chosen to

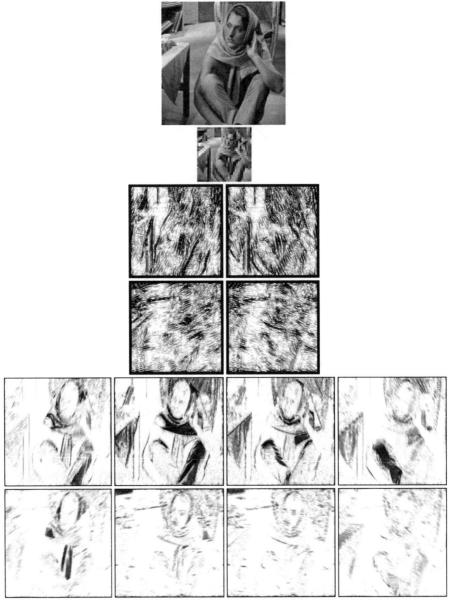

Fig. 1 An illustration of the subband decomposition obtained using the discrete shearlet transform. The *top image* is the original *Barbara* image. The image below the top image is of the coarse-scale reconstruction. Images of the subband reconstructions for levels $j = 0$ and $j = 1$ are given below with an inverted grayscale for presentation purposes. As explained in the text, there are four directional subbands corresponding to $j = 0$ and eight directional subbands corresponding to $j = 1$

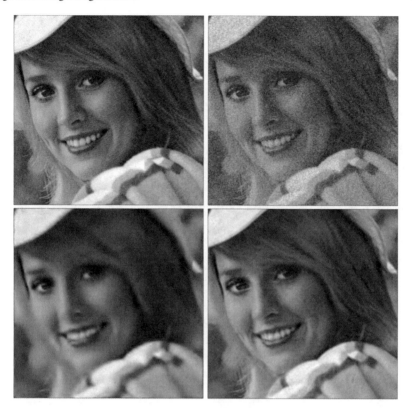

Fig. 2 Image denoising results of a piece of the *Elaine*. From *top left*, clockwise: Original image, noisy image (SNR = 10.46 dB), shearlet transform (SNR = 16.47 dB), and wavelet transform (SNR = 14.00 dB)

be $T_{j,\ell} = \frac{\sigma_{j,\ell}^2}{\sigma_{j,\ell,m}}$, where $\sigma_{j,\ell}$ is the standard deviation of the noise for the subband indexed by the scaling parameter j and the directional parameter ℓ, and $\sigma_{j,\ell,m}$ denotes the standard deviation of the mth coefficient of the image at scale j and direction ℓ.

A numerical demonstration of this shearlet-based denoising approach is given in Fig. 2, where this method is compared against a similar scheme based on a nonsubsampled wavelet transform. To assess the performance of the algorithm, we have used the standard *signal-to-noise* ratio to measure how much noise is present before and after the estimate is made. Recall that the signal-to-noise ratio (SNR) is given by

$$\mathrm{SNR}(f, f_{\mathrm{est}}) = 10\log_{10}\left[\frac{\mathrm{var}(f)}{\mathrm{mean}(f - \tilde{f})}\right],$$

where $\mathrm{var}(f)$ is the variance of the image, and is measured in decibels.

We refer to [32] for additional numerical tests and details about this version of the shearlet thresholding algorithm.

2.3 Denoising Using Shearlet-Based Total Variation Regularization

An alternative approach for the application of shearlets to problems of image denoising consists in combining the ideas of wavelet thresholding described above with other classical methods for denoising.

Indeed, there is another very successful philosophy to image denoising which is based on the theory of the partial differential equations and variational methods, such as diffusion equations and total variation (TV) minimization (cf. [10, 74, 89]). Intuitively, the idea of diffusion equations is to model a noisy image as a function $\tilde{f}$ on $\Omega \subset \mathbb{R}^2$ and to computed its denoised version as the solution of a suitable diffusion process (isotropic and anisotropic) with $\tilde{f}$ as initial condition. It is clear that this produces an image which is more "regular" than the original one. Alternatively, one can produce a similar regularization process by minimizing an energy functional of the form

$$E(f; \lambda, \tilde{f}) = \frac{\lambda}{2} \int_{\Omega} (f - \tilde{f})^2 \, dx \, dy + P(f),$$

where the first term, called *fidelity term*, encourages the similarity between $\tilde{f}$ and its denoised version f, and the second term $P(f, \nabla f)$, called *penalty term*, controls the regularity of the solution. Indeed, it is known that there are strong relations between regularisation methods and diffusion filters [75].

In particular, let us consider a classical version of a regularization method based on *Total Variation (TV) regularization*, which consists in minimizing the functional

$$\int_{\Omega} \phi(\|\nabla f\|) \, dx \, dy + \frac{\lambda}{2} \int_{\Omega} (f - \tilde{f})^2 \, dx \, dy,$$

where $\phi \in C^2(\mathbb{R})$ is an even regularization function (cf. [4]). In the above expression, the penalty term involves the *total variation* of f, which, for a function $f \in W^{1,1}(\Omega)$, is defined as

$$TV(f) = \int_{\Omega} \|\nabla f\| \, dA,$$

where $\nabla f = \left(\frac{\partial u}{\partial x_1}, \frac{\partial u}{\partial x_2} \right)$ and $\| \ \|$ is the standard Euclidean norm. That is, the penalty term ensures that the solution of the regularization method minimizes the total variation of f (cf. [62] for a discussion of the role of TV in imaging applications). The minimizer can be found by solving the equation solution of

$$\frac{\partial f}{\partial t} = \nabla \cdot \left(\frac{\phi'(\|\nabla f\|)}{\|\nabla f\|} \nabla f \right) - \lambda (f - f_0),$$

subjected to the von Neumann boundary condition. In the case when $\lambda = 0$ and $\lim_{x \to \infty} \phi'(x)/x = 0$, this equation is considered a special case of the Perona and Malik diffusion equation (cf. [72]):

$$\frac{\partial f}{\partial t} = \nabla \cdot (\rho(\|\nabla f\|) \nabla f),$$

where $\rho(x) = \phi'(x)/x$. There are several other aspects of these problems which go beyond the space limitations of this chapter, and we refer the interested reader to the references mentioned at the beginning of this section.

Fig. 3 Detail images of experimental results. From the *top, clockwise*: Original image, noisy image (SNR $= 11.11$ dB), diffusion-based estimate using 53 iterations (SNR $= 14.78$ dB), shearlet-based diffusion estimate using six iterations (SNR $= 16.15$ dB), shearlet-based TV estimate using two iterations ($L = 7$, SNR $= 16.29$ dB), TV-based estimate using 113 iterations (SNR $= 14.52$ dB)

The regularization methods described above can be very effective in image denoising, and generally provide superior denoising performance especially when applied to images with negligible texture and fine-scale features. Yet a drawback is that they can result in estimates that are reminiscent of oil-paintings, with loss of important detail when applied to images that contain complex textures and shading. To improve upon these methods, combinations of such techniques with sparse representations have recently been proposed (e.g., [7, 15, 28, 58, 81, 90]). A similar combination has been proposed using shearlets in [31].

The idea of the shearlet-based Total Variation (TV) Regularization approach is rather simple yet very effective. Assuming a denoised estimate is found by thresholding a shearlet representation (using the method from Sect. 2.2), let P_S be the projection operator that retains the non-threshold shearlet coefficients of f. The shearlet-based TV method is then described as essentially solving

$$\frac{\partial f}{\partial t} = \nabla \cdot \left(\frac{\phi'(\|\nabla P_S(f)\|)}{\|\nabla P_S(f)\|} \nabla P_S(f) \right) - \lambda_{x,y}(f - \tilde{f})$$

with the boundary condition $\frac{\partial f}{\partial n} = 0$ on $\partial \Omega$ and the initial condition $f(x,y,0) = \tilde{f}(x,y)$ for $x,y \in \Omega$. Note that the fidelity parameter $\lambda_{x,y}$ is spatially varying. It is based on a measure of local variances that is updated after a number of iterations L or progressions of artificial time steps (see [40] for more details).

Another diffusion variant based on shearlets has been to solve

$$\frac{\partial f}{\partial t} = \nabla \cdot (\rho(\|\nabla P_S f\|)\nabla P_S f)$$

with the Neumann boundary condition $\frac{\partial f}{\partial n} = 0$ on $\partial \Omega$ and the initial condition $f(x,y,0) = f_0(x,y)$ for $x,y \in \Omega$.

Ilustrations of these techniques, including a comparison with standard TV, are done using an image of flowers. Close-ups of the results are shown in Fig. 3.

2.4 Complex-Valued Denoising

Another variant of the shearlet denosing algorithm was developed to deal with the problem of reducing complex-valued noise [68]. This problem arises in synthetic aperture radar (SAR) interferometry where interferometric phase noise reduction is a main challenge.

Recall that an SAR image is a complex-valued two-dimensional array and is often displayed in terms of its magnitude without any phase information. Interferograms are obtained by multiplying a SAR image by the complex conjugate of a second SAR image obtained from a slightly different location. These interferograms contain information about topographic height and are used to produce digital elevation maps (DEM). A typical problem is that the complex-valued noise in these phase estimates cause errors with the phase unwrapping needed for the height information to be formed.

The *n-looks complex image* is defined as

$$f = \frac{1}{n} \sum_{k=1}^{n} f_1(k) f_2^*(k) = |f|e^{j\psi}, \tag{4}$$

where f_1 and f_2 are a pair of 1-look complex SAR images. The phase quality depends on the amplitude of the correlation coefficient and is given as

$$\rho = \frac{E\left[f_1 f_2^*\right]}{\sqrt{E\left[|f_1|^2\right] E\left[|f_1|^2\right]}} = |\rho|e^{j\theta}, \tag{5}$$

where $|\rho|$ is the coherence and θ is the phase of the complex correlation coefficient. By using an appropriate phase noise model, a shearlet coefficient shrinkage method can be derived which adapts the one presented in Sect. 2.2. Figure 4 gives an illustration of this method and compares it to a wavelet-based method. In this example a single look image is given. The difference between the ideal phase image and the estimate is given by counting the number of residues.

Fig. 4 Noisy interferometric phase filtering methods. From *left to right*: Noisy interferogram with coherence $|\rho| = 0.5$ (number of residues is 14, 119), wavelet-based estimate (number of residues is 80), shearlet-based estimate (number of residues is 20)

2.5 Other Shearlet-Based Denoising Techniques

Additional methods of image denoising based on shearlets were recently presented in the literature [9, 13, 14, 18, 46, 83, 85, 93, 94] but discussing those in detail would go beyond the scope of this chapter. Many of these methods use variants of the shearlet shrinkage strategy described above, or they introduce some form of adaptivity in the thresholding.

3 Inverse Problems

In many scientific and industrial applications, the objects or the features of most interest cannot be observed directly, but must be inferred from other observable quantities. The aim of inverse problems is to reconstruct the cause for such observed variables. An especially important class of inverse problems, for example, concerns the determination of the structural properties of an object from measurements of the absorbed or scattered radiation taken at an array of sensors, which occurs in applications such as Computerized Tomography (CT) or Synthetic Aperture Radar (SAR). Another example is the removal of image degradation due to optical distortion or motion blur.

In most of these problems, the relationship between the observed data y and the feature of interest f is approximately linear, and can be modeled mathematically as

$$y = Kf + z, \tag{6}$$

where K is a linear operator and z is Gaussian noise [3]. Since the operator K is typically not invertible (i.e., K^{-1} is unbounded), some "regularization" is needed to invert the problem. Unfortunately, traditional regularization methods (e.g., Tikhonov regularization or truncated Singular Value Decomposition [63, 64, 86]) have the undesirable effect that important features to be recovered are lost, as evident in imaging applications where the regularized reconstructions are blurred versions of the original. This phenomenon is of particular concern since, in many situations, the most relevant information to be recovered is indeed contained in edges or other sharp transitions. To address this issue, a number of different methods have been proposed, including hidden Markov models, Total Variation regularization and Anisotropic Diffusion [38, 73, 84, 88]. However, while these methods produce visually appealing results, their rationale is essentially heuristic and they offer no sound theoretical framework to assess the ultimate method performance.

As will be described in the following, recent ideas from sparse representations can be applied to develop a theoretical and computational framework for the regularized inversion of a large class of inverse problems. Specifically, in contrast with more traditional regularization techniques and the other methods mentioned above, the shearlet representation provides a rigorous theoretical framework which is very effective at dealing with a large class of inverse problems and which is especially effective in the recovery of information associated with edges and other singularities.

3.1 Inverting the Radon Transform

By taking advantage of the ability of the shearlet representation to represent some important classes of operators, a novel method for the regularized inversion of the Radon transform was introduced in [17, 29]. This is a problem of great interest since the Radon transform is the underlying mathematical framework of Computerized

Tomography (CT), which has become an essential tool in medical diagnostics and preventive medicine.[2] The Radon transform maps a Lebesgue integrable function f on $\mathbb{R}^2$ into the set of its line integrals

$$Rf(\theta,t) = \int_{\ell(\theta,t)} f(x)\,dx,$$

where $\ell(\theta,t)$, with $t \in \mathbb{R}, \theta \in S^1$, are the lines $\{x \in \mathbb{R}^2 : x \cdot \theta = t\}$.

Since the shearlet-based approach to invert the Radon transform adapts a number of ideas from the Wavelet–Vaguelette Decomposition (WVD) introduced by Donoho in [21], let us start by briefly recalling the main ideas of the WVD.

Suppose that the operator K in (6) maps the space $L^2(\mathbb{R}^2)$ into the Hilbert space Y. The WVD consists in selecting a well localized orthonormal wavelet basis $\{\psi_{j,m}\}$ of $L^2(\mathbb{R}^2)$ and an appropriate orthonormal basis $U_{j,k}$ of Y so that any $f \in L^2(\mathbb{R}^2)$ can be expressed as

$$f = \sum_{j,k} c_{j,k} [Kf, U_{j,k}] \psi_{j,m}, \qquad (7)$$

where $c_{j,k}$ are known scalars and $[\cdot,\cdot]$ is the inner product in Y (we refer to [21] for more detail). It follows that f can be recovered from the observed data Kf and, consequently, an estimate of f can be obtained from the decomposition (7) applied to the noisy data $Kf + z$ using a wavelet thresholding algorithm like those described in Sect. 2. The main advantage of this approach is that, unlike the classical Singular Value Decomposition (SVD), the basis functions employed in the decomposition formula (7) do not derive entirely from the operator K, but can be chosen to capture most efficiently the features of the object f to be recovered. Indeed this approach turns out to outperform SVD and other standard methods. For functions f in a certain class (specifically, if f is in a certain family of Besov spaces), then the WVD method converges to f with the optimal rate, provided that the thresholding parameters are properly selected [21, 54].

By adapting the WVD approach within the shearlet framework, it is shown in [17, 29] that a function $f \in L^2(\mathbb{R}^2)$ is recovered from the Radon data Rf using the expansion

$$f = \sum_{j,\ell,m} 2^j [Rf, U_{j,\ell,m}] \psi_{j,\ell,m}, \qquad (8)$$

where $\{\psi_{j,\ell,m}\}$ is a Parseval frame of shearlets of $L^2(\mathbb{R}^2)$, and $\{U_{j,\ell,m}\}$ is a related system which is obtained by applying an operator closely related to the Radon transform to the shearlet system. The advantage of this representation is that the shearlet system $\{\psi_{j,\ell,m}\}$ is optimally suited to represent images f containing edges. Thus, if f is a cartoon-like image and the observed Radon data Rf are corrupted by additive Gaussian noise, it is proved that the shearlet-based estimator obtained from shearlet thresholding provides an *optimal* Mean Square Error for the recovery of f. In particular, this approach outperforms the standard WVD as well as other traditional methods, in which cases the MSE has a slower decay rate. With respect

---

[2] More than 72 millions of medical CT scans were performed in the USA, in 2007.

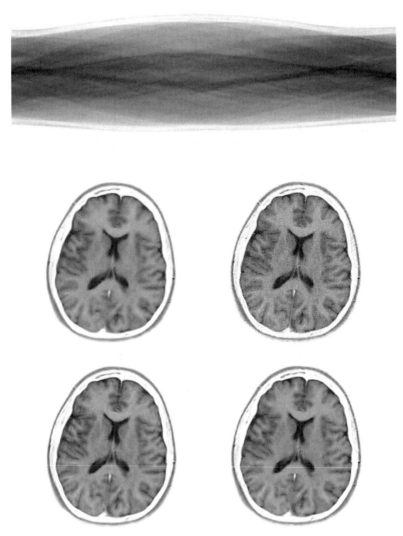

Fig. 5 From the *top, clockwise*: noisy Radon projections (SNR = 34.11 dB); unfiltered reconstruction (SNR = 11.19 dB); shearlet-based estimate (SNR = 21.68 dB); curvelet-based estimate
(SNR = 21.26 dB); wavelet-based estimate (SNR = 20.47 dB)

to a somewhat similar result based on the curvelet approach [6], the shearlet-based
method provides a simpler and more flexible mathematical construction which leads
to a an improved numerical implementation and performance. A typical application
of the shearlet-based regularized reconstruction algorithm is reported in Fig. 5 where
the method is compared against the curvelet-based algorithm and the wavelet-based
one (corresponding to WVD). We refer to [17, 29] for additional numerical tests
and details on the algorithm.

A related method of using shearlets to control noise amplification when inverting the Radon transform was presented in [1] in the case when the sampled data is compressed.

3.2 Deconvolution

When image degradations include the blurring introduced by camera motion as well as the noise introduced by the electronics of the system, the model of the degradation can be given as a convolution operation. The process of undoing this convolution operation is commonly known as deconvolution and is known to be an ill-posed inverse problem. To regularize the ill-posed problem, the sparse representation properties of shearlets can be utilized.

The idea of using a sparse representation to regularize deconvolution as well as other inverse problems has been suggested before (see for example [21] and [6]). However, unique to this shearlet approach is the ability for a multi-scale and anisotropic regularization inversion to be done before noise shrinkage [70]. An additional benefit is the use of a cross-validation function to adaptively select the thresholding values.

A digitally recorded image is a finite discrete data set, so an image deconvolution problem can be formulated as a matrix inversion problem. Without loss of generality, we assume the recorded images/arrays are of size $N \times N$. Let γ denote an $N \times N$ array of samples representing a zero mean additive white Gaussian noise with variance σ^2. Let y denote the observed image and x is assumed to represent the image to be estimated. Then, the deconvolution problem can be formulated as

$$\mathbf{y} = H\mathbf{f} + \gamma,$$

where $\mathbf{y}$, $\mathbf{f}$, and γ are $N^2 \times 1$ column vectors representing the arrays y, f, and γ lexicographically ordered, and H is the $N^2 \times N^2$ matrix that models the blur operator. When the assumption of periodic boundary is made, the problem can be described as

$$y(n_1, n_2) = (f * h)(n_1, n_2) + \gamma(n_1, n_2), \qquad (9)$$

where $0 \leq n_1, n_2 \leq N - 1$, $*$ denote circular convolution, and h denotes the point spread function (PSF). In the discrete Fourier transform (DFT) domain, (9) reduces to

$$\widehat{y}(k_1, k_2) = \widehat{h}(k_1, k_2)\widehat{f}(k_1, k_2) + \widehat{\gamma}(k_1, k_2), \qquad (10)$$

where $\widehat{y}(k_1, k_2), \widehat{h}(k_1, k_2), \widehat{f}(k_1, k_2)$ and $\widehat{\gamma}(k_1, k_2)$ are the discrete Fourier transforms of y, h, f, and γ, respectively, for $-N/2 \leq k_1, k_2 \leq N/2 - 1$. In this formulation, it is evident that if there exist indices (k_1, k_2) where $|\widehat{h}(k_1, k_2)|$ contains values at or near zero, then the system will be ill-conditioned.

Using the regularized inverse operator

$$H'_\alpha(k_1,k_2) = \frac{\overline{\overline{h}}(k_1,k_2)}{|\widehat{h}(k_1,k_2)|^2 + \alpha}$$

for some regularizing parameter $\alpha \in \mathbb{R}^+$, an estimate in the DFT domain is given by

$$\widehat{f_\alpha}(k_1,k_2) = \widehat{y}(k_1,k_2)H'_\alpha(k_1,k_2),$$

for $N/2 \leq k_1, k_2 \leq N/2 - 1$. Applying the multi-channel implementation of shear-lets, we can adaptively control the regularization parameter to be the best suited for each frequency supported trapezoidal region. Let $g_{j,\ell}$ denote the shearlet filter that will correspond to a given scale j and direction ℓ. The shearlet coefficients of an estimate for a given regularization parameter α can be computed in the DFT domain as

$$\widehat{c}(f_\alpha)^S_{j,\ell}(k_1,k_2) = \widehat{y}(k_1,k_2)\,\widehat{g}_{j,\ell}(k_1,k_2)\,H'_\alpha(k_1,k_2),$$

for $N/2 \leq k_1, k_2 \leq N/2 - 1$.

The regularization parameters $\{\alpha\}$ will act to suppress a noise amplification, yet it is desirable to allow a noise amplification as long as the remaining noise level can be adequately controlled by shearlet shrinkage methods.

Shearlet threshold values can be adaptively found by using a generalize cross val-idation (GCV) as follows: Let $\mathbf{y}$, $\mathbf{f}$, and γ denote the observed noisy image, original image, and the colored noise so that $\mathbf{y} = \mathbf{f} + \gamma$. The noise is assumed to be second-order stationary (i.e., the mean is constant and the correlation between two points depends only on the distance between them).

The soft thresholding function $T_\tau(c)$ is defined to be equal to $c - \tau\text{sign}(c)$ if $|c| > \tau$ and zero otherwise for a given threshold parameter τ. Assuming the noise process γ is stationary and that $\langle \gamma, \psi_{j,\ell,m} \rangle$ represents a shearlet coefficient of a ran-dom vector γ at scale j, direction ℓ, and location m, the following lemma is obtained.

Lemma 1 ([70]). $E\left[|\langle \gamma, \psi_{j,\ell,m} \rangle|^2\right]$ *depends only on the scale j and direction ℓ.*

This means the shearlet transform of stationary correlated noise is stationary within each scale and directional component. Let $\mathbf{y}_{j,\ell}$ denote the vector of shearlet coefficients of $\mathbf{y}$ at scale j and direction ℓ. $L_{j,\ell}$ is the number of shearlet coefficients on scale j and direction ℓ, and L is the total number of shearlet coefficients. Given

$$R_{j,\ell}(\tau_{j,\ell}) = \frac{1}{L_{j,\ell}}\|T_{\tau_{j,\ell}}(\mathbf{y}_{j,\ell}) - \mathbf{f}_{j,\ell}\|^2, \tag{11}$$

the total risk is

$$R(\tau) = \sum_j \sum_\ell \frac{L_{j,\ell}}{L} R_{j,\ell}(\tau_{j,\ell}). \tag{12}$$

This means the minimizing the mean squared error or risk function R can be achieved by minimizing $R_{j,\ell}(\tau_{j,\ell})$ for all j and ℓ. Assuming $L_{j,\ell,0}$ is the total number

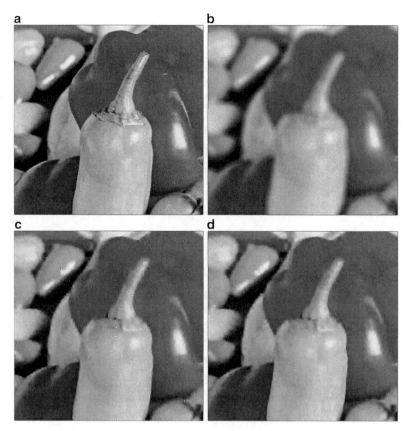

Fig. 6 Details of the image deconvolution experiment with a *Peppers* image. (**a**) Original image. (**b**) Noisy blurred image, BSNR = 30 dB. (**c**) ForWaRD estimate, ISNR = 4.29 dB. (**d**) Shearlet-based estimate, ISNR = 5.42 dB

of shearlet coefficients that were replaced by zero after threshold, we now have the following:

Theorem 1 ([70]). *The minimizer of*

$$GCV_{j,\ell}(\tau_{j,\ell}) = \frac{\frac{1}{L_{j,\ell}} \| T_\tau(\mathbf{y}_{j,\ell}) - \mathbf{y}_{j,\ell} \|^2}{\left[\frac{L_{j,\ell,0}}{L_{j,\ell}}\right]^2}, \tag{13}$$

is asymptotically optimal for the minimum risk threshold $R_{j,\ell}(\tau_{j,\ell})$ for scale j and directional component ℓ.

This means finding the values $\tau_{j,\ell}$ that minimize the cross validation function $GCV_{j,\ell}$ for each j and ℓ, leads to a shearlet-based estimate that will likely be close to the ideal noise-free image.

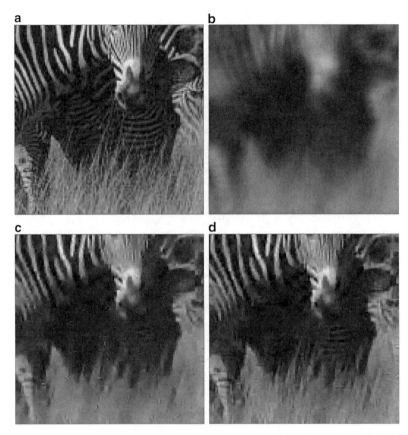

Fig. 7 Details of the image deconvolution experiment with a *Zebra* image. (**a**) Original image. (**b**) Noisy blurred image, BSNR = 30 dB. (**c**) ForWaRD estimate, ISNR = 5.53 dB. (**d**) Shearlet-based estimate, ISNR = 6.03 dB

Define $\widehat{c}(y)^S_{j,\ell}(k_1,k_2)$ as $\widehat{y}(k_1,k_2)\widehat{g}_{j,\ell}(k_1,k_2)$ for $-N/2 \le k_1,k_2 \le N/2 - 1$, and $\tilde{c}(f_\alpha)^S_{j,\ell}$ to be the estimate of $c(f_\alpha)^S_{j,\ell}$ after thresholding the coefficients by using the GCV formula given in (13). That is, for a given α, set $\tilde{c}(f_\alpha)^S_{j,\ell} = T_{\tau'_{j,\ell}}(c(f_\alpha)^S_{j,\ell})$ for $\tau'_{j,\ell} = \arg\min_{\tau_{j,\ell}} GCV_{j,\ell}(\tau_{j,\ell})$. Then, one option for finding the optimal α for each thresholded set of shearlet coefficients $\tilde{c}(f_\alpha)^S_{j,\ell}$ can be found by minimizing the cost function

$$\sum_{k_1}\sum_{k_2} \frac{|\widehat{h}(k_1,k_2)|}{|\widehat{h}(k_1,k_2)|^2+\eta} \left|\widehat{h}(k_1,k_1)\widehat{\tilde{c}}(f_\alpha)^S_{j,\ell}(k_1,k_2) - \widehat{c}(y)^S_{j,\ell}(k_1,k_2)\right|^2,$$

where $\eta = N^2\sigma^2/\|c(y)^S_{j,\ell} - \mu(c(y)^S_{j,\ell})\|^2_2$, $\mu(y)$ denotes the mean of y, and σ is the estimated standard deviation of the noisy blurred image. For other variances and options on finding α, we refer to [70].

We use the improvement in signal-to-noise-ratio (ISNR) to measure the success of the routines and the blurred signal-to-noise ratio (BSNR) to give in understanding of the problem setup. The ISNR is given as

$$\text{ISNR} = 10\log_{10}\left(\frac{\|f-y\|_2^2}{\|f-\tilde{f}\|_2^2}\right),$$

and the BSNR is given as

$$\text{BSNR} = 10\log_{10}\left(\frac{\|(f*h)-\mu(f*h)\|_2^2}{N^2\sigma^2}\right),$$

for an $N \times N$ image. Notice that both are measured in decibels.

Applications of the shearlet-based deconvolution algorithm are shown in Figs. 6 and 7, where it is compared against the highly competitive wavelet-based deconvolution algorithm known as *Fourier-Wavelet Regularized Deconvolution* (ForWaRD) [65]. Figures 6 and 7 display close-ups of some experiments results where the blur was a 9×9 boxcar blur [65].

3.3 Inverse-Halftoning

Halftoning is a process of rendering an image into a binary (black-and-white) image. Halftoning techniques include error diffusion methods such as those by Floyd-Steinburg and Jarvis et al. [87, 47, 36, 48]. At times these halftoned images may need resizing, enhancement, or removal of aliasing artifacts. There may also be a need for these images to be restored to their original gray-scale images for other reasons such as compression or for digital achieving of old newspapers and articles.

Given the Floyd-Steinberg filter

$$h_{\text{FS}} = \frac{1}{16}\begin{bmatrix} 0 & \bullet & 7 \\ 3 & 5 & 1 \end{bmatrix},$$

or the Jarvis error filter

$$h_{\text{J}} = \frac{1}{48}\begin{bmatrix} 0 & 0 & \bullet & 7 & 5 \\ 3 & 5 & 7 & 5 & 3 \\ 1 & 3 & 5 & 3 & 1 \end{bmatrix},$$

the quantization error at $\bullet$ is diffused over a causal neighborhood according to the matrix values. Specifically, each pixel is identified in a raster-scan indexing scheme and the pixel's gray-scale value is made into a binary number by thresholding (1, if the value is greater than or equal to $1/2$, and 0 otherwise). The quantization error is then diffused on neighboring pixels using the weights from h_{FS} or h_{J}. Let p and q denote the impulse responses determined by the error diffusion model. The relation between the original $N \times N$ gray-scale image f and the resultant halftone y image can be approximately modeled as

$$y(n_1,n_2) = (p*f)(n_1,n_2) + (q*v)(n_1,n_2) \tag{14}$$

where $0 \leq n_1, n_2 \leq N - 1$ and υ are considered an additive white Gaussian noise even though the process does not involve randomness. In the DFT domain, (14) can be written as

$$\widehat{y}(k_1, k_2) = \widehat{p}(k_1, k_2)\widehat{f}(k_1, k_2) + \widehat{q}(k_1, k_2)\widehat{v}(k_1, k_2),$$

for $-N/2 \leq k_1, k_2 \leq N/2 - 1$. Assuming $\widehat{h}$ denotes the DFT of the diffusion filters h_{FS} or h_J, the transfer functions $\widehat{p}$ and $\widehat{q}$ are given by

$$\widehat{p}(k_1, k_2) = \frac{C}{1 + (C-1)\widehat{h}(k_1, k_2)}$$

and

$$\widehat{q}(k_1, k_2) = \frac{1 - \widehat{h}(k_1, k_2)}{1 + (C-1)\widehat{h}(k_1, k_2)}.$$

where the constant $C = 2.03$ when $h = h_{FS}$ or $C = 4.45$ when $h = h_J$ [48].

To approximately invert the halftoning process, we use the regularized inverse operator

$$P'_\alpha(k_1, k_2) = \frac{\overline{\widehat{p}}(k_1, k_2)}{|\widehat{p}(k_1, k_2)|^2 + \alpha^2 |\widehat{q}(k_1, k_2)|^2} \tag{15}$$

for some regularizing parameter $\alpha \in R^+$. This gives an image estimate in the Fourier domain as

$$\widehat{f}_\alpha(k_1, k_2) = \widehat{y}(k_1, k_2) P'_\alpha(k_1, k_2),$$

for $-N/2 \leq k_1, k_2 \leq N/2 - 1$.

This regularization process can be separated into a shearlet domain as done previously. Assuming $g_{j,\ell}$ denotes the shearlet filter for scale j and direction ℓ, the shearlet coefficients of an estimate of the image for a given regularization parameter α can be computed in the DFT domain as

$$\widehat{c}(f_\alpha)^S_{j,\ell}(k_1, k_2) = \widehat{y}(k_1, k_2)\widehat{g}_{j,\ell}(k_1, k_2) P'_\alpha(k_1, k_2)$$

for $-N/2 \leq k_1, k_2 \leq N/2 - 1$.

The remaining aspect of this problem is transformed into a form of a denoising problem which can be dealt with by thresholding the estimated shearlet coefficients using the GCV formulation determined previously.

We illustrate the performance of the shearlet inverse halftoning algorithm (see [34] for more details) by using the *Barbara* image. The image was halftoned using the Floyd-Steinberg algorithm and comparisons were done with a wavelet-based method as well as a LPA-ICI-based method [66, 37]. Close-ups of the results are shown in Fig. 8.

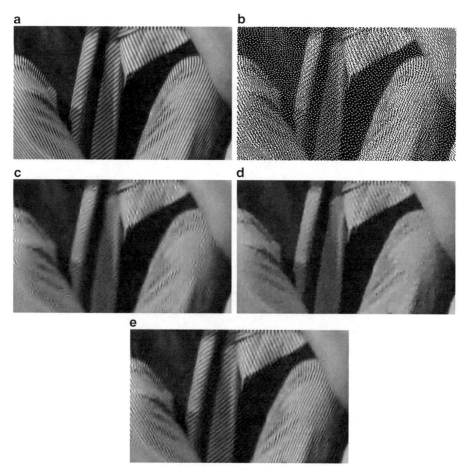

Fig. 8 Close-ups of inverse halftoning experiment with the *Barbara* image. (**a**) Original image. (**b**) Floyd-Steinberg halftone. (**c**) Wavelet-based estimate (SNR $= 19.30\,$dB). (**d**) LPA-ICI-based estimate (SNR $= 19.25\,$dB). (**e**) Shearlet-based estimate (SNR $= 22.02\,$dB)

4 Image Enhancement

Image enhancement is a term describing an improvement of the visual properties of an image for the purpose of interpretation or perception either for human or computer vision systems. Mathematically, given an image whose pixel values are described as the array $y(k_1,k_1)$, it produces an altered image $y_e(k_1,k_2)$ by an application of an enhancement transformation En. That is $y_e(k_1,k_2) = En(y(k_1,k_2))$ for $-N/2 \leq k_1, k_2 \leq N/2 - 1$. If we assume that y is a grayscaled image whose pixels are integers ranging from 0 to 255, then a fairly simple enhancement map is $En(t) = 255 - t$, which was used in Fig. 1 to improve the contrast in the images of the shearlet coefficients. Other simple enhancement transforms are based on logarithm, exponential, or piece-wise linear functions. Another powerful and widely used enhancement technique is *histogram equalization* which produces a transformation En such that the histogram of the its pixel values of the enhanced image y_e is evenly distributed. More recently, several image enhancement techniques have been introduced which are based on ideas from multiscale analysis [52, 53, 56, 80].

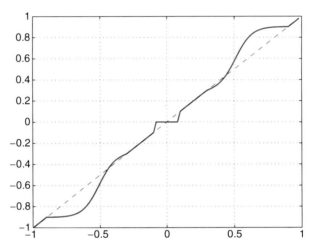

Fig. 9 Enhancement map: $b = 0.35$, $\beta = 10$, $T1 = 0.1$, $T2 = 0.3$, $T3 = 0.9$

Image enhancement is frequently used in medical imaging, where it can be helpful to emphasize visual features which are important for medical diagnostic. For example, the enhancement of mammography images can be very useful to improve the visibility of small tumors for early detection. To this goal, many image processing techniques based on multiscale analysis have been found effective, such as the approach of Laine et al. [52], which investigates mammography feature analysis using the dyadic wavelet transform, the approach by Strickland et al. [82], which uses the undecimated wavelet transform for detecting and segmenting calcifications in mammograms, and the enhancement method of Chang et al. [11], which is based on overcomplete multiscale representations.

In this section, we adapt some of the ideas proposed in the literature for image enhancement to construct an image enhancement algorithm based on the shearlet representation. The intuitive idea behind this approach is that, since shearlet coefficients are closely related to edges and other essential geometric features, it is possible to enhance such features by controlling the magnitude of the shearlet coefficients. Thus, we introduce an appropriate adaptive nonlinear mapping function on the shearlet coefficients with the goal to amplify weak edges, while keeping the other strong edges intact. We define this nonlinear operator as follows, using the notation $\operatorname{sigm}(t) = (1 + e^{-t})^{-1}$:

$$\text{En}(t) = 0 \text{ if } |t| < T_1$$
$$\text{En}(t) = \text{sign}(t)T_2 + \bar{a}(\text{sigm}(\beta(\bar{t}-b)) - \text{sigm}(-\beta(\bar{t}+b)))$$
$$\text{if } T_2 \leq |y| \leq T_3$$
$$\text{En}(t) = t \quad \text{otherwise} \tag{16}$$

where $t \in [-1,1]$, $\bar{a} = a(T_3 - T_2)$, $\bar{t} = \text{sign}(t)\frac{|t|-T_2}{T_3-T_2}$, $0 \leq T_1 \leq T_2 < T_3 \leq 1$, $b \in (0,1)$, and a, dependent on the gain factor β and b, is defined as

$$a = \frac{1}{\text{sigm}(\beta(1-b)) - \text{sigm}(-\beta(1+b))}.$$

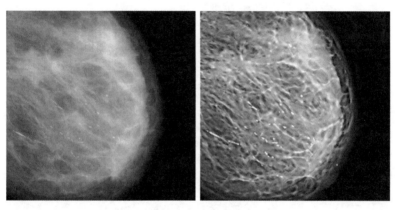

Fig. 10 Image enhanced of mammogram. From *left to right*: Original mammogram with spiculated masses and calcifications, Image enhanced mammogram using shearlets

In this formulation, the parameters T_1, T_2, and T_3 are selected threshold values, and b and β control the threshold and rate of enhancement, respectively. The interval $[T_2, T_3]$ serves as a sliding window for feature selectivity. It can be adjusted to emphasize important features within a specific range. These parameters can be adaptively selected by using the standard deviation of the pixel values for each scale j and direction ℓ. Using this nonlinear function, the shearlet coefficients are pointwise modified for image enhancement by

$$\langle y_e, \psi_{j,\ell,m} \rangle = \max_m(|\langle y, \psi_{j,\ell,m} \rangle|) En\left(\frac{\langle y, \psi_{j,\ell,m} \rangle}{\max_m(|\langle y, \psi_{j,\ell,m} \rangle|)}\right),$$

where $\max_m(|\langle y, \psi_{j,\ell,m} \rangle|)$ is the maximum absolute amplitude of $|\langle y, \psi_{j,\ell,m} \rangle|$ as a function of position m and $\langle y_e, \psi_{j,\ell,m} \rangle$ denotes the shearlet coefficients of the enhanced image y_e. The resultant enhanced image y_e is found by simply inverting the transform. Figure 9 shows an enhancement map curve representing the enhanced coefficients versus the original coefficients.

Results of the shearlet-based image enhancement method were presented in [69], where they have been applied to enhanced mammogram images. Examples of the

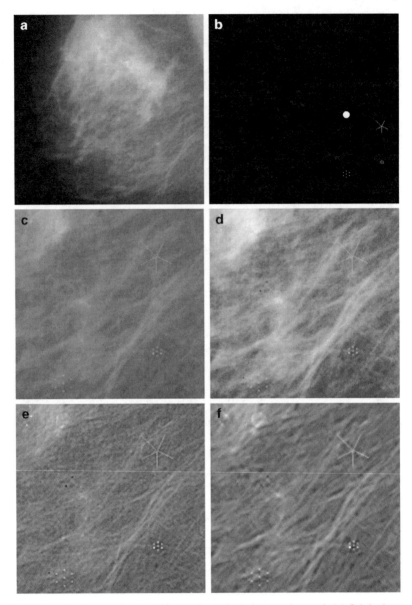

Fig. 11 Enhanced of mammogram with a mathematical phantom inserted. (**a**) Original mammogram. (**b**) Mathematical phantom. (**c**) Region of interest (ROI) image. (**d**) Enhanced ROI image using histogram equalization. (**e**) Enhanced ROI image using NSWT. (**f**) Enhanced ROI image using shearlets

application of this algorithm are shown in Figs. 10 and 11, where the enhanced mammograms obtained using shearlets are compared with those obtained using the nonsubsampled wavelet transform (NSWT) and a standard histogram equalization method. In the experiment Fig. 11, we created mathematical models of phantoms to validate our enhancement methods against any false alarms arising from our enhancement techniques. This phantom is a good model for features such as micro-calcifications, masses, and spicular objects which occur in real data.

In our experiments, we used 1, 8, 8, 16, 16 directions in the scales from coarser to finer, respectively, as done in [33] for the shearlet decomposition. The standard deviation of pixel values were used to adaptively select the values for T_1, T_2, and T_3. We choose $b = 0.23, 0.14, 0.10, 0.10$ and $\beta = 20, 35, 45, 35$ for the directions in the scales from coarser to finer, respectively. In the first experiment, we enhanced a mammogram image using shearlets as shown in Fig. 10. In the second experiment, we blended a normal mammogram with the phantoms and compared our enhance-ment method with that of the histogram equalization and the NSWT as shown in Fig. 11. As in Fig. 11, enhancement by shearlets provided a significant improvement in contrast for each feature included in the blended mammogram; whereas features such as mass (white disc) is hard to see in the enhanced ROI images obtained by the NSWT and histogram equalization methods.

5 Edge Analysis and Detection

The detection and analysis of edges is a primary task in a variety of image pro-cessing and computer vision applications. In fact, since edges are usually the most prominent features in natural images and scientific data, the localization of edges is a fundamental low level task for higher level applications such as shape recognition, 3D reconstruction, data enhancement and restoration.

Edges can be formally characterized as those points of a function u, defined on a domain $\Omega \subset \mathbb{R}^2$, for which the gradient is noticeably large, that is,

$$\{x \in \Omega \subset \mathbb{R}^2 : |\nabla u(x) > p\},$$

where p is some suitable chosen threshold. It is clear that this simple characteri-zation of edges does not translate directly into an effective edge detection scheme, since images are usually affected by noise and the differential operator is extremely sensitive to noise. As a consequence, in the most common edge detector schemes, to watch out for the interference of noise, the image is first smoothed out or mollified. For example, in the classical Canny edge detection algorithm [8] the image is first convolved with a scalable Gaussian function as

$$u_a = u * G_a,$$

where $G_a(x) = G(a^{-1}x)$, $a > 0$, $x \in \mathbb{R}^2$ and $G(x) = \frac{1}{\pi} e^{-x^2}$. Next, the edge points are identified as the local maxima of the gradient of u_a. Notice that this approach involves a scaling parameter a: as a decreases, the detection of the edge location becomes more accurate; however, as a decreases, also the detector's sensitivity to noise increases. As a result, the performance of the edge detector depends heavily on the scaling factor a (as well as the threshold).

There is an interesting and useful relationship between edge detection and wavelet analysis which was first observed by Mallat, Hwang and Zhong in [60, 61] and can be summarized as follows. Given an image $u \in L^2(\mathbb{R}^2)$, a simple computation shows that its continuous wavelet transform with respect to an admissible real and even function ψ can be written as

$$W_\psi u(a,x) = \int_{\mathbb{R}^2} u(y) D_a \psi(y - x) \, dy = u * D_a \psi(x).$$

where $D_a \psi(x) = a^{-1} \psi(a^{-1}x)$. In particular, if $\psi = \nabla G$, then

$$\nabla u_a(x) = u * \nabla G_a(x) = u * D_a \psi(x) = W_\psi u(a,x). \tag{17}$$

This shows that the maxima of the magnitude of the gradient of the smoothed image u_a correspond precisely to the maxima of the magnitude of the wavelet transform $W_\psi u(a,x)$. This observation provides a natural mathematical framework for the multiscale analysis of edges which was successfully developed in [60, 61].

5.1 Edge Analysis Using Shearlets

The main limitation of the Canny edge detector or the wavelet method described above is that both methods are essentially isotropic and, as a result, are not very efficient at dealing with the anisotropic nature of the edges. The difficulty in accurately identifying the location of edges is particularly evident in the presence of noise and when several edges are close together or cross each other, such as the situation of 2-dimensional projections of 3-dimensional objects [95]. In such cases, the following limitations of traditional edge detectors is particularly evident:

- *Difficulty in distinguishing close edges.* The isotropic Gaussian filtering causes edges running close together to be blurred into a single curve.
- *Poor angular accuracy.* In the presence of sharp changes in curvature or crossing curves, the isotropic Gaussian filtering leads to an inaccurate detection of the edge orientation. This affects the detection of corners and junctions.

To better deal with the edge information, a number of methods were introduced which replace the scalable collection of isotropic Gaussian filters G_a, $a > 0$, in (17) with a family of steerable and scalable anisotropic Gaussian filters such as

$$G_{a_1,a_2,\theta}(x_1,x_2) = a_1^{-1/2} a_2^{-1/2} R_\theta \, G(a_1^{-1}x_1, a_2^{-1}x_2),$$

where $a_1, a_2 > 0$ and R_θ is the matrix of rotation by the angle θ (see [71, 88, 39]). Unfortunately, the design and implementation of such filters are computationally involved and there is no theoretical setting to decide how to design such family of filters to best capture the edges.

The shearlet framework has the advantage of providing a well justified mathematical setting for efficiently representing the edge information. In fact, as discussed in

"Introduction to Shearlets" and "Analysis and Identification of Multidimensional Singularities using the Continuous Shearlet Transform" of this volume, the continuous shearlet transform can be applied to precisely characterize the geometric information associated with the edges through its asymptotic behavior at fine scales. The results can be summarized as follows.

Let an image u be modeled as piecewise smooth function in $\Omega = [0,1]^2$. That is, we assume that u is smooth everywhere on Ω, except for a collection of finitely many piecewise smooth curves, denoted by Γ, where jump discontinuities may occur. Then, the asymptotic decay properties of the continuous shearlet transform $\mathscr{SH}$ of u are as follows [44]:

- If $p \notin \Gamma$, then $|\mathscr{SH}_\psi u(a,s,p)|$ *decays rapidly*, as $a \to 0$, for each $s \in \mathbb{R}$.
- If $p \in \Gamma$ and Γ is smooth near p, then $|\mathscr{SH}_\psi u(a,s,p)|$ *decays rapidly*, as $a \to 0$, for each $s \in \mathbb{R}$ unless $s = s_0$ is the normal orientation to Γ at p. In this last case, $|\mathscr{SH}_\psi u(a,s_0,p)| \sim a^{\frac{3}{4}}$, as $a \to 0$.
- If p is a corner point of Γ and $s = s_0$, $s = s_1$ are the normal orientations to Γ at p, then $|\mathscr{SH}_\psi u(a,s_0,p)|, |\mathscr{SH}_\psi u(a,s_1,p)| \sim a^{\frac{3}{4}}$, as $a \to 0$. For all other orientations, the asymptotic decay of $|\mathscr{SH}_\psi u(a,s,p)|$ is faster (even if not necessarily "rapid").

Here by "rapid decay", we mean that, given any $N \in \mathbb{N}$, there is a $C_N > 0$ such that $|\mathscr{SH}_\psi u(a,s,p)| \leq Ca^N$, as $a \to 0$. It is also useful to observe that spike-type singularities produce a very different behavior than jump discontinuities on the decay of the continuous shearlet transform. Consider, for example, a Dirac delta distribution centered at t_0. In this case a simple calculation (see [49]) shows that

$$|\mathscr{SH}_\psi \delta_{t_0}(a,s,t_0)| \asymp a^{-3/4}, \quad \text{as } a \to 0,$$

that is, the continuous shearlet transform of δ_{t_0}, at $t = t_0$ increases at fine scales. The decay is rapid for $t \neq t_0$.

These observations show that the continuous shearlet transform precisely describes the geometric information of the edges and the other singular points of an image. This is in contrast with the wavelet transform which cannot provide any information about the edge orientation.

5.2 Edge Detection Using Shearlets

An algorithm for edge detection based on shearlets was introduced in [91, 92], where a discrete shearlet transform was described with properties specifically designed for this task. In fact, the discrete shearlet transform which was presented above for image denoising, produces large sidelobes around prominent edges[3] which interfere with the detection of the edge location. By contrast, the special discrete shearlet

---

[3] The same problem occurs if one uses a standard discrete wavelet or curvelet transform.

transform introduced in [91, 92] is not affected by this issue since the analysis filters are chosen to be consistent with the theoretical results in [44, 45], which require that the shearlet generating function ψ satisfies certain specific symmetry properties in the Fourier domain (this is also discussed in "Analysis and Identification of Multidimensional Singularities Using the Continuous Shearlet Transform" of this volume).

The first step of the *shearlet edge detector algorithm* consists in selecting the edge point candidates of a digital image $u[m_1, m_2]$. They are identified as those points $(\overline{m_1}, \overline{m_2})$ which, at fine scales j, are local maxima of the function

$$M_j u[m_1, m_2]^2 = \sum_\ell (\mathscr{SH} u[j, \ell, m_1, m_2])^2.$$

Here $\mathscr{SH} u[j, \ell, m_1, m_2]$ denotes the discrete shearlet transform. According to the properties of the continuous shearlet transform summarized above, we expect that, if $(\overline{m_1}, \overline{m_2})$ is an edge point, the discrete shearlet transform of u will behave as

$$|\mathscr{SH} u[j, \ell, \overline{m_1}, \overline{m_2}]| \sim C 2^{-\beta j},$$

where $\beta \geq 0$. If, however, $\beta < 0$ (in which case the size of $|\mathscr{SH} u|$ increases at finer scales), then $(\overline{m_1}, \overline{m_2})$ will be recognized as a spike singularity and the point will be classified as noise. Using this procedure, edge point candidates for each of the oriented components are found by identifying the points for which $\beta \geq 0$. Next, a *non-maximal suppression routine* is applied to these points to trace along the edge in the edge direction and suppress any pixel value that is not considered to be an edge. Using this routine, at each edge point candidate, the magnitude of the shearlet transform is compared with the values of its neighbors along the gradient direction (this is obtained from the orientation map of the shearlet decomposition). If the magnitude is smaller, the point is discarded; if it is the largest, it is kept.

Extensive numerical experiments have shown that the shearlet edge detector is very competitive against other classical or state-of-the-art edge detectors, and its performance is very robust in the presence of noise. An example is displayed in Fig. 12, where the shearlet edge detector is compared against the wavelet edge detector (which is essentially equivalent to the Canny edge detector) and the Sobel and Prewitt edge detectors. Notice that both the Sobel and Prewitt filters are 2D discrete approximations of the gradient operator. The performance of the edge detectors is assessed using the Pratt's Figure of Merit, which is a fidelity function ranging from 0 to 1, where 1 is a perfect edge detector. This is defined as

$$\text{FOM} = \frac{1}{\max(N_e, N_d)} \sum_{k=1}^{N_d} \frac{1}{1 + \alpha d(k)^2},$$

where N_e the number of actual edge points, N_d the number of detected edge points, $d(k)$ the distance from the kth actual edge point to the detected edge point and α is a scaling constant typically set to 1/9. The numerical test reported in the figures show that the shearlet edge detector consistently yields the best value for FOM.

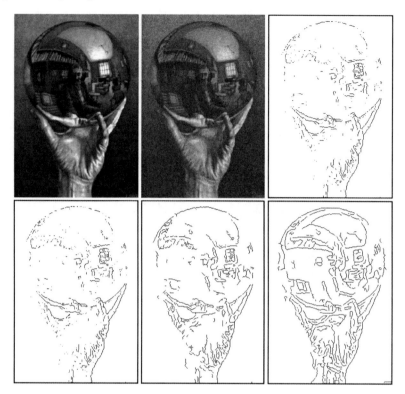

Fig. 12 Results of edge detection methods. From *top left, clockwise*: Original image, noisy image (PSNR = 25.94 dB), Prewitt result (FOM = 0.31), shearlet result (FOM = 0.94), wavelet result (FOM = 0.59), and Sobel result (FOM = 0.32)

5.3 Edge Analysis Using Shearlets

As observed above, the continuous shearlet transform has the ability to precisely characterize the geometry of the edges. These properties lead directly to a very effective algorithm for the estimation of the edge orientation, which was originally introduced in [92]. Specifically, by taking advantage of the parameter associated with the orientation variable in the shearlet transform, the edge orientations of an image u, can be estimated by searching for the value of the shearing variable s which maximizes $\mathscr{SH}_{\psi}u(a,s,p)$ at an edge point p, when a is sufficiently small. Discretely, this is obtained by fixing a sufficiently fine scales (i.e., $a = 2^{-2j}$ sufficiently "small") and computing the index $\tilde{\ell}$ which maximizes the magnitude of the discrete shearlet transform $\mathscr{SH}u[j,\ell,m]$ as

$$\tilde{\ell}(j,m) = \arg\max_{\ell} |\mathscr{SH}u[j,\ell,m]|. \tag{18}$$

Once this is found, the corresponding angle of orientation $\theta_{\tilde{\ell}}(j,m)$ associated with the index $\tilde{\ell}(j,m)$ can be easily computed. As illustrated in [92], this approach leads to a very accurate and robust estimation for the local orientation of the edge curves.

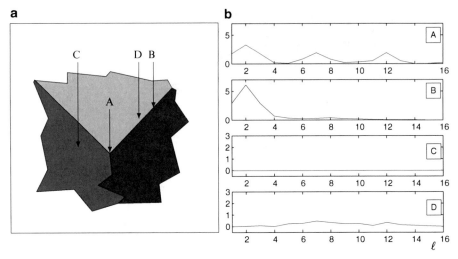

Fig. 13 (**a**) Test image and representative points A (junction), B (regular edge point), C (smooth region), D (near edge). (**b**) Magnitude of the Discrete Shearlet Transform, as a function of the orientation parameter ℓ at the locations $m_0 = A, B, C, D$ indicated in (**a**). Notice the different scaling factor used in the y-axis, for the plots of points C and D

Indeed, the sensitivity of the shearlet transform to the edge orientation is useful for the extraction of landmarks, another imaging application, which is important in problems of classification and retrieval. To illustrate the general principle, consider the simple image in Fig. 13 consisting of large smooth regions separated by piecewise smooth curves. The junction point A, where three edges intersect, is certainly the most prominent object in the image, and this can be easily identified by looking at values of the shearlet transform. In fact, if one examines the discrete shearlet transform $\mathscr{SH}u[j_0, l, m_0]$, at a fixed (fine) scale j_0 and locations m_0, as a function of the shearing parameter l, the plot immediately identifies the local geometric properties of the image. Specifically, as illustrated in Fig. 13b, one can recognize the following four classes of points inside the image. At the junction point $k_0 = A$, the function $|\mathscr{SH}u[j_0, \ell, m_0]|$ exhibits three peaks corresponding to the orientations of the three edge segments converging into A; at the point $m_0 = B$, located on a smooth edge, $|\mathscr{SH}u[j_0, \ell, m_0]|$ has a single peak; at a point $m_0 = D$, inside a smooth region, $|\mathscr{SH}u[j_0, \ell, m_0]|$ is essentially flat; finally, at a point $m_0 = C$ "close" to an edge, $|\mathscr{SH}u[j_0, \ell, m_0]|$ exhibit two peaks, but they are much smaller in amplitude than those for the points A and B. A similar behavior was observed, as expected, for more general images, even in the presence of noise.

Based on these observations, a simple and effective algorithm for classifying smooth regions, edges, corners and junction points of an image was proposed and validated in [92].

6 Image Separation

Blind source separation is a classical problem in signal processing whose object is the separation of a set of signals from a set of mixed signals, with very little information about the source signals and the mixing process. The traditional techniques for addressing this problem rely on the assumption that the source signals are essentially decorrelated, so that they can be separated into additive subcomponents which are statistically independent. The main weakness of these techniques is that they are very sensitive to noise. On the other hand, recent results have shown that sparse representations such as shearlets can be applied to design extremely robust source separation algorithms [25, 26, 79].

In the following, we describe a very effective algorithm for image separation, recently proposed in [25, 26] which takes advantage of the ability of the shearlet representation in dealing with edge curves and other elongated features. This approach is especially tailored to deal with the situation of images, such as astronomical or biological images, where it is important to separate pointlike objects from curvelike ones.

6.1 Image Model

The class of the images of interest, denoted by J, are modeled as a composition of point- and curve-like objects. That is, a point-like object is a function P which is smooth except for finitely many point singularities and has the form

$$P(x) = \sum_{i=1}^{m} |x - x_i|^{-\frac{3}{2}}.$$

A curve-like object is a distribution C with delta singularity along a closed curve $\tau : [0,1] \to \mathbb{R}^2$. Hence, an image in J will be of the form

$$f = P + C.$$

The goal of the Geometric Separation Problem is to recover P and C from the observed signal f.

The basic idea is to choose a redundant dictionary containing two representations systems Φ_1, Φ_2, that is

$$\mathscr{D} = \Phi_1 + \Phi_2,$$

where each system sparsely represents only one of the different components of $f \in J$. Specifically, Φ_1 is chosen to be a Parseval frame of shearlets which, as discussed above, provides optimally sparse approximations of functions which are smooth apart from curve singularities; Φ_2 is chosen to be a smooth wavelet orthonormal basis, which, as known, provides optimally sparse approximations of functions which are smooth apart from point singularities.

6.2 Geometric Separation Algorithm

In the algorithmic approach to the image separation problem devised in [50], an image $f \in J$ is examined at various resolution levels, denoted by f_j, $j \in \mathbb{Z}$, where $f_j = f * F_j$ and F_j is a bandpass filter associated with the frequency band centered at 2^j. Hence, for each $j \in \mathbb{Z}$,

$$f_j = P_j + C_j,$$

where P_j and C_j are the point-like and curve-like components of f_j, respectively, at scale j. At the resolution level j, the following optimization problem is defined:

$$(\widehat{W}_j, \widehat{S}_j) = \mathrm{argmin}_{W_j, S_j} ||\Phi_1^T S_j||_1 + ||\Phi_2^T W_j||_1 \quad \text{subject to } f_j = S_j + W_j, \quad (19)$$

where $\Phi_1^T S_j$ and $\Phi_2^T W_j$ are the shearlet and wavelet coefficients of the signals S_j and W_j, respectively.

The following theoretical result from [26] ensures the convergence of the geometric separation problem at fine scales.

Theorem 2. *Let* $(\widehat{W}_j, \widehat{S}_j)$ *be the solutions to the optimization problem* (20) *for each scale* j. *Then*

$$\lim_{j \to \infty} \frac{||\widehat{W}_j - P_j||_2 + ||\widehat{S}_j - C_j||_2}{||P_j||_2 + ||C_j||_2} = 0.$$

This shows that the components P_j and C_j of f_j are recovered with asymptotically arbitrarily high precision at fine scales.

In practice, an image f is not purely a sum of a point-like and a curve-like components, and contains an additional part that can be modeled as a noise term. In this situation, one can modify the optimization problem (20) as follows

$$(\widehat{W}_j, \widehat{S}_j) = \mathrm{argmin}_{W_j, S_j} ||\Phi_1^T S_j||_1 + ||\Phi_2^T W_j||_1 + \lambda ||f_j - W_j - S_j||_2, \quad (20)$$

subject to $f_j = S_j + W_j$. Notice that this type of expression is sometimes called an *infimal convolution* in the image processing literature. In this modified form, the additional noisy component in the image is characterized by the property that it cannot be represented sparsely by either one of the two representation systems and, thus, will be allocated to the residual term $(f_j - W_j - S_j)$.

An example of the application of the shearlet-based geometric separation algorithm to a noisy image is illustrated in Fig. 14, where the result is compared to

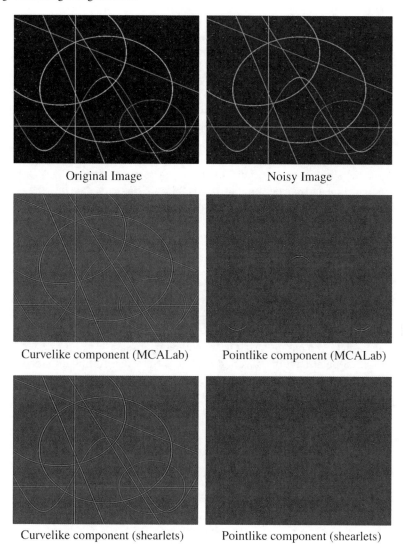

Fig. 14 Example of geometric separation on a noisy synthetic image. The shearlet-based geometric separation algorithm is compared against MCALab

the MCALab algorithm [35], another separation algorithm which employs a combination of curvelet and wavelet representations. The figure shows that the shearlet-based approach is very effective at separating the pointlike and curvelike components of the image, and it produces significantly less artifacts than MCALab. We refer to [26] for a more detailed discussion.

7 Shearlets Analysis of 3D Data

A number of results have recently appeared dealing with the application of shearlet-based methods for the analysis and processing of 3D data sets. Similar to the 2D case, the nearly optimally sparse approximation properties of 3D shearlet representations can be exploited for data denoising and feature extraction. As expected, dealing with 3D data sets entails more challenges in terms of memory storage so that particular attention is required to devise numerical efficient implementations.

A 3D Discrete Shearlet Transform (3D DST) was proposed in [51] and tested on video denoising. The algorithm follows essentially the ideas of the 2D discrete shearlet algorithm and can be summarized as follows. First, the data in the frequency domain are divided into three pyramidal regions, each one aligned with one of the orthogonal axis. The directional filtering stage is based on computing the DFT in the pseudopolar domain. In particular, in the first pyramidal region, this is defined as $(u, v, w) = (\xi_1, \frac{\xi_2}{\xi_1}, \frac{\xi_3}{\xi_1})$. Hence, at each fixed resolution level, the 3D DST algorithm proceeds as follows.

- The multiscale filter stage decomposes f_d^{j-1} into a low-pass f_a^j and high-pass f_d^j array.
- $\hat{f}_d^j$ is rearranged onto a pseudo-polar grid.
- A directional band pass filtering is applied on the pseudo-polar data.
- The pseudo-polar data is converted back to a Cartesian formulation and the inverse DFT is computed.

The algorithm runs in $O(N^3 \log(N))$ operations.

A shearlet thresholding routine based on the 3D DST algorithm (3DSHEAR) was applied to a problem of video denoising and the performance was compared against the following state-of-the-art algorithms: the Dual Tree Wavelet Transform (DTWT) and Surfacelets (SURF). We also compared against the 2D discrete shearlet transform (2DSHEAR), which was applied frame by frame, to illustrate the benefit of using a 3D transform, rather than a 2D transform acting on each frame. Figure 15 shows a side-by-side comparison of the denoising algorithm performance on a typical frame extracted from the video sequence Mobile. Additional comparisons and discussion can be found in [67, 51].

A shearlet-based method of analyzing videos from multiple views to autonomously estimate the kinematic state of an object was developed in [76]. In particular, a target's kinematics state was parametrized as a vector

$$\mathbf{X} = \begin{bmatrix} \mathbf{s} & \mathbf{r} \end{bmatrix}$$

whose respective corresponding elements were spatial position $\mathbf{s} = [x, y, z]$ and rotational orientation $\mathbf{r} = [h, p, r]$. Information from a Bayesian filter was merged to stabilize 2D recognition and tracking so that observation and object were concurrent. In this application, the shearlet transform was used to extract image features reliably. It particular, it was used to determine two-dimensional locations in the midst of illumination changes and discontinuities.

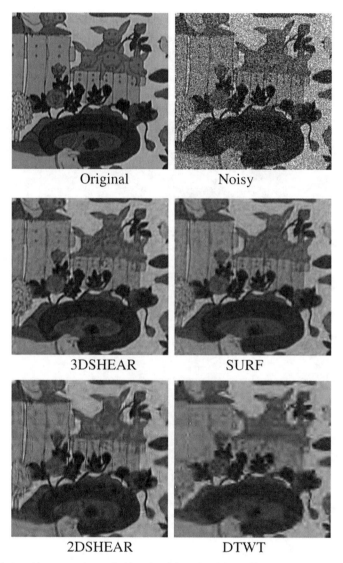

Fig. 15 Side-by-side comparison of video denoising algorithms, illustrated on a frame extracted from a video sequence. The 3D Discrete Shearlet Tranform (3DSHEAR) is compared against the Dual Tree Wavelet Transform (DTWT), the Surfacelet Tranform (SURF) and the 2D Siscrete Shearlet Transform (2DSHEAR)

In an effort to improve the state estimation routine, a continuous type of 3D shearlet transform was developed to analyze video data. In this case, the 3D shearlet transform was being used for detecting surface boundaries [77]. An illustration of the power of using a 3D shearlet surface/edge detector routine over a slice-by-slice detection of the 2D shearlet edge detector is given in Fig. 16. In this example, a solid spherical harmonic of order 2 and degree 7 is generated with a gradient shading applied to each slice. The same slices are analyzed by the 2D shearlet edge detector for comparisons. The images show the contour surface plot of this spherical harmonic and image slices through the center aligned with the x,y, and z axis.

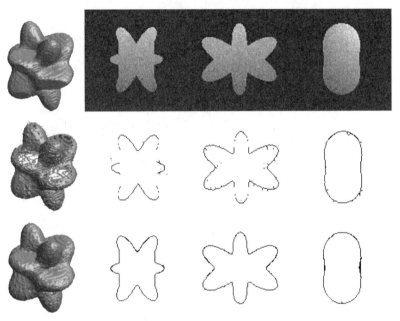

Fig. 16 Illustrations of 3D edge detection of a *solid spherical harmonic* of order 2 and degree 7 with a 2D gradient shadding applied. The images on the left display the 3D contour surface plots and the images on the left display the slices of sphere through the center

8 Additional Applications

Among the other areas of image processing that benefit from the use of the shearlet representation, we also recall image fusion and inpainting. In image fusion, the goal is to process and synthesize information provided by various sensors. A novel image fusion algorithm based on shearlets and local energy was recently proposed in [57], where it was shown that this approach outperforms traditional methods by preserving more details in the source images and further improving the subjective quality of fused image. In [19], another shearlet-based image fusion method was developed for panchromatic high-resolution images and multispectral images, and it was shown that it provides superior performance in terms of spatial resolution and preservation of spectral information. For the applications of inpainting, which can be described as an interpolation or estimation problem with missing data, a shearlet-based techique was recently presented in [41].

Acknowledgements

D.L. acknowledges support from NSF grants DMS 1008900 and DMS (Career) 1005799.

References

1. J. Aelterman, H. Q. Luong, B. Goossens, A. Pizurica, W. Philips, *Compass: a joint framework for Parallel Imaging and Compressive Sensing in MRI*, Image Processing (ICIP), 17th IEEE International Conference on (2010), 1653–1656.
2. A. Averbuch, R. R. Coifman, D. L. Donoho, M. Israeli, and Y. Shkolnisky, *A framework for discrete integral transformations I - the pseudo-polar Fourier transform*, SIAM Journal on Scientific Computing **30**(2) (2008), 764–784.
3. M. Bertero, *Linear inverse and ill-posed problems*, Advances in Electronics and Electron Physics (P.W. Hawkes, ed.), Academic Press, New York, 1989.
4. L. Blanc-Feraud, P. Charbonnier, G. Aubert, and M. Barlaud, *Nonlinear image processing: modelling and fast algorithm for regularization with edge detection*, Proc. IEEE ICIP-95, **1** (1995), 474–477.
5. P. J. Burt, E. H. Adelson, *The Laplacian pyramid as a compact image code*, IEEE Trans. Commun. **31** (4) (1983), 532–540.
6. E. J. Candès, and D. L. Donoho, *Recovering edges in ill-posed inverse problems: optimality of curvelet frames*, Annals Stat. **30**(3) (2002), 784–842.
7. E. J. Candès and F. Guo, *New multiscale transforms, minimum total variation synthesis: applications to edge-preserving image reconstruction*, Signal Proc. **82**(11) (2002), 1519–1543.
8. F. J. Canny, *A computational approach to edge detection*, IEEE Trans. Pattern Anal. Machine Intell. **8**(6) (1986), 679–698.
9. H. Cao, W. Tian, C. Deng, *Shearlet-based image denoising using bivariate model*, Progress in Informatics and Computing (PIC), 2010 IEEE International Conference on **2** (2010), 818–821.
10. T. Chan, J. Shen, *Image Processing And Analysis: Variational, PDE, Wavelet, And Stochastic Methods*, SIAM, Philadelphia (2005).
11. C. Chang, A. F. Laine, *Coherence of Multiscale Features for Contrast Enhancement of Digital Mammograms*, IEEE Trans. Info. Tech. in Biomedicine **3**(1) (1999), 32–46.
12. G. Chang, B. Yu and M. Vetterli, *Adaptive Wavelet Thresholding for Image Denoising and Compression*, IEEE Trans. Image Processing, **9** (2000), 1532–1546.
13. X. Chen, C. Deng, S. Wang, *Shearlet-Based Adaptive Shrinkage Threshold for Image Denoising*, E-Business and E-Government (ICEE), 2010 International Conference on (2010), 1616–1619.
14. X. Chen, H. Sun, C. Deng, *Image Denoising Algorithm Using Adaptive Shrinkage Threshold Based on Shearlet Transform*, Frontier of Computer Science and Technology, 2009, Fourth International Conference on (2009), 254–257.
15. R. R. Coifman and A. Sowa, *Combining the calculus of variations and wavelets for image enhancement*, Appl. Comput. Harmon. Anal., **9** (2000), 1–18.
16. F. Colonna, G. R. Easley, *Generalized discrete Radon transforms and their use in the ridgelet transform*, Journal of Mathematical Imaging and Vision, **23** (2005), 145–165.
17. F. Colonna, G. R. Easley, K. Guo, and D. Labate, *Radon Transform Inversion using the Shearlet Representation*, Appl. Comput. Harmon. Anal. **29**(2) (2010), 232–250.
18. C. Deng, H. Sun, X. Chen, *Shearlet-Based Adaptive Bayesian Estimator for Image Denoising*, Frontier of Computer Science and Technology, 2009, Fourth International Conference on (2009), 248–253.

19. C. Deng, S. Wang, X. Chen, *Remote Sensing Images Fusion Algorithm Based on Shearlet Transform*, Environmental Science and Information Application Technology, 2009, International Conference on **3** (2009), 451–454.

20. D. L. Donoho, *Unconditional bases are optimal bases for data compression and for statistical estimation*, Appl. Comput. Harmon. Anal. **1**(1) (1993), 100–115.

21. D. L. Donoho, *Nonlinear solution of linear inverse problems by wavelet-vaguelette decomposition*, Appl. Comput. Harmon. Anal. **2** (1995), 101–126.

22. D. L. Donoho, *De-noising by soft thresholding*, IEEE Trans. Info. Theory **41** (1995), 613–627.

23. D. L. Donoho and I. M. Johnstone, *Ideal spatial adaptation via wavelet shrinkage*, Biometrika **81** (1994), 425–455.

24. D. L. Donoho and I. M. Johnstone, *Adapting to unknown smoothness via wavelet shrinkage*, J. Amer. Stat. Assoc. **90**(432) (1995), 1200–1224.

25. D. L. Donoho and G. Kutyniok, *Geometric Separation using a Wavelet-Shearlet Dictionary*, SampTA-09 (Marseille, France, 2009), Proc., 2009.

26. D. L. Donoho and G. Kutyniok, *Microlocal analysis of the geometric separation problem*, Comm. Pure Appl. Math., to appear.

27. D. L. Donoho, M. Vetterli, R. A. DeVore, and I. Daubechies, *Data compression and harmonic analysis*, IEEE Trans. Inform. Theory, **44** (1998), 2435–2476.

28. S. Durand and J. Froment, *Reconstruction of wavelet coefficients using total variation minimization*, SIAM J. Sci. Comput., **24**(5) (2003), 1754–1767.

29. G. R. Easley, F. Colonna, and D. Labate, *Improved Radon Based Imaging using the Shearlet Transform*, Proc. SPIE, Independent Component Analyses, Wavelets, Unsupervised Smart Sensors, Neural Networks, Biosystems, and Nanoengineering VII, **7343**, Orlando, April 2009.

30. G. R. Easley, D. Labate, *Critically sampled composite wavelets*, Signals, Systems and Computers, 2009 Conference Record of the Forty-Third Asilomar Conference on (2009), 447–451.

31. G. R. Easley, D. Labate, and F. Colonna, *Shearlet-Based Total Variation for Denoising*, IEEE Trans. Image Processing, **18**(2) (2009), 260–268.

32. G. R. Easley, D. Labate, and W-Q Lim, *Sparse Directional Image Representations using the Discrete Shearlet Transform*, Appl. Comput. Harmon. Anal. **25**(1) (2008), 25–46.

33. G. R. Easley, V. Patel, D. M. Healy, Jr., *An M-channel Directional Filter Bank Compatible with the Contourlet and Shearlet Frequency Tiling*, Wavelets XII, Proceedings of SPIE, San Diego, CA (2007), 26–30.

34. G. R. Easley, V. M. Patel, and D. M. Healy, Jr., *Inverse halftoning using a shearlet representation*, Proc. of SPIE *Wavelets XIII*, **7446**, San Diego, August 2009.

35. M. J. Fadilli, J. L Starck, M. Elad, and D. L. Donoho, *MCALab: reproducible research in signal and image decomposition and inpainting*, IEEE Comput. Sci. Eng. Mag. **12**(1) (2010), 44–63.

36. R. W. Floyd and L. Steinberg, *An adaptive algorithm for spatial grayscale*, Proc. Soc. Image Display **17**(2) (1976), 75–77.

37. A. Foi, V. Katkovnik, K. Egiazarian, and J. Astola, *Inverse halftoning based on the anisotropic LPA-ICI deconvolution*, Proc. Int. TICSP Workshop Spectral Methods Multirate Signal Processing, (Vienna, Austria) (2004), 49–56.

38. D. Geman and C. Yang, *Nonlinear image recovery with half-quadratic regularization*, IEEE Trans. Image Proc. **4** (1995), 932–946.

39. J. Geusebroek, A. W. M. Smeulders, and J. van de Weijer, *Fast anisotropic Gauss filtering*, IEEE Trans. Image Proc. **8** (2003), 938–943.

40. G. Gilboa, Y. Y. Zeevi, and N. Sochen, *Texture preserving variational denoising using an adaptive fidelity term*, Proc. VLSM, Nice (2003), 137–144.

41. R. Gomathi and A. Kumar, *An efficient GEM model for image inpainting using a new directional sparse representation: Discrete Shearlet Transform*, Computational Intelligence and Computing Research (ICCIC), 2010 IEEE International Conference on (2010), 1–4.

42. B. Goossens, J. Aelterman, H. Luong, A. Pizurica, and W. Philips, *Efficient design of a low redundant Discrete Shearlet Transform*, Local and Non-Local Approximation in Image Processing, 2009, International Workshop on (2009), 112–124.
43. K. Guo and D. Labate, *Optimally sparse multidimensional representation using shearlets*, SIAM J. Math. Anal. **39** (2007), 298–318.
44. K. Guo and D. Labate, *Characterization and analysis of edges using the continuous shearlet transform*, SIAM J. Imaging Sciences **2** (2009), 959–986.
45. K. Guo, D. Labate and W. Lim, *Edge analysis and identification using the continuous shearlet transform*, Appl. Comput. Harmon. Anal. **27** (2009), 24–46.
46. Q. Guo, S. Yu, X. Chen, C. Liu, and W. Wei, *Shearlet-based image denoising using bivariate shrinkage with intra-band and opposite orientation dependencies*, Computational Sciences and Optimization, 2009, International Joint Conference on, **1** (2009), 863–866.
47. J. Jarvis, C. Judice, and W. Ninke, *A survey of techniques for the display of continuous tone pictures on bilevel displays*, Comput. Graph and Image Proc. **5** (1976), 13–40.
48. T. D. Kite, B. L. Evans, and A. C. Bovik, *Modeling and quality assessment of halftoning by error diffusion*, IEEE Trans. Image Proc. **9** (2000), 909–922.
49. G. Kutyniok and D. Labate, *Resolution of the wavefront set using continuous shearlets*, Trans. Amer. Math. Soc. **361** (2009), 2719–2754.
50. G. Kutyniok and W. Lim, *Image separation using shearlets*, in: Curves and Surfaces (Avignon, France, 2010), Lecture Notes in Computer Science 6920, Springer, 2012.
51. D. Labate and P. Negi, *3D Discrete shearlet transform and video denoising*, Wavelets and Sparsity XIV (San Diego, CA, 2011), SPIE Proc. 8138, SPIE, Bellingham, WA, 2011.
52. A. F. Laine, S. Schuler, J. Fan, and W. Huda, *Mammographic feature enhancement by multiscale analysis*, IEEE Trans. Med. Imag. **13**(4) (1994), 725–752.
53. A. F. Laine and X. Zong, *A multiscale sub-octave wavelet transform for de-noising and enhancement*, Wavelet Applications, Proc. SPIE, Denver, CO, August 6-9, 1996, **2825**, 238–249.
54. N. Lee and B J Lucier, *Wavelets methods for inverting the Radon transform with noisy data*, IEEE Trans. Image Proc. **10**(1) (2001), 79–94.
55. W. Q. Lim, *The discrete shearlet transform: a new directional transform and compactly supported shearlet frames*, Image Proc. IEEE Transactions on **19**(5) (2010), 1166–1180.
56. J. Lu and D. M. Healy, Jr., *Contrast enhancement via multi-scale gradient transformation*, Wavelet Applications, Proc. SPIE, Orlando, FL, April 5-8, 1994.
57. L. Lü , J. Zhao, and H. Sun, *Multi-focus image fusion based on shearlet and local energy*, Signal Processing Systems (ICSPS), 2010 2nd International Conference on, **1** (2010), V1-632–V1-635.
58. J. Ma and M. Fenn, *Combined complex ridgelet shrinkage and total variation minimization*, SIAM J. Sci. Comput., **28**(3) (2006), 984–1000.
59. S. Mallat, *A Wavelet Tour of Signal Processing*, Academic Press, San Diego, 1998.
60. S. Mallat and W. L. Hwang, *Singularity detection and processing with wavelets*, IEEE Trans. Inf. Theory **38**(2) (1992), 617–643.
61. S. Mallat and S. Zhong, *Characterization of signals from multiscale edges*, IEEE Trans. Pattern Anal. Mach. Intell. **14**(7) (1992), 710–732.
62. Y. Meyer, *Oscillating Patterns in Image Processing and Nonlinear Evolution Equations*, AMS, Providence, 2001.
63. F. Natterer, *The Mathematics of Computerized Tomography*, Wiley, New York, 1986.
64. F. Natterer and F. Wübbeling, *Mathematical Methods in Image Reconstruction*, SIAM Monographs on Mathematical Modeling and Computation, Philadelphia, 2001.
65. R. Neelamani, H. Choi, and R. G. Baraniuk, *ForWaRD: Fourier-wavelet regularized deconvolution for ill-conditioned systems*, IEEE Trans. Image Proc. **52**(2) (2004), 418–433.
66. R. Neelamani, R. Nowak, and R. Baraniuk, *Model-based inverse halftoning with Wavelet-Vaguelette Deconvolution*, Proc. IEEE Int. Conf. Image Proc. (2000), 973–976.
67. P. Negi and D. Labate, *3D discrete shearlet transform and video processing*, IEEE Trans. Image Proc., in press 2012.

68. V. M. Patel, G. R. Easley, and R. Chellappa, *Multiscale directional filtering of noisy InSAR phase images*, Proc. SPIE, Independent Component Analyses, Wavelets, Neural Networks, Biosystems, and Nanoengineering VII **7703**, Orlando, April 2010.

69. V. M. Patel, G. R. Easley, and D. M. Healy, Jr., *A new multiresolution generalized directional filter bank design and application in image enhancement*, Proc. IEEE International Conference on Image Proc., San Diego, October 2008, 2816–2819.

70. V. M. Patel, G. R. Easley, and D. M. Healy, Jr., *Shearlet-based deconvolution*, IEEE Trans. Image Proc. **18**(12) (2009), 2673–2685.

71. P. Perona, *Steerable-scalable kernels for edge detection and junction analysis*, Image Vis. Comput. **10** (1992), 663–672.

72. P. Perona and J. Malik, *Scale-space and edge detection using anisotropic diffusion*, IEEE Trans. Pattern Anal. Mach. Intel. **12** (1990), 629–639.

73. L. Rudin, S. Oscher, and E. Fatemi, *Nonlinear total variation based noise removal algorithms*, Phys. D **60** (1992), 259–268.

74. O. Scherzer, M. Grasmair, H. Grossauer, M. Haltmeier and F. Lenzen, *Variational Methods in Imaging*, Springer, Applied Mathematical Sciences 167, 2009.

75. O. Scherzer and J. Weickert, *Relations between regularization and diffusion filtering* Journal of Mathematical Imaging and Vision **12**(1) (2000), 43–63.

76. D. A. Schug and G. R. Easley, *Three dimensional Bayesian state estimation using shearlet edge analysis and detection*, Communications, Control and Signal Processing (ISCCSP), 2010 4th International Symposium on (2010), 1–4.

77. D. A. Schug, G. R. Easley, and D. P. O'Leary, *Three-dimensional shearlet edge analysis*, Proc. SPIE, *Independent Component Analyses, Wavelets, Neural Networks, Biosystems, and Nanoengineering IX*, Orlando, April 2011.

78. J. L. Starck, E. J. Candès, and D. L. Donoho, *The curvelet transform for image denoising*, IEEE Trans. Im. Proc. **11** (2002), 670–684.

79. J. L Starck, M. Elad, and D. L. Donoho, *Image decomposition via the combination of sparse representation and a variational approach*, IEEE Trans. Image Proc. **14** (2005), 1570–1582.

80. J. L. Starck, F. Murtagh, E. J. Candès, and D. L. Donoho, *Gray and color image contrast enhancement by the curvelet transform*, IEEE Trans. Imag. Proc. **12**(6) (2003), 706–717.

81. G. Steidl, J. Weickert, T. Brox, P. Mrázek, and M. Welk, *On the equivalence of soft wavelet shrinkage, total variation diffusion, total variation regularization, and SIDEs*, SIAM J. Numer. Anal. **42** (2004), 686–713.

82. R. N. Strickland and H. I. Hahn, *Wavelet Transforms for Detecting Microcalcifications in Mammograms*, IEEE Trans. on Med. Imag. **15**(2) (1996), 218–229.

83. H. Sun and J. Zhao, *Shearlet Threshold Denoising Method Based on Two Sub-swarm Exchange Particle Swarm Optimization*, Granular Computing (GrC), 2010 IEEE International Conference on (2010), 449–452.

84. S. Teboul, L. Blanc-Feraud, G. Aubert, and M. Barlaud, *Variational approach for edge-preserving regularization using coupled PDEs*, IEEE Trans. Image Proc. **7** (1998), 387–397.

85. W. Tian, H. Cao, and C. Deng, *Shearlet-based adaptive MMSE estimator for image denoising*, Intelligent Computing and Intelligent Systems (ICIS), 2010 IEEE International Conference on **2** (2010), 689–692.

86. A. N. Tikhonov, *Solution of incorrectly formulated problems and the regularization method*, Soviet Math. Doklady **4** (1963), 1035–1039.

87. R. Ulichney, *Digital Halftoning*, MIT Press, Cambridge, MA, 1987.

88. J. Weickert, *Foundations and applications of nonlinear anisotropic diffusion filtering*, Z. Angew. Math. Mechan. **76** (1996), 283–286.

89. J. Weickert, *Anisotropic Diffusion in Image Processing*, Teubner, Stuttgart, 1998.

90. M. Welk, G. Steidl and J. Weickert, *Locally analytic schemes: A link between diffuusion filtering and wavelet shrinkage*, Appl. and Comput. Harmon. Anal. **24** (2008), 195–224.

91. S. Yi, D. Labate, G. R. Easley, and H. Krim, *Edge detection and processing using shearlets*, Proc. IEEE Int. Conference on Image Proc., San Diego, October 12-15, 2008.

92. S. Yi, D. Labate, G. R. Easley, and H. Krim, *A Shearlet approach to edge analysis and detection*, IEEE Trans. Image Proc. **18**(5) (2009), 929–941

93. X. Zhang, X. Sun, L. Jiao, and J. Chen, *A Non-Local Means Filter with Translating Invariant Shearlet Feature Descriptors*, Wireless Communications Networking and Mobile Computing (WiCOM), 2010 6th International Conference on (2010), 1–4.
94. X. Zhang, Q. Zhang, and L. Jiao, *Image Denoising with Non-Local Means in the Shearlet Domain*, Multi-Platform/Multi-Sensor Remote Sensing and Mapping (M2RSM), 2011 International Workshop on (2011), 1–5.
95. D. Ziou and S. Tabbone, *Edge Detection Techniques An Overview*, Internat. J. Pattern Recognition and Image Anal. **8**(4) (1998), 537–559.

Index

G. Kutyniok and D. Labate (eds.), *Shearlets: Multiscale Analysis for Multivariate Data*,
Applied and Numerical Harmonic Analysis, DOI 10.1007/978-0-8176-8316-0,
© Springer Science+Business Media, LLC 2012

Applied and Numerical Harmonic Analysis

A.I. Saichev and W.A. Woyczyński: *Distributions in the Physical and Engineering Sciences* (ISBN 978-0-8176-3924-2)

R. Tolimieri and M. An: *Time-Frequency Representations* (ISBN 978-0-8176-3918-1)

G.T. Herman: *Geometry of Digital Spaces* (ISBN 978-0-8176-3897-9)

A. Procházka, J. Uhliř, P.J.W. Rayner, and N.G. Kingsbury: *Signal Analysis and Prediction* (ISBN 978-0-8176-4042-2)

J. Ramanathan: *Methods of Applied Fourier Analysis* (ISBN 978-0-8176-3963-1)

A. Teolis: *Computational Signal Processing with Wavelets* (ISBN 978-0-8176-3909-9)

W.O. Bray and C.V. Stanojević: *Analysis of Divergence* (ISBN 978-0-8176-4058-3)

G.T Herman and A. Kuba: *Discrete Tomography* (ISBN 978-0-8176-4101-6)

J.J. Benedetto and P.J.S.G. Ferreira: *Modern Sampling Theory* (ISBN 978-0-8176-4023-1)

A. Abbate, C.M. DeCusatis, and P.K. Das: *Wavelets and Subbands* (ISBN 978-0-8176-4136-8)

L. Debnath: *Wavelet Transforms and Time-Frequency Signal Analysis* (ISBN 978-0-8176-4104-7)

K. Gröchenig: *Foundations of Time-Frequency Analysis* (ISBN 978-0-8176-4022-4)

D.F. Walnut: *An Introduction to Wavelet Analysis* (ISBN 978-0-8176-3962-4)

O. Bratteli and P. Jorgensen: *Wavelets through a Looking Glass* (ISBN 978-0-8176-4280-8)

H.G. Feichtinger and T. Strohmer: *Advances in Gabor Analysis* (ISBN 978-0-8176-4239-6)

O. Christensen: *An Introduction to Frames and Riesz Bases* (ISBN 978-0-8176-4295-2)

L. Debnath: *Wavelets and Signal Processing* (ISBN 978-0-8176-4235-8)

J. Davis: *Methods of Applied Mathematics with a MATLAB Overview* (ISBN 978-0-8176-4331-7)

G. Bi and Y. Zeng: *Transforms and Fast Algorithms for Signal Analysis and Representations* (ISBN 978-0-8176-4279-2)

J.J. Benedetto and A. Zayed: *Sampling, Wavelets, and Tomography* (ISBN 978-0-8176-4304-1)

E. Prestini: *The Evolution of Applied Harmonic Analysis* (ISBN 978-0-8176-4125-2)

O. Christensen and K.L. Christensen: *Approximation Theory* (ISBN 978-0-8176-3600-5)

L. Brandolini, L. Colzani, A. Iosevich, and G. Travaglini: *Fourier Analysis and Convexity* (ISBN 978-0-8176-3263-2)

W. Freeden and V. Michel: *Multiscale Potential Theory* (ISBN 978-0-8176-4105-4)

O. Calin and D.-C. Chang: *Geometric Mechanics on Riemannian Manifolds* (ISBN 978-0-8176-4354-6)

J.A. Hogan and J.D. Lakey: *Time-Frequency and Time-Scale Methods* (ISBN 978-0-8176-4276-1)

C. Heil: *Harmonic Analysis and Applications* (ISBN 978-0-8176-3778-1)

K. Borre, D.M. Akos, N. Bertelsen, P. Rinder, and S.H. Jensen: *A Software-Defined GPS and Galileo Receiver* (ISBN 978-0-8176-4390-4)

Applied and Numerical Harmonic Analysis (Cont'd)

T. Qian, V. Mang I, and Y. Xu: *Wavelet Analysis and Applications* (ISBN 978-3-7643-7777-9)

G.T. Herman and A. Kuba: *Advances in Discrete Tomography and Its Applications* (ISBN 978-0-8176-3614-2)

M.C. Fu, R.A. Jarrow, J.-Y. J. Yen, and R.J. Elliott: *Advances in Mathematical Finance* (ISBN 978-0-8176-4544-1)

O. Christensen: *Frames and Bases* (ISBN 978-0-8176-4677-6)

P.E.T. Jorgensen, K.D. Merrill, and J.A. Packer: *Representations, Wavelets, and Frames* (ISBN 978-0-8176-4682-0)

M. An, A.K. Brodzik, and R. Tolimieri: *Ideal Sequence Design in Time-Frequency Space* (ISBN 978-0-8176-4737-7)

B. Luong: *Fourier Analysis on Finite Abelian Groups* (ISBN 978-0-8176-4915-9)

S.G. Krantz: *Explorations in Harmonic Analysis* (ISBN 978-0-8176-4668-4)

G.S. Chirikjian: *Stochastic Models, Information Theory, and Lie Groups, Volume 1* (ISBN 978-0-8176-4802-2)

C. Cabrelli and J.L. Torrea: *Recent Developments in Real and Harmonic Analysis* (ISBN 978-0-8176-4531-1)

M.V. Wickerhauser: *Mathematics for Multimedia* (ISBN 978-0-8176-4879-4)

P. Massopust and B. Forster: *Four Short Courses on Harmonic Analysis* (ISBN 978-0-8176-4890-9)

O. Christensen: *Functions, Spaces, and Expansions* (ISBN 978-0-8176-4979-1)

J. Barral and S. Seuret: *Recent Developments in Fractals and Related Fields* (ISBN 978-0-8176-4887-9)

O. Calin, D. Chang, K. Furutani, and C. Iwasaki: *Heat Kernels for Elliptic and Sub-elliptic Operators* (ISBN 978-0-8176-4994-4)

C. Heil: *A Basis Theory Primer* (ISBN 978-0-8176-4686-8)

J.R. Klauder: *A Modern Approach to Functional Integration* (ISBN 978-0-8176-4790-2)

J. Cohen and A. Zayed: *Wavelets and Multiscale Analysis* (ISBN 978-0-8176-8094-7)

D. Joyner and J.-L. Kim: *Selected Unsolved Problems in Coding Theory* (ISBN 978-0-8176-8255-2)

J.A. Hogan and J.D. Lakey: *Duration and Bandwidth Limiting* (ISBN 978-0-8176-8306-1)

G. Chirikjian: *Stochastic Models, Information Theory, and Lie Groups, Volume 2* (ISBN 978-0-8176-4943-2)

G. Kutyniok and D. Labate: *Shearlets* (ISBN 978-0-8176-8315-3)

*For a fully up-to-date list of ANHA titles, visit **http://www.springer.com/series/4968?detailsPage=titles** or **http://www.springerlink.com/content/t7k8lm/**.*

Printed by Publishers' Graphics LLC USA
MO20120424-009
2012